MEDICAL DEVICES

MEDICAL DEVICES
SURGICAL AND IMAGE-GUIDED TECHNOLOGIES

Edited by

Martin Culjat
Rahul Singh
Hua Lee

WILEY

A JOHN WILEY & SONS, INC., PUBLICATION

Cover Design: Michael Rutkowski
Cover Image: Flexible rectoscope, courtesy of KARL STORZ Endoscopy-America, Inc.

Published by John Wiley & Sons, Inc., Hoboken, New Jersey
Published simultaneously in Canada

For general information on our other products and services or for technical support, please contact our Customer Care Department within the United States at (800) 762-2974, outside the United States at (317) 572-3993 or fax (317) 572-4002.

Wiley also publishes its books in a variety of electronic formats. Some content that appears in print may not be available in electronic formats. For more information about Wiley products, visit our web site at www.wiley.com.

Library of Congress Cataloging-in-Publication Data:

Culjat, Martin, 1977–
  Medical Devices: Surgical and Image-Guided Technologies / Martin Culjat, Rahul Singh, Hua Lee.
    p. cm.
  Includes bibliographical references.
  ISBN 978-0-470-54918-6 (hardback)
  1. Medical instruments and apparatus–Textbooks. 2. Surgical instruments and apparatus–Textbooks. I. Singh, Rahul, 1975– II. Lee, Hua. III. Title.
  R856.C85 2012
  610.28′4–dc23

                                  2012020590

Printed in the United States of America

ISBN: 9780470549186

10 9 8 7 6 5 4 3 2 1

Jean-Jacques Lemaire, Eric J. Behnke, Andrew J. Frew, and
Antonio A. F. DeSalles

Recent decades have seen considerable advances in the development of medical devices and technologies. Innovations in instrumentation, implantable devices, and imaging systems have led to new diagnostic and therapeutic techniques and even new medical disciplines. Because of these and other advances in medicine, an increasing number of conditions can now be treated and patient outcomes continue to improve. Researchers, engineers, and clinicians in the biomedical engineering field are now developing the next generation of technologies that will enable procedures never imagined and make modern medicine accessible to more people worldwide. A challenge is to realize these innovations while reducing rather than increasing the cost of health care.

This book is intended primarily for the growing number of undergraduates, graduate students, medical students, and researchers who are interested in medical device design. Currently, there is a lack of concise, modern, device-focused texts that are written for such an audience. As the complexity of medical technologies continues to increase, there will be an acute need for individuals with the knowledge and skills necessary to lead this growing field.

The content of this text was inspired by research activities at the UCLA Center for Advanced Surgical and Interventional Technology (CASIT). To gauge a preliminary assessment of the effectiveness of this book's technical coverage, the editors and several of the authors participated in a one-quarter seminar course at the UC, Santa Barbara during the fall of 2008, receiving superb ratings and reviews. The class attracted students from all engineering majors, as well as the pre-med program, with a breadth of audience and interest level that this book carries through gracefully.

The technical content in this book is presented in a comprehensive manner, consistent with junior/senior undergraduate and first-year graduate students' background level in mathematics, physics, chemistry, and biology. The chapters are written and organized in the form of independent modules, such that lectures can be configured with a high degree of flexibility from year to year. Each chapter was written by one or more clinical or engineering experts, primarily from the fields of biomedical engineering, electrical engineering, mechanical engineering, computer science, surgery, and radiology.

The book is organized into five sections, each with a separate focus. The first section *Introduction to Medical Devices* features two chapters. Chapter 1 provides a brief introduction on the history, future, and terminology related to medical devices, and Chapter 2 provides a thorough overview of factors to consider during the medical device design process, including topics such as regulatory affairs and manufacturing. The second section focuses on *Minimally Invasive Devices and Techniques* and features four chapters. Chapter 3 discusses principles and tools of laparoscopic surgery, Chapter 4 describes minimally invasive techniques in ophthalmology, Chapter 5 discusses surgical robotics and their application to minimally invasive surgery, and Chapter 6 describes interventional applications of catheters and catheter technologies. *Energy Delivery Devices and Systems* are described in the third section. This section contains chapters on electrosurgical tools used for cautery and coagulation of tissues (Chapter 7), devices used to ablate tissues such as tumors (Chapter 8), and lasers and their application to medicine (Chapter 9). The fourth section, *Implantable Devices and Systems* features chapters on implantable devices for vascular and cardiovascular procedures (Chapter 10), circulatory assist devices for heart failure (Chapter 11), and orthopedic implants, such as hip replacements and spinal fusion devices (Chapter 12). The final section covers *Imaging and Image-Guided Techniques* and includes four chapters. Chapter 13 focuses on endoscopic devices and systems for minimally invasive procedures; Chapter 14 on ultrasound devices used for both imaging and therapy; Chapter 15 on X-ray imaging technologies, including fluoroscopy, mammography, and computed tomography (CT); and Chapter 16 on techniques for image fusion and image-guided navigation of instruments during neurosurgery.

This book does not attempt to cover all of the medical devices and technologies in use today. Instead, the chapters were carefully selected such that a broad spectrum of representative topics in biomedical engineering could be discussed comprehensively. These topics are highly relevant to the state-of-the-art minimally invasive, image-guided, and interventional techniques that are used today.

The editors would like to thank everyone at the CASIT for their input into the development of this project. Additional thanks goes to Ms. Susan Ly for her assistance with copy editing.

MARTIN CULJAT
RAHUL SINGH
HUA LEE

▰▰▰▰ CONTRIBUTORS

Robert M. Beardsley, Jules Stein Eye Institute, University of California, Los Angeles, Los Angeles, CA, USA

Eric J. Behnke, Department of Neurosurgery and Radiology Oncology, University of California, Los Angeles, Los Angeles, CA, USA

Axel Boese, INKA - Intelligente Katheter, Otto-von-Guericke University Magdeburg, Magdeburg, Germany

G. Bryan Cornwall, Research and Clinical Resources, NuVasive, Inc., San Diego, CA, USA

Martin Culjat, Departments of Bioengineering and Surgery, Center for Advanced Surgical and Interventional Technology (CASIT), University of California, Los Angeles, Los Angeles, CA, USA

Antonio A. F. DeSalles, Department of Neurosurgery and Radiation Oncology, University of California, Los Angeles, Los Angeles, CA, USA

Michael Douek, Department of Radiological Sciences, University of California, Los Angeles, Santa Monica, CA, USA

Erik Dutson, Department of Surgery-General, Center for Advanced Surgical and Interventional Technology (CASIT), University of California, Los Angeles, Los Angeles, CA, USA

Edward Ebramzadeh, Biomechanical Engineering and Surgical Research Facility, Santa Monica UCLA Medical Center and Orthopaedic Hospital, Santa Monica, CA, USA

Andrew J. Frew, Department of Neurosurgery and Radiation Oncology, University of California, Los Angeles, Los Angeles, CA, USA

Warren Grundfest, Departments of Bioengineering, Electrical Engineering, and Surgery, Center for Advanced Surgical and Interventional Technology (CASIT), University of California, Los Angeles, Los Angeles, CA, USA

John Ho, Department of Pediatric Cardiology, University of California, Los Angeles, Los Angeles, CA, USA

Allen Y. Hu, Retina Division, Jules Stein Eye Institute, University of California, Los Angeles, Los Angeles, CA, USA

Jean-Pierre Hubschman, Retina Division, Jules Stein Eye Institute, University Of California Los Angeles, Los Angeles, CA, USA

Todd S. Johnson, Global Extremities Development, Reconstructive Division, Zimmer, Inc., Warsaw, IN, USA

Colin Kealey, Department of Surgery, Center for Advanced Surgical and Interventional Technology (CASIT), University of California, Los Angeles, Los Angeles, CA, USA

Murray Kwon, Department of Surgery-Cardiothoracic, University of California, Los Angeles, Los Angeles, CA, USA

Jean-Jacques Lemaire, Department of Neurosurgery, Auvergne University, Centre Hospitalier et Universitaire, Clermont-Ferrand, France

Dan Levi, Department of Pediatric Cardiology, University of California, Los Angeles, Los Angeles, CA, USA

David Lu, Department of Radiological Sciences, University of California, Los Angeles, Los Angeles, CA, USA

Justin McWilliams, Department of Radiological Sciences, University of California, Los Angeles, Los Angeles, CA, USA

Jon Moseley, Implant Technology, Wright Medical Technology, Inc., Arlington, TN, USA

Amit P. Mulgaonkar, Biomedical Engineering IDP, Center for Advanced Surgical and Interventional Technology (CASIT), University of California, Los Angeles, Los Angeles, CA, USA

Gregory Nighswonger, KARL STORZ Endoscopy-America, Inc., El Segundo, CA, USA

Asael Papour, Department of Electrical Engineering, University of California, Los Angeles, Los Angeles, CA, USA

Camellia Racu-Keefer, Southern California Permanente Medical Group, Department of General Surgery, San Diego, CA, USA

Paymon Rahgozar, Department of Surgery-Cardiothoracic, University of California, Los Angeles, Los Angeles, CA, USA

David Rigberg, Department of Surgery-Vascular, University of California, Los Angeles, Los Angeles, CA, USA

Mark Roden, Medical Vision Systems, LLC., Playa Del Rey, CA, USA

Jacob Rosen, Department of Computer Engineering, Baskin School of Engineering, University of California, Santa Cruz, Santa Cruz, CA, USA

Sophia N. Sangiorgio, Biomechanical Engineering and Surgical Research Facility, Santa Monica UCLA Medical Center and Orthopaedic Hospital, Santa Monica, CA, USA

Rahul Singh, Department of Bioengineering and Surgery, Center for Advanced Surgical and Interventional Technology (CASIT), University of California, Los Angeles, Los Angeles, CA, USA

Oscar Stafsudd, Department of Electrical Engineering, University of California, Los Angeles, Los Angeles, CA, USA

Zachary Taylor, Department of Bioengineering, Center for Advanced Surgical and Interventional Technology (CASIT), University of California, Los Angeles, Los Angeles, CA, USA

Allan Tulloch, Department of Surgery, Center for Advanced Surgical and Interventional Technology (CASIT), University of California, Los Angeles, Los Angeles, CA, USA

Scott Um, Minimally Invasive Surgery/Bariatric Surgery, Center for Advanced Surgical and Interventional Technology (CASIT), University of California, Los Angeles, Los Angeles, CA, USA

PART I

INTRODUCTION TO MEDICAL DEVICES

CHAPTER 1

Introduction

MARTIN CULJAT

Center for Advanced Surgical and Interventional Technology (CASIT), University of California,
Los Angeles, CA

1.1 HISTORY OF MEDICAL DEVICES

The fields of medicine and surgery are as old as the origin of man. There are
surviving records of medical procedures and theories dating back thousands of
years, from the ancient Egyptians to the Babylonians, Hebrews, Indians, Chinese,
and Greeks. Internal diseases were poorly understood and often blamed on super-
natural beings and treated by medicine men during religious ceremonies. External
injuries and diseases, on the other hand, were often treated surgically with techniques developed independently by multiple civilizations, some using concepts
similar to those used currently. Application of dressings to wounds has been nearly
universal throughout the history of man, with recorded evidence of the usage of
leaves, clay, tar, bark, snow, sand, down feathers, and animal hides (Bishop, 1960).
Both cobwebs and heated cautery irons have been used to control bleeding in multiple cultures, and suture needles have been developed using tools such as bone
splinters and thorns. Insect jaws have been used as sutures in at least three continents, typically by encouraging a termite to bite through the wound with its
powerful jaws and subsequently removing its body. Fracture fixation has been
practiced by many civilizations, using materials such as hardened animal hides,
clay, and wood. Many other early surgical tools have been discovered across the
world, such as bark and feather quills for wound drainage and cleaning by the North
American Lakota Indians, as well as bamboo, shells, sharks' teeth, and bones as
surgical scalpels in New Britain in the South Pacific. Relatively modern embodiments of surgical tools were used as far back as Roman times, with metal forceps,
scalpels, speculas, surgical needles, urinary catheters, and cautery irons discovered
in the buried ruins of Pompeii, dating from the first century CE (Greenhill, 1875).

Medical Devices: Surgical and Image-Guided Technologies, First Edition.
Edited by Martin Culjat, Rahul Singh, and Hua Lee.
© 2013 John Wiley & Sons, Inc. Published 2013 by John Wiley & Sons, Inc.

The Middle Ages was a relatively slow period in the advancement of medical tools and interventions. More significant was the increasing study and understanding of anatomy, physiology, pharmacology, surgery, and other fields relevant to medicine. These advances were particularly evident in the Middle East and Europe, where comprehensive texts were written on these topics, often to be lost, rediscovered, and translated. Persian and Arabic physicians were credited with many significant achievements in medicine during the Islamic Golden Age from the eighth century CE until the Mongol invasions in the thirteenth century. In Europe, the birth of universities in the twelfth century and the Renaissance in the fourteenth century likewise encouraged the study and advancement of medicine. Consequently, surgical and patient care techniques slowly began to improve, leading to better patient outcomes. These improvements began to accelerate in the nineteenth century.

Before the nineteenth century, surgery was largely performed without anesthesia, in nonsterile environments, and without the benefit of preoperative visualization of internal anatomy of the patient. The introduction of analgesics such as ether and chloroform in the 1840s was a major advance, as patients were no longer subjected to tremendous pain while conscious, and surgeons were able to perform longer and more complex procedures. Before this time, the best surgeons were often those who could operate the fastest, and surgery was mostly limited to a few procedures, such as bladder stone removal, vessel ligation, leg amputation, and excision of superficial tumors (Tilney, 2011). The acceptance of asepsis, or sterilization, in the late nineteenth century significantly reduced postoperative deaths, which had often occurred at hospitals in as many as 80% of patients. The discovery of the X-ray in 1895 and the subsequent birth of radiology enabled physicians for the first time to study the anatomy of the body. Together, these achievements transformed the surgical discipline and led to a rapid expansion of interventions that could be successfully performed.

Throughout history, war was a major catalyst for advances in medicine and surgery, allowing physicians and surgeons to practice and popularize experimental tools, drugs, and techniques, many of which still benefit the global population at present. The introduction of firearms and canons in the Battle of Crécy in 1346 and machine guns in the 1870 Franco-Prussian War underscored the need for improved battlefield care, as these weapons led to more severe wounds, more rapid infections, and deaths due to tetanus (Tilney, 2011). The Crimean War in the 1850s highlighted the poor conditions on and off the battlefield, where large armies were used and the wounded were cared for in overcrowded, unsanitary conditions. During this war, five out of six deaths resulted from cholera, dysentery, and malaria, and above-knee amputations had a 90% mortality rate because of infection. Typhus, typhoid, and dysentery caused two-thirds of deaths in the American Civil War, and many others died from wound infections. However, this period of warfare also saw the introduction of nursing teams, the foundation of the Red Cross, improved surgical techniques, and the occasional use of analgesics and antisepsis. World War I saw significant improvements in asepsis, successful abdominal and plastic surgery techniques, the introduction of blood transfusions, and large-scale immunization of soldiers against typhoid. World War II led to improved burn management, use of blood banking and intravenous fluids, a better understanding of pharmaceuticals, and the standardization of care. Introduction of plastic fluid bags, tubing, and tools following World War II was a major development in asepsis, as contamination from patient to patient was virtually eliminated.

In the twentieth century, advancements in antibiotics, pharmaceuticals, and anesthesia improved outcomes for patients worldwide and brought modern medicine to a more global population. At the same time, innovations in materials, manufacturing, electronics, and computing accelerated the use of technology in medicine and led to the birth of the medical device industry. Some notable technological advances in the twentieth century include electrocardiology (ECG) (1903), stereotactic surgery (1908), endoscopy (1910), electroencephalography (EEG) (1929), dialysis machines (1943), disposable catheters (1944), defibrillators (1947), ventilators (1949), hip replacements (1962), artificial heart (1963), diagnostic ultrasound (1965), balloon catheters (1969), cochlear implants (1969), laser eye surgery (1973), positron emission tomography (PET) (1976), magnetic resonance imaging (MRI) (1980), surgical robotics (1985), and intravascular stents (1988) (Challoner, 2009).

Recent technological innovations have spawned entirely new approaches to surgery. For example, traditional open surgeries that are associated with large incisions and extensive patient trauma have recently began to give way to minimally invasive techniques, such as laparoscopy. In these procedures, small "keyhole" incisions are made on the patient's body, significantly reducing trauma and recovery times. Precision tools have been developed to operate through these small openings and allow clinicians to perform an array of tasks from outside of the body. Interventional devices, such as catheters, are now commonly fed deep into the vasculature from needle incisions in the skin to deliver medication, measure pressure, or widen obstructed blood vessels. Some techniques, such as tissue ablation, can now be performed either laparoscopically, with a catheter, or even noninvasively from outside of the skin surface. Many minimally invasive procedures have already began to move toward robotic control or automation.

Implantable devices have also continued to mature. The use of new biocompatible and nonthrombogenic materials and coatings has led to vast improvements in a range of implant technologies, from coronary stents to hip replacements and cardiac assist devices. Computer-aided design, finite-element modeling, and precision machining have enabled implant designers to better customize implantable devices for individual patients or conditions. Likewise, miniaturization of electronics, improved battery technologies, and advanced telemetry systems have enabled the implantation of robust sensing and stimulation systems, such as cardiac pacemakers, deep brain stimulators, and cochlear implants.

Some of the greatest changes in modern medicine have occurred in the field of imaging. Improvements in 3D imaging techniques, such as computed tomography (CT) and MRI, have given clinicians an unprecedented view of patients' internal anatomic and pathophysiologic processes. In addition to facilitating diagnosis, imaging techniques have been adapted to guide interventions, facilitating more targeted and less invasive delivery of therapy. A current trend is an increased use of image fusion, or the combination of multiple imaging technologies. These techniques merge data from disparate sources such as ultrasound, MRI, CT, and PET, to give a more complete picture of a patient's disease state. Image fusion can be used as a preoperative, intraoperative, or postoperative tool in fields such as neurosurgery and prostate surgery.

All of these technological advances have led to rapid worldwide growth in the medical device industry. At present, the global medical device industry has estimated worldwide sales greater than $300 billion (Zack's Equity Research, 2011). The US medical device market is the world's largest market, with an estimated value greater than $105 billion in 2011 (Espicom Healthcare Intelligence, 2011). Both the US and worldwide markets are expected to continue to grow rapidly because of the development of new products, aging populations, geographic expansion, and emerging markets.

The next two decades will likely see the introduction of a new generation of medical technologies. Some potential advances include micro- and nanoscale implantable devices and coatings, tissue-engineered organs, increased use of portable and wearable devices, devices enabling new minimally invasive and noninvasive surgical and therapeutic techniques, advances in real-time image fusion, automation of surgical tasks, and telesurgery. Medical devices and technologies will be increasingly available to the growing worldwide population through the use of lower cost materials and manufacturing techniques, as well as by increased competition from manufacturers in countries with lower labor costs. Additional cost savings and improved worldwide access can be achieved by developing simpler devices or techniques that can be used by a broader range of clinicians with lower skill sets or less training. Medical simulation and telementoring technologies may further broaden access to clinicians in rural or underserved areas, by providing these clinicians with high quality training tools and remote guidance from expert surgeons and physicians.

1.2 MEDICAL DEVICE TERMINOLOGY

The US Food and Drug Administration (FDA), which governs the manufacture and distribution of medical technology, describes a medical device as "an instrument, apparatus, implement, machine, contrivance, implant, *in vitro* reagent, or other similar or related article, including a component part, or accessory, that is intended for use in the diagnosis of disease or other conditions or in the cure, mitigation, treatment, or prevention of disease, in man or other animals or intended to affect the structure or any function of the body of man or other animals. Also, it does not achieve any of its primary intended purposes through chemical action within or on the body of man or other animals and is not dependent on being metabolized for the achievement of any of its primary intended purposes" (FDA, 2010).

In other words, a medical device is any product, not including drugs or vaccines, used in the care or treatment of patients (or animals) for disease prevention, or for diagnosis of a disease or condition. This definition includes products such as bandages, bedpans, and *in vitro* laboratory kits, as well as prosthetic limbs, pacemakers, and X-ray systems.

A convenient way to categorize today's medical devices is with the North American Industry Classification System (NAICS) (U.S. Census Bureau, 2007). This system assigns products with a six-digit code, based on the industry and/or business sector that the product falls within. According to the International Trade Association (ITA) of the US Department of Commerce, all medical devices fall within one of eight NAICS categories (ITA, 2009). These categories are as follows:

- *In Vitro Diagnostic Substances*: Includes chemical, biological, or radioactive substances used for diagnostic tests performed in test tubes, petri dishes, machines, and other diagnostic test-type devices (NAIC 325413). This group comprised roughly 10% of the US medical device market in 2007, measured by the value of shipments (ITA, 2009).
- *Surgical Appliances and Supplies*: Includes a wide range of products such as orthopedic devices, prosthetic appliances, surgical implants, surgical dressings, crutches, surgical sutures, personal industrial safety devices (except protective eyewear), hospital beds, and operating room tables (NAIC 339113). These products made up 28% of the US medical device market in 2007 (ITA, 2009).
- *Ophthalmic Goods*: Includes prescription eyeglasses, contact lenses, sunglasses, eyeglass frames, reading glasses made to standard powers, and protective eyewear (NAIC 339115). Ophthalmic goods made up approximately 5% of the total market in 2007 (ITA, 2009).
- *Dental Equipment and Supplies*: Includes dental equipment and supplies used by dental laboratories and offices, such as dental chairs, dental instrument delivery systems, dental hand instruments, dental impression material, and dental cements (NAIC 339114). This group accounted for 5% of the market in 2007 (ITA, 2009).
- *Dental Laboratories*: Includes dentures, crowns, bridges, and orthodontic appliances customized for individual application (NAIC 339116). These comprised about 4% of the market in 2007 (ITA, 2009).
- *Surgical and Medical Instruments*: Includes medical, surgical, ophthalmic, and veterinary instruments and apparatus, except for electromedical, electrotherapeutic, and irradiation products (defined later). Examples include syringes, hypodermic needles, anesthesia apparatus, blood transfusion equipment, catheters, surgical clamps, and medical thermometers (NAIC 339112). This group was 26% of the medical device market in 2007 (ITA, 2009).
- *Electromedical and Electrotherapeutic Apparatuses*: Includes MRI equipment, medical ultrasound equipment, pacemakers, hearing aids, electrocardiographs, electromedical endoscopic equipment, and other related products (NAIC 334510). This group comprised roughly 19% of the medical device market in 2007 (ITA, 2009).
- *Irradiation Apparatuses*: Includes X-ray devices, CT systems, fluoroscopy systems, and other diagnostic or therapeutic products using β-rays, γ-rays,

X-rays, or other ionizing radiation (NAIC 334517). These devices made up about 8% of the medical device market in 2007 (ITA, 2009).

Clearly, there are various types of products that are classified by NAICS and ITA as medical devices. However, many medical devices are not necessarily considered *devices* in the engineering sense of the word. Some, including dental cements and acoustic scanning gels, are *materials*. Others may be considered *supplies*, such as surgical masks, surgical drapes, and test kits, or general *equipment*, such as hospital beds and surgical tables. The bulk of the remaining products are used directly for diagnostic, therapeutic, or surgical action on patients and can be classified as *tools*, *instruments*, *devices*, or *systems*. These types of products are mostly confined within the latter three NAICS categories—surgical and medical instruments, electromedical and electrotherapeutic apparatus, and irradiation apparatus. There are notable exceptions—implantable devices, such as breast implants and stents, fall within the surgical appliances and supplies group and surgical instruments used for dental applications, such as dental drills, fall within the dental equipment and supplies group.

The terms *tools*, *instruments*, *devices*, and *systems* are used frequently in this book and can be thought to fit into a technological spectrum of varying complexity (Fig. 1.1). In general, *tools* are the simplest and usually do not have moving parts. The term *instrument* can be interpreted broadly, but often has moving parts and may use electric power. Both tools and instruments are usually controlled by a clinician to perform an action on a patient. In engineering, the term *device* is usually reserved for more complex *active* products that are powered or that are capable of transduction of energy from one form into another. However, in medicine, many *passive* products are also considered as devices, especially those that are implantable. *Systems* are the most complex, are usually powered, and

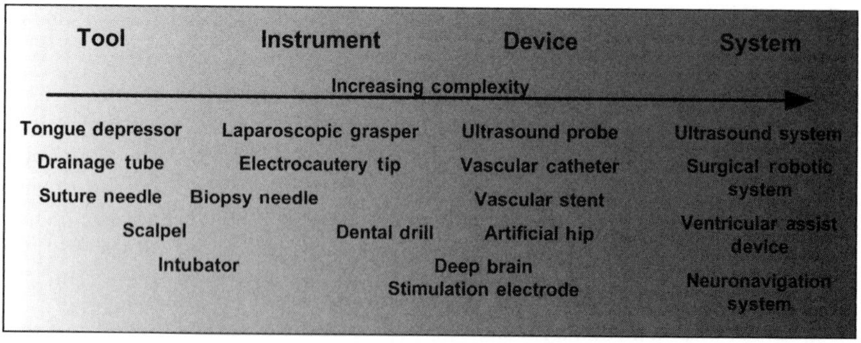

Tool	Instrument	Device	System
Increasing complexity →			
Tongue depressor	Laparoscopic grasper	Ultrasound probe	Ultrasound system
Drainage tube	Electrocautery tip	Vascular catheter	Surgical robotic system
Suture needle	Biopsy needle	Vascular stent	
Scalpel	Dental drill	Artificial hip	Ventricular assist device
Intubator		Deep brain Stimulation electrode	Neuronavigation system

Figure 1.1 Examples of tools, instruments, devices, and systems that fall within the spectrum of medical technologies. Complexity increases from left to right. All of these technologies are often included under the umbrella of the term *medical device*. Note that passive implantable devices, such as vascular stents and artificial hips, are usually considered devices rather than supplies, tools, or instruments, because of their high complexity and usage.

sometimes include one or more tools, instruments, or devices. Many instruments or devices cannot operate or be controlled without a full system in place. However, it should be acknowledged that these four terms are often used interchangeably, and the fact that any of these four terms can be used synonymously with the term *medical device* adds further confusion. A spectrum representing these terms is illustrated in Figure 1.1, along with a few examples that fall within each category, or across multiple categories. Note that the term *apparatus*, which is used by the NAICS, is a general term that also encompasses all four terms but is not used commonly in engineering or in this book.

This book focuses on medical instruments, devices, and systems used for the most advanced techniques practiced at present in medicine and surgery, including minimally invasive, image-guided, and interventional procedures. These terms are also often confused and should be clarified as well. *Minimally invasive procedures* are those in which tools, instruments, or devices are inserted through small incisions to perform procedures with minimal patient trauma. Some examples of minimally invasive instruments or devices are biopsy needles, laparoscopic instruments, microsurgical instruments in ophthalmology, and catheters. *Image-guided procedures* are procedures performed with the assistance of an imaging technique, such as ultrasound, fluoroscopy, CT, or MRI. Real-time image-guided techniques are performed with the imaging device or system capturing images during the procedure, or intraoperatively. Image-guided procedures may also be performed with the images captured before the procedure, or preoperatively. *Interventional procedures* are typically performed with both minimally invasive and image-guidance techniques. A good example of an interventional device is a vascular catheter, as most are inserted through small incisions made with a needle, and under the guidance of ultrasound or fluoroscopy. Another example is a probe used during stereotactic neurosurgery, which is inserted into the brain through a small incision in the skull under the guidance of 3D CT or MRI image data. Devices that are placed using interventional techniques, such as stents that are placed using catheters, are considered interventional devices.

Finally, some clarification is given regarding the study of medical devices. Bioengineering and biomedical engineering are broad fields that focus on the design and development of medical devices and technologies, used for any of a number of medical applications. Both are multidisciplinary, in that they include or overlap with many other fields. In bioengineering/biomedical engineering graduate programs, students can have backgrounds in many disciplines such as mechanical engineering, electrical engineering, chemical engineering, computer science, materials science, physics, physiology, biology, and chemistry. Some have backgrounds in business, law, or medicine.

Bioengineering is often used synonymously with biomedical engineering. However, bioengineering is a broader term that encompasses biomedical engineering and is defined as the investigation of biological topics using engineering principles. Bioengineering includes not only medical devices as defined by the FDA but also fields such as biotechnology, which is related to the development of pharmaceuticals and pharmaceutical-based products. Biomedical engineering, on the other hand, is the application of engineering principles directly to human health and focuses more on medical devices than on bioengineering. The distinction between the two fields is perhaps best illustrated by the development of artificial organs. Artificial organs such as electromechanical artificial hearts, ventricular assist devices, or prosthetic limbs use mechanical- or electrical-based technologies and, therefore, fall more specifically under the realm of biomedical engineering. The growth of artificial organs and tissue using biological materials, referred to as *tissue engineering*, is primarily a bioengineering discipline and has not yet been commercialized or categorized by NAICS. This book deals mostly with the biomedical engineering and traditional medical devices, as defined by the FDA and categorized by NAICS, but is also highly relevant to the greater bioengineering field.

1.3 PURPOSE OF THE BOOK

It is not possible to thoroughly describe the full spectrum of medical devices in a single textbook. We have therefore attempted to focus on the most advanced technologies, particularly those that are applicable to minimally invasive, image-guided, and interventional techniques. The aim of this book is to introduce readers to the instruments, devices, and systems used in various medical disciplines and to provide a brief historical context and glimpse into the future within each field.

CHAPTER 2

Design of Medical Devices

GREGORY NIGHSWONGER

KARL STORZ Endoscopy-America, Inc., El Segundo, CA

2.1 INTRODUCTION

From mankind's earliest attempts at the practice of medicine, in China, Greece, Egypt, India, Europe, and elsewhere, records show evidence of primitive instruments having been developed to assist in some of those efforts. Vestiges of a few of those ancient implements can still be recognized in some modern surgical tools. Materials have become more sophisticated, and design principles have certainly evolved to become more highly formalized and structured. Yet, the fundamental purpose of medical devices has remained the same: to serve as a tool for physicians and surgeons, aiding them in research, diagnosis, and treatment of disease and injury.

The purpose of this chapter is to outline some of the essential steps in taking a device concept through development to manufacture and preparation for market, with emphasis on those steps that are unique to medical technology as opposed to nonmedical products. Many of these steps are taken for virtually all medical devices, but certain processes will differ in scale and complexity. For example, the development of new, improved surgical forceps will entail far fewer members of the development team and less sophisticated processes than will the development of a full magnetic resonance imaging (MRI) system or video endoscope.

2.2 THE MEDICAL DEVICE DESIGN ENVIRONMENT

The design and development of medical devices take place within a rather special environment, very unlike that of most other types of products. This environment encompasses a complex array of government (state and federal in the United States, as well as international agencies) regulation and oversight, applicable standards

Medical Devices: Surgical and Image-Guided Technologies, First Edition.
Edited by Martin Culjat, Rahul Singh, and Hua Lee.
© 2013 John Wiley & Sons, Inc. Published 2013 by John Wiley & Sons, Inc.

(again, US and international) and similar influences that guide and define the design process. Of course, the impact of this environment will be more or less significant, depending on the nature of the device once again and the potential market envisioned.

As the design process begins, consideration must be given to the specific regulatory bodies that will eventually provide oversight of the path to market. Similar consideration must be given to the standards that will be relevant to the intended device.

2.2.1 US Regulation

Government regulation and oversight of medical and surgical devices are considered as necessities because these devices are used to diagnose and treat patients, while also protecting the safety of both patients and caregivers. In the United States, medical devices are regulated by the Center for Devices and Radiological Health (CDRH) of the Food and Drug Administration (FDA). The agency's mandate is to promote and protect the public health by making safe and effective medical devices available in a timely manner. The CDRH is responsible for premarket and postmarket regulation of medical devices. The standard for demonstrating safety and effectiveness is determined in part by the risk associated with the device in question. Devices are classified according to their perceived risk using a three-tiered system (class I, II, or III) (Kaplan et al., 2004).

Class I devices represent the lowest risk and are subject to general controls, which are published standards pertaining to labeling, manufacturing, postmarket surveillance, and reporting and generally do not require formal FDA review. Examples of class I medical devices include elastic bandages, arm slings, tongue depressors, and medical thermometers.

Class II devices are higher risk devices. General controls alone have been found to be insufficient to provide reasonable assurance of safety and effectiveness; however, there is adequate information available to establish special controls. Special controls may include performance standards, design controls, and postmarket surveillance programs. Most class II devices also require FDA clearance of a premarket notification application, either a Premarket Notification 510(k) submission or a PreMarket Approval (PMA) submission before being marketed. Data must be provided in the 510(k) to demonstrate that the new device is "substantially equivalent" to a legally marketed predicate device. A PMA is submitted to the FDA to demonstrate that a new device, or the one that has been modified, is both safe and effective. This represents a higher standard than is required for 510(k) submissions and generally requires inclusion of data on human use from a formal clinical study in addition to laboratory studies. Examples of class II devices include physiologic monitors, powered wheelchairs, infusion pumps, and surgical drapes.

Class III devices include those that are believed to pose the highest potential risk. These devices either are life sustaining/supporting, of substantial importance in preventing impairment of human health, or present a high risk of illness or injury, and so general and special controls alone are inadequate to provide reasonable assurance of safety and effectiveness. Before being legally marketed, most class III devices require agency approval of a PMA, generally requiring clinical data demonstrating reasonable assurance that the device is safe and effective in the target population. Examples of class III medical devices include replacement heart valves, implantable pacemakers, and blood vessel stents.

The FDA/CDRH also provides special pathways to the approval of class III products that may be in development to treat diseases or other conditions among very small populations of patients. For example, a Humanitarian Use Device (HUD) is one that is intended to benefit patients by treating and diagnosing diseases or conditions affecting fewer than 4000 individuals in the United States each year. The Office of Orphan Products Development determines if a device meets requirements to be designated as an HUD. In addition, a Humanitarian Device Exemption (HDE) application is submitted to the FDA. The HDE is similar to a PMA, but an HUD is exempt from the effectiveness requirements of a PMA, so does not require results of clinical investigations demonstrating that the device is effective for its intended purpose. Nevertheless, the HDE must demonstrate that the device is safe, and the probable benefits outweigh the probable risks.

2.2.2 Differences in European Regulation

In many respects, the regulatory process in the United States is similar to those of countries within the European Union (EU); however, there are a few significant differences. First and foremost, the EU system makes ample use of notified bodies (NBs) to implement regulatory control over medical devices. Acting as independent commercial organizations that perform many functions that are similar to FDA's CDRH, NBs are authorized to issue the CE mark, which is necessary for marketing certain medical devices (Kaplan et al., 2004).

In the EU, device marketing has been regulated by a series of medical device directives established initially in the 1990s to describe the essential requirements for obtaining European clearance. More recently, the directives have been modified to provide fundamental requirements for clinical evaluation and postmarket surveillance. The directives define four classifications of medical devices, based on risk and specifying levels of testing and evidence required for approval. Class I includes low risk devices, ranging from stethoscopes to wheelchairs; little evaluation is generally needed before the device is marketed. Class IIa includes low to medium risk devices, such as hearing aids, electrocardiographs, and ultrasonic diagnostic equipment and certain other imaging equipment. Class IIb encompasses medium to high risk devices, including radiology equipment, such as X-ray machines, as well as surgical lasers, nonimplantable infusion pumps, ventilators, and intensive care monitoring equipment. Class III is the highest risk group and includes balloon catheters and prosthetic heart valves.

Manufacturers of new devices in classes IIa, IIb, and III must work with one or more NBs to demonstrate safety and conformity with criteria established in the directives. This is done most often by referring to technical standards of relevant international organizations. The manufacturer must also demonstrate that the device performs effectively for its intended purpose as defined by the manufacturer. The clinical data used to do this may include a critical evaluation of the relevant scientific literature available at the time the device is demonstrated to be equivalent to another device that already complies with relevant essential requirements and for which there are data. Alternatively, the manufacturer may present a critical evaluation of the results of all reported clinical investigations that have addressed residual safety concerns.

As to the approval criteria for high risk devices, there is a significant difference between those of the CDRH and the EU. Marketing approval for class III high risk (and some class II) devices in the United States is predicated on the manufacturer demonstrating the device to be reasonably safe and effective, which typically requires a prospective, randomized controlled clinical trial. To receive marketing approval for a similar device in the EU, the manufacturer must conduct some human clinical investigations, but it is not compulsory for these clinical trials to be randomized (Fraser et al., 2011). This arguably subtle difference has a profound impact on the size and scope of the clinical studies for regulatory approval, generally making the approval process more rapid and less costly to manufacturers, and has shaped the handling of some early device testing related to products intended for eventual marketing in the United States (Kaplan et al., 2004).

2.2.3 Standards

Medical device standards are applicable to a range of device-related issues, including design, testing, labeling, active implantable devices, sterilization, packaging, risk management, safety, software, biological evaluation, quality management, and various other topics. Published standards are the result of collaboration between industry representatives and regulatory agencies. And, while the various standards organizations develop and publish multiple standards related to similar issues, there are continuing attempts to harmonize standards among different nations and organizations. Among the major standards organizations, a few are listed as follows:

- Association for the Advancement of Medical Instrumentation (AAMI): It is a nonprofit organization founded in 1967 and is an alliance of more than 6000 members from around the world. Located in Arlington, Virginia, AAMI is a primary resource for the industry, the professions, and government for national and international standards covering medical instruments and devices.
- ASTM International: Formerly known as the *American Society for Testing and Materials* (ASTM), the society is a globally recognized leader in the development and delivery of international voluntary consensus standards for many commercial and technical fields, including healthcare and medical products. ASTM international standards are the result of contributions by its members, including more than 30,000 technical experts and business professionals representing 135 countries. The ASTM is located in West Conshohocken, Pennsylvania.

- The British Standards Institution (BSI): As the National Standards Body of the United Kingdom, the BSI develops private, national, and international standards in more than 150 countries. With headquarters in London, the organization publishes standards on a range of medical devices, including dental equipment, neurosurgical implants, prosthetics, anesthetic, and respiratory equipment.
- International Electrotechnical Commission (IEC): As one of the world's leading organizations for the preparation and publication of international standards for all electrical, electronics, and related technologies, the IEC uses a number of technical committees and related groups to address issues related to diagnostic imaging systems, electromedical equipment, and other medical electronics products. The IEC is located in Geneva, Switzerland.
- International Organization for Standardization (ISO): With its Central Secretariat in Geneva, Switzerland, the ISO is the world's largest developer and publisher of international standards. The ISO comprises a network of the national standards institutes of 162 countries. Although the majority of its standards are specific to particular products, materials, or processes, the ISO also developed generic management system standards that can be applied to any organization, regardless of product or service. These include the ISO 9001 quality standard, which is the basis of implementation of quality management systems by the medical device industry.

2.3 BASIC DESIGN PHASES

The fundamental steps of medical device design and development include the following:

- Feasibility research
- Planning and organization
- Conceptualizing and Review
- Testing and refining
- Proving
- Pilot testing and release to manufacturing

2.3.1 Feasibility

Among others, the first question to be asked in the initial design process is whether there is a need for the device. To answer this, the following must be considered:

- What will the device do, and how will the device do it?
- Who will use the proposed device? (physicians, surgeons, nurses, consumers, etc.)

- Why is the device needed? Which generally means, how is its need being met now?
- Can it be done?

The answers to these questions are generally found through research, which most often means actually going to the physician's office or visiting the surgeon in the operating room. The focus is on observing how they are currently doing their work and listening to their suggestions of what they actually need. Discussion with them about what they would look for in a new device, as well as what they would not want, is also important. Identifying a new potential market for a device will often benefit from meeting with key opinion leaders, as they are needed and available.

2.3.2 Planning and Organization — Assembling the Design Team

Today's medical devices make use of sophisticated mechanical technologies, as well as advanced electronics and software, new materials and coating processes, advanced machining and fabrication technologies, and much more. So the design and development of new devices most often requires a team of individuals with diverse skill sets, and often groups of teams, with specialists representing each of the technologies involved to provide expertise and knowledge in those specific areas.

Because good communication is required between all members of the team, a production manager is needed to oversee it. Depending on the type of device to be developed, the appropriate project manager may be one with a relatively broad technical background, who can operate across the various pertinent technologies. On the other hand, it may be one with stronger administrative expertise, who make certain that projects are completed on budget and on time.

The team itself will generally include members whose functional areas of expertise will take in various facets of mechanical, electrical, compliance, systems, manufacturing, and quality engineering. Often, tasks in several of these areas will be performed by the same individual or team.

Designers who are experts in the use of computer-aided design (CAD) software have a critical role to play on the design team. Use of CAD has had a profound influence on the development of today's medical devices. High end CAD systems enable designs for highly complex medical devices to be generated in a fraction of the time once required to render finished drawings. Equally important, drawings created with CAD can be changed to reflect refinements in the design concepts or as a result of testing, for example. Use of CAD, especially when combined with rapid prototyping systems, offers important advantages for streamlining the device design and development process.

Similarly, the role of electrical and electronics engineers has a significant bearing on the design process and their work must be in correlation with other design functions. While many companies purchase what are essentially off-the-shelf electronic components and design other parts of the device around these, some companies also develop proprietary electronic components. For example, in developing a new high definition camera head for endoscopic use, the software tools of the engineering team will work closely with the mechanical design team to create a circuit design that will not only fit perfectly into the housing but also allow necessary testing and facilitate assembly. This can be done by creating circuit board models that can be physically assembled with a three-dimensional prototype.

2.3.3 When to Involve Regulatory Affairs

Requirements for regulatory approval can have an obvious impact on the product design process when very complex devices are involved. The need for device approval, in fact, can ultimately force a decision not to go ahead with a given project. If the project requires a PMA, and the market is insufficiently developed, few, if any, companies would proceed. At a university, such a project could conceivably proceed to development of a new product—if only for the exercise. But the private sector would most likely not, because the return on investment may simply not be there.

Because of the nature of these decisions, regulatory affairs should be involved early, at least in making a determination of what will be required as far as regulatory compliance is concerned. If the decision is that the new device will require a standard 510(k) procedure, then it is quite simple. A 510(k) could actually be started before any product is actually finished because it requires only the idea of a product to begin the process. In the case of having to file a PMA application, the regulatory department's real involvement with the PMA comes after the project is completed but before the device is a tangible finished product.

For example, the FDA requires medical manufacturers to create and maintain a design history file (DHF) for each device developed. The DHF is a compilation of records that describe the design history of the finished product and encompasses all design activities related to its development, as well as accessories, major components, labeling, packaging, and production processes. The FDA also requires manufacturers of class II and class III medical devices to implement design controls, which are essentially a development and control plan used to manage the development of a new product. The DHF is where these activities are documented.

2.3.4 Conceptualizing and Review

At this early stage, and depending on the complexity of the proposed device systems involved, the first steps involve successive meetings to deal strictly with rough concepts on paper, just ideas (Fig. 2.1). This is often a period of brainstorming, for not only the total concept but also components or subsystems that may be involved. Several design concepts may be initially developed, often just rough sketches for consideration. Conceptualizing usually takes into account the specific design requirements of the device, as well as specifications, features, characteristics, and nature of the end use environment and applications (Fig. 2.2).

After the selection of the initial concept is made, any parts or components that involve principles that have not been tested previously are identified, and then

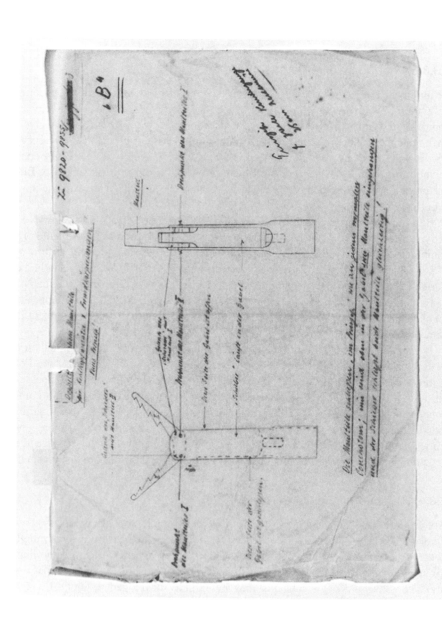

Figure 2.1 Initial concepts of medical device design generally begins with ideas put to paper, such as this early sketch by Karl Storz showing

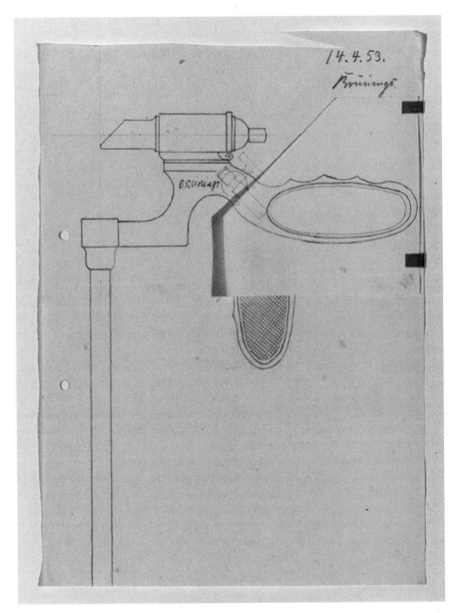

Figure 2.2 Shop drawings have often reflected changes in device concepts, such as the change in handle configuration shown in the figure.

specific narrow functions have to be separated from the greater concept—the big picture—for testing and review. This continues until the product concept takes shape and is ready to flesh out.

Once the high level configuration of this concept or system is more or less visualized, then it has to be broken down into its individual components or technologies and assigned, not for immediate start of work, but for discussion. Discussion would focus on finding out what are the possibilities and the limitations for the finished device, as well as what it can and cannot do. Consideration would also have to be given to whether the concept of the device would adequately address the issue identified during the feasibility assessment phase described above. Costs also have to be kept in mind at all times.

At the same time, it is not too early to have team members looking at materials selection, software requirements, if any, and so on. Findings in these areas can, at times, assist in providing direction even in concept phases, as can early involvement by regulatory and even manufacturing at times, which ensure compliance with quality system requirements and maintain the DHF that will also be used in post-product release phases. Other factors, such as ergonomics and human factors, may also be considered and can ultimately influence functional design characteristics.

2.3.5 Testing and Refinement

The individual teams continue working on their assigned areas of the concept. The pertinent principles and mechanisms are usually tested just on their own in order to prove validity, completely separate from any systems functions. This determines whether it can be incorporated into the whole picture. If not, one needs to start over and consider another configuration, because what was envisioned will either be too costly or will not function as intended.

Once each of these subparts has been proven to work as predicted, then one has to start looking at the entire picture, taking a systems-based approach. So, each individual would be designing more than his or her own part, leading to some form of breadboard testing. The breadboard approach, not to be confused with prototyping, is where a number of mechanisms are physically fitted together as a system and are tested for function toward a desired outcome.

Some parts of this process can be accomplished through computer simulation and testing (Fig. 2.3). That is, some of the relationships defined at the subsystem level can be mechanical simulations. Nevertheless, there are systems that consist of or contain various parts that cannot be simulated but, in the end, must still be tested and verified.

2.3.6 Proving the Concept

The next step, once everything is verified, would be to build a prototype that would actually resemble the final product (Fig. 2.4). Functionally, it would work as anticipated.

The prototype has several virtues. Because the subject is medical devices, actual patients are the intended subjects of the completed device. Unlike breadboards, the prototype can potentially be used with patients. For example, a prototype is likely to be used, under very special conditions of course, in live animal laboratory research or with patients. It depends in part on how close to a finished good that prototype is.

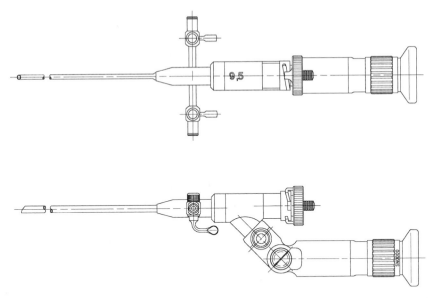

Figure 2.3 Computer simulations, such as this view of a multifunctional device handle, offer an important tool for developing and refining device designs.

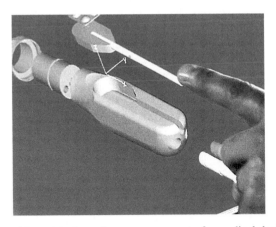

Figure 2.4 Assembly and testing of every component of a medical device are essential.

The prototype is intended to verify that all the separate components work together (Fig. 2.5). That is, each of the subsystems must provide the function it is required to provide and must behave as predicted. More importantly, the prototype must function as intended, not with a simulated input, but with a real input, providing the correct diagnostic or therapeutic effect to the patient being tested. It must meet the needs of both the practitioner and the patient.

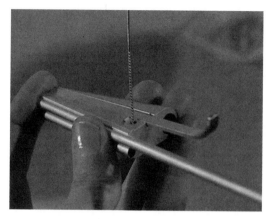

Figure 2.5 Each device component (electrical, mechanical, and others) must be tested, both individually and as part of the functioning unit.

In some cases, of course, a series of prototypes may be necessary. The first iteration is the α-prototype, which is generally created using rapid prototyping methods such as stereolithography or CNC modeling. Such methods are capable of producing functioning prototypes quickly and at a relatively low cost. While the α-prototype is created as a functioning unit that meets all design specifications, it may not perform as predicted. Testing of the α-prototype and assessment of performance will determine where changes and refinements are needed. These will be incorporated into the β-prototype, which would be the next step in device development. In addition, functions that were not included initially may be incorporated into the β-prototype or subsequent iterations for testing and evaluation.

2.3.7 Pilot Testing and Release to Manufacturing

During the developmental process, every individual responsible for a part is expected to consider not only how their components could be prototyped but also manufacturability issues. Because medical devices are not just one-of-a-kind products, manufacturability is a significant aspect of design. Decisions must be made regarding materials to be used, special coatings, machining, packaging, sterilization, and other issues. In most cases, each person who is responsible for a part or subsystem has to think about such issues in terms of manufacturability.

Once the prototype device is evaluated, considered final, and signed off by all the individuals involved, it goes to manufacturing. It is necessary to have the manufacturing team to see if all the assumptions that have been made about manufacturability are correct. If each one of the individuals working on these parts has consulted with manufacturing specialists and verified those assumptions, most issues will have been addressed by the time the prototype is complete. In that case, the role of the manufacturing team is to identify areas that can be simplified or made easier for manufacturing.

The next step in the process is that the manufacturing team works more closely with the people who developed the product to make sure that the device gets manufactured according to the way it was planned. Typically, the manufacturing team will produce new sets of drawings or documents that will provide information to the people who are responsible for finalizing the manufacturing process. This would be followed in most cases by performing a short pilot run. That would be the first step in manufacturing.

During these final stages, of course, procedures would also be followed to ensure regulatory compliance. Activities would also include developing appropriate instructions for use of the device, sterilization procedures, training plans for end users, labeling, and other such issues.

It is important to note that sterilization and processing are critical design considerations for medical devices. Simply put, these procedures are essential to ensuring that devices or instruments do not transmit infectious organisms to surgeons, staff, or patients. Also, the methods used for processing must not adversely affect the devices themselves. Device design, therefore, must consider whether the materials used are compatible with common modes of processing and sterilization and that the design is not so complex as to compromise processing or make it excessively difficult.

To comply with FDA requirements under its quality system regulation, manufacturers must also develop and follow quality systems to help ensure that their products consistently meet a range of applicable requirements and specifications (Figs. 2.6–2.8). The quality systems for FDA-regulated medical devices are known as current good manufacturing practices, or cGMPs. Because the quality system regulation must apply to such a broad range of devices, it does not provide a detailed description of how a specific device must be manufactured. Instead, a framework is offered for all manufacturers to comply with by developing and following procedures and adding any details that are appropriate to a given device according to

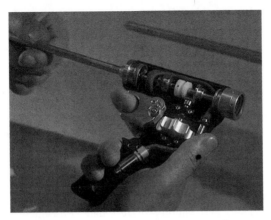

Figure 2.6 Device testing can include evaluation in a test rig to assess device performance when mounted near similar or related devices to ensure functional compatibility.

Figure 2.7   Computer-controlled quality testing of device components aid compliance with FDA quality system regulations.

Figure 2.8   Assessment of new electronic device designs often entails electromagnetic radiation testing to ensure that its operation will not interfere with other devices.

the current state-of-the-art manufacturing for that specific device. Despite the regulation's flexibility, each manufacturer is responsible for establishing requirements for each type of device that will result in devices that are safe and effective (FDA, 2011a).

In a related area, FDA's Good Laboratory Practice (GLP) regulations deal with the organization, process, and conditions under which laboratory studies are planned, performed, monitored, recorded, and reported. These practices are intended to promote the quality and validity of test data. Preclinical studies should be performed under GLP conditions if they are intended to support submission of research or marketing submission to the FDA (FDA, 2011b).

2.4 POSTMARKET ACTIVITIES

Once the device is on the market, feedback from end users may prompt further changes, which may also require the involvement of regulatory affairs. And the manufacturer must comply with appropriate regulations and guidelines for post-market surveillance, which may require a portion of the verification and validation processes to be repeated. All design changes must be accompanied by an updated risk assessment.

2.5 FINAL NOTE

The introduction of new products to the market is an important aspect of any business strategy, and no less so for medical device manufacturers. The goal of the device design and development process is ultimately to launch a safe and effective product to the appropriate market, at the right time and at the right price. While formalizing an efficient strategy for achieving this goal is important, it is equally critical to bear in mind that the same process must also be sufficiently flexible to accommodate numerous variations such as market conditions, new technology developments, and so on.

This aspect of the development process also brings to mind another issue: reliability engineering. Reliability engineering is an often undervalued facet of the design process for medical devices. And yet, it should be considered a fundamental part of product development, beginning at the earliest stages. Device manufacturers may also overlook reliability testing as an essential method for developing reliable products in a shorter period of time and at the lower cost. And yet, trying to add processes to ensure reliability late in the product development cycle often results in costly redesigns and delays. Clearly, a thorough product reliability process should parallel the product development process. (Giuntini, 2000).

PART II

MINIMALLY INVASIVE DEVICES AND TECHNIQUES

CHAPTER 3

Instrumentation for Laparoscopic Surgery

CAMELLIA RACU-KEEFER

Southern California Permanente Medical Group, San Diego, CA

SCOTT UM, MARTIN CULJAT, and ERIK DUTSON

Center for Advanced Surgical and Interventional Technology (CASIT), University of California, Los Angeles, CA

3.1 INTRODUCTION

The early beginnings of laparoscopic surgery date back to the early twentieth century when Han Christian Jacobaeus performed the first laparoscopy and thoracoscopy on a patient in Stockholm, Sweden in 1910. This approach at the time was promoted as a method for evaluating patients with fluid accumulation in the abdomen (Ballantyne, 1995). Minimally invasive surgery has rapidly evolved, particularly over the past 20 years, and now serves as an alternative, and often, as the preferred approach to traditional open surgery (Fig. 3.1). This method employs a camera and surgical instruments that are introduced into the body through small incisions (Fig. 3.2). This historically evolved from the use of endoscopes for visualization of internal organs, as with colonoscopy, which has been performed for many centuries. Antonin Jean Desormeaux, known as the father of endoscopy, developed a device he called the endoscope in 1853, which used mirrors and lenses with a lamp flame as a light source, and was primarily used in urologic cases (Berci, 1976). Also known as endosurgery, the minimally invasive approach led to many fields, including arthroscopy, angioscopy, and laparoscopy.

Medical Devices: Surgical and Image-Guided Technologies, First Edition.
Edited by Martin Culjat, Rahul Singh, and Hua Lee.
© 2013 John Wiley & Sons, Inc. Published 2013 by John Wiley & Sons, Inc.

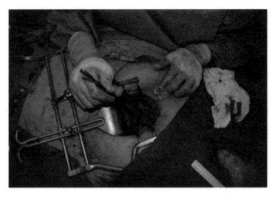

Figure 3.1   In traditional open surgery, a large incision is made in the body for insertion of instruments and visualization of the procedure.

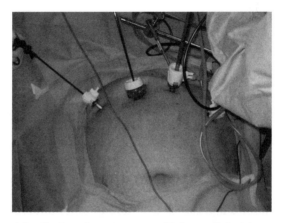

Figure 3.2   In minimally invasive surgery, multiple small incisions are made for the introduction of the camera and instruments.

Laparoscopy refers to surgery in the abdominal cavity using a laparoscope, or camera, and minimally invasive instruments. Laparoscopy was first developed by gynecologists and since then, it has been adopted by general surgeons for many operations. Laparoscopic cholecystectomy is recognized as the hallmark of this evolution for gallstone disease and has replaced the open cholecystectomy since 1987 when Mouret pioneered this technique in Lyon, France.

Minimally invasive surgery has been rapidly embraced and adopted clinically because it provides shorter recovery time, lower risk of infection, reduced postoperative pain and trauma to the patient, and reduction in hospital length of stay (Faraz and Payandeh, 2000). Along with its benefits, this approach carries with it unique risks that are directly related to the operative technique, such as during the initial step of entry into the abdomen. Understanding the indicated uses and limitations is critical for optimal use of laparoscopic instruments and prevention and treatment of potential injuries.

Laparoscopic surgery is challenging. Many procedures require reverse hand movements, haptic sensation from the instrument tips to the hands is limited, and straining body posture is often incurred on the surgeon. Selection of the correct instruments can allow the surgeon to maximize dexterity and mitigate many of these challenges. Ergonomics and the success of the procedure can be further improved through the design of new laparoscopic instruments, particularly those that focus on dexterity enhancement and remote manipulation (Faraz and Payandeh, 2000).

This chapter aims to introduce laparoscopic surgery, describe the design features of standard laparoscopic instruments that facilitate laparoscopic surgery, identify the current limitations of such instruments, describe innovative technology, and discuss future applications.

3.2 BASIC PRINCIPLES

During laparoscopic surgery, multiple small incisions are made in the abdomen. A hollow, cylindrical device, called a *trocar* is inserted into each incision. A laparoscope is inserted into one of the trocars to allow for visualization into the abdominal cavity, and a video display of the abdominal cavity is provided to the surgeons on an external monitor (Figs. 3.3 and 3.4). Long surgical instruments, such as graspers, scissors, or staplers, are inserted directly through the remaining trocars, allowing the surgeon to operate with his or her hands outside of the body and with the instrument tips inside the abdomen. These instruments are used to grasp tissues, cut tissues, clamp tissues, tie sutures, and perform other surgical tasks.

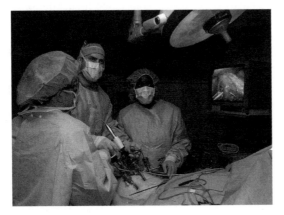

Figure 3.3 In laparoscopic surgery, surgeons control surgical instruments with their hands outside of the body. Retrieved from http://en.wikipedia.org/wiki/File:Laparoscopic_stomach_surgery.jpg.

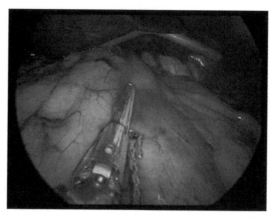

Figure 3.4   Instrument tips are located within the abdomen, and the surgical field is displayed via laparoscope on an external monitor.

Laparoscopic surgery would be challenging, both visually and mechanically, without sufficient working space between the abdominal wall and the internal organs. Space is created by insufflating the abdominal cavity, or filling it with carbon dioxide, throughout the procedure. CO_2 has the advantages that it can be absorbed by tissues in the body and removed by the respiratory system, and that it is nonflammable. The CO_2-filled working space in the abdomen is termed as *pneumoperitoneum*. In order to maintain the pneumoperitoneum, the trocars must also function as air seals.

During surgery, the trocar is positioned at the abdominal wall and serves as a pivot point about which the instrument can rotate (Fig. 3.5). Most laparoscopic instruments have four degrees of freedom (DOF), meaning that there are four translational or rotational movements that can be made to vary the instrument position in space. These standard DOF include translation in and out of the abdominal cavity, rotation of the shaft within the trocar and the abdominal cavity, and tilt about the pivot point in two axes. Some specialized laparoscopic instruments have additional degrees of freedom, which allow for improved access and dexterity within the abdominal cavity.

Many different types of instruments have been adapted from open surgery for use during laparoscopic surgery. These instruments can be used for a variety of surgical tasks, including dissection, cutting, suturing, clipping, and stapling. Dissecting, which is the act of physically separating tissues, is performed using graspers and hooks. Cutting is done using endoscopic scissors, and suturing is done using needle holders. Clipping is the application of a metal clip to blood vessels or ducts for the occlusion of fluid flow and is performed using a clip applier. Specialized laparoscopic staplers have been developed to rapidly staple

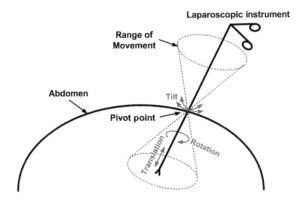

Figure 3.5 During laparoscopic surgery, a surgical instrument pivots about the abdominal wall through a trocar. A typical instrument has four degrees of freedom (rotation, translation, and tilt in two axes) that define its range of movement.

tissues together. Retrieval bags can be used to remove tissues, such as tumors, through the trocars. All of these instruments and tools are described in further detail in the following text.

Powered laparoscopic instruments are also available for more complex tasks such as imaging, cautery, and ablation. As mentioned earlier, laparoscopes are used for visualization of the surgical field. Some laparoscopes are nonpowered telescopic instruments, based on a rod-lens system that is fed into a video camera. However, these have largely been replaced by digital laparoscopes, with charge-coupled devices (CCDs) and fiber optics. Laparoscopic ultrasound probes are commonly used and allow for real-time imaging within organs and tissues. Most have long shafts with articulated tips for additional degrees of freedom for improved access to tissues in the abdominal cavity. Electrosurgery instruments, electrocautery instruments, and ultrasonic shears are electrically powered devices that can be used to rapidly cut or coagulate tissues during laparoscopic surgery. These are all common general surgical instruments that have been adapted for laparoscopy. Devices are also available for ablation of tumors and other tissues during laparoscopy, including radio-frequency (RF) ablation and cryoblation devices. Laparoscopes (Chapter 13), ultrasound probes (Chapter 14), cautery devices (Chapter 7), and ablation devices (Chapter 8) are described in more detail elsewhere in this book. This chapter focuses on nonpowered laparoscopic tools and instruments, including trocars, graspers, scissors, hooks, needle holders, clip appliers, staplers, and specimen retrieval tools.

3.3   LAPAROSCOPIC INSTRUMENTATION

3.3.1   Trocars

To begin a laparoscopic procedure, the surgeon must first be able to gain access to the intraabdominal cavity. A trocar is used to enter the abdomen via an incision and serves as the working channel that allows the surgeon to introduce instruments across the abdominal wall and to work within the abdominal cavity (Fig. 3.6). In general, access techniques can be divided into open and closed techniques. The open technique involves placing the first trocar under direct visualization. In this case, an incision is made larger than the size of the trocar such that the surgeon can see inside the abdominal cavity as the trocar is inserted, and the incision closes around the trocar after the placement. This technique was introduced for abdominal access by Hasson in 1978 and has been described as the safest technique for insertion of the first trocar (Bonjer et al., 1997). Once the initial trocar is placed, a laparoscope is inserted into the abdomen and the subsequent trocars are placed under direct camera visualization. The closed technique involves placing the first trocar blindly and is generally faster and easier to perform than the open technique. However, this technique is contraindicated in patients who have scar tissues inside the abdomen from prior surgeries because of the risk of injury to those intraabdominal organs and blood vessels.

For almost a decade, gynecologists had incorporated laparoscopy in their practice, but general surgeons remained skeptical. After the introduction of the open Hasson technique, general surgeons adopted laparoscopic surgery more readily,

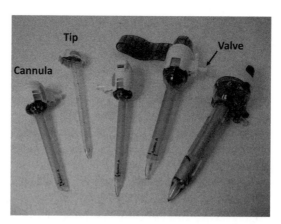

Figure 3.6   Trocars, shown here with varying diameters, are used to gain access into the abdominal cavity during laparoscopic surgery. The two parts at left are disassembled components of an identical trocar at the center. The cannula is the outer component (far left), which makes contact with the abdominal wall, and the trocar tip is the inner component (second from left) that slides into the cannula and is used only during the initial trocar insertion into the abdominal wall. The trocar tip component is removed after insertion, and laparoscopic tools are then inserted through the cannula.

because an open approach reduced concerns of injury to the bowel and vascular structures (Hasson, 1978). No individual access technique can be characterized as the safest or best technique, as each has its own particular set of complications. Laparoscopic complications of 20–40% are related to the initial steps of the operation, specifically during placement of the initial trocar (Hashizume and Sugimachi, 1997). These access complications have traditionally been thought to be rare, with a range of 5/10,000 up to 3/1000 (Champault et al., 1996). However, more recent literature suggests that access injuries are more severe and likely more common than initially identified (Chandler et al., 2001). Increasingly, open techniques are used and recommended by general surgeons for the introduction of the first trocar.

A trocar, whether used with the open or closed technique, has several notable features, including the cannula, valve, and trocar tip (Fig. 3.6). The cannula is the hollow cylindrical portion of the trocar, the outer portion of which is anchored in the abdominal wall, and the inner portion of which is used to introduce laparoscopic instruments into the abdomen. Once the cannula is placed through the abdominal wall, it should remain fixed in the desired position and should have a mechanism in place to prevent it from sliding out with the passage of instruments. Some cannulas have smooth outer surfaces, and others have threaded surfaces or other mechanisms for fixation to the abdominal wall (Fig. 3.7).

Because the gas pressure inside the pneumoperitoneum is higher than atmospheric pressure, the trocar must also serve as an air seal. It, therefore, must be designed such that air does not escape each time an instrument is introduced, during its use, or on withdrawal of the instrument. An internal air seal is achieved to sustain the pneumoperitoneum by incorporating a valve within the trocar (Fig. 3.6). A variety of different valve systems have been designed, including flap valves (oblique or transverse), piston (or trumpet) valves, duckbill valves, or magnetic

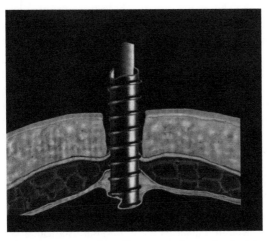

Figure 3.7 Thread pattern on used to fix a cannula at the incision site. *Source*: © 1999–2012 WeBSurg® IRCAD®, All Rights Reserved.

Figure 3.8 Trocars with blunt tip (top) and sharp tip with spring-loaded blade (bottom).

ball valves. The valves are made to retract either manually or automatically with the introduction of an instrument. The outer diameter of the cannula must also maintain an air seal through fixation to the abdominal wall.

Most trocars that are used at present have a separate tip component that fits into the cannula and is used to create the entry point into the abdomen and abdominal wall layers (Fig. 3.6). The trocar tip and cannula are together used to access the abdomen, and the inner trocar tip is removed once the trocar is in place. Some trocars do not have a separate trocar tip, and the tip is simply incorporated onto the end of the cannula.

Tips can be either sharp or blunt (Fig. 3.8). Sharp tips cut through the tissues to create a path through the abdominal wall using a spring-loaded blade; noncutting tips spread and stretch the tissue apart to allow the trocar to traverse the abdominal wall. In obese patients with thick abdominal walls, optical trocars are often used to gain access to the abdominal cavity. These trocars have glass tips that allow a laparoscope to be inserted into the trocar to visualize each layer of the abdomen during the penetration of the abdominal wall. The choice of trocar tip depends on the surgeon's training and comfort level (Leroy et al., 2005).

Trocars are made with different sizes, shapes, and materials. Trocars are designed with different outer diameters, varying from 3 mm to 30 mm, and different lengths. The diameter is determined based on the type and size of instrument to be used, and length is selected based on the individual patient's abdominal wall thickness. The material is typically plastic or metal, each with various properties that can impact clinical outcomes. Plastic trocars are designed such that they minimize light reflection from the laparoscope and also have the advantage that they are insulators. However, during electrosurgery, in which electrical current flows across the body to cut or coagulate tissues near the tip of an electrosurgical instrument, metal, plastic, and hybrid metal–plastic trocars are all used. All three can result in tissue damage because of capacitive coupling, but use of conductive metal trocars can minimize potential tissue injury by allowing energy to dissipate over greater surface areas, thus decreasing the localized current density.

A variation of the traditional trocar design is the balloon trocar. Balloon trocars provide a favorable design solution when there is difficulty in stabilizing the trocar-cannula unit within the abdominal wall. Balloon trocars have a blunt tip with an inflatable balloon attached to the distal end of the cannula and a thick foam more proximal to the balloon. The balloon is inflated once the trocar tip has been inserted through the abdominal wall, providing a sealing mechanism on the inside of the anterior abdominal wall so that gas does not escape the abdomen from the incision site, and the cannula does not move from its desired position. The external foam acts to create a tight seal to prevent escape of pneumoperitoneum.

3.3.2 Standard Laparoscopic Instruments

Laparoscopic instruments are used to perform the bulk of the tasks during laparoscopic surgery. Laparoscopic instruments all have long, cylindrical shafts that allow a surgeon to operate deep within the abdominal cavity through a trocar. Instrument lengths vary from 18 to 45 cm, with most ranging from 34 to 37 cm long. Shorter instruments (18 to 25 cm) are adapted for cervical and pediatric surgery, and for specific procedures in adults, such as gallbladder and antireflux procedures. Longer 45 cm instruments are best suited to obese or tall patients. With the increase in the obese population, the use of long instruments is becoming very common (Mutter et al., 2005).

Instruments vary in diameter from 1.8 to 12 mm, but most instruments are designed to pass through 5 or 10 mm trocars. Each instrument has varying rigidity, with longer and thinner instruments generally more flexible and the length a function of the force it can deliver at the tip of the instrument. Smaller diameter instruments allow for smaller incisions but are more fragile. Most instruments, such as graspers, scissors, and hooks, and articulated or angulated instruments, are available in the 5 mm diameter size. The 12 mm diameter size is usually reserved for large instruments such as laparoscopes, specimen retrieval bags, and staplers.

Most laparoscopic instruments such as graspers and scissors have basic opening and closing functions. Many instruments can also be rotated 360° using a built-in knob near the handle, which can be turned with a finger. This may facilitate tissue retraction but can be a disadvantage if the goal is to rotate the tissue within the grasper.

A certain number of instruments offer articulation at their tip, in addition to the usual four degrees of freedom (Fig. 3.9). Articulation is useful during cutting or stapling of tissues at a certain angle. When trocars are placed in a suboptimal position, introducing a straight instrument and achieving a certain angle to the tissue can be difficult. Articulation of the instrument helps to improve the alignment of the instrument and the working tissues.

3.3.2.1 Graspers Graspers are used for holding or dissecting tissues. The variety of graspers is distinguished by the design of the tip. For instance, graspers can be pointed, flat, curved, or fenestrated (Fig. 3.10). Fenestrated graspers, or those

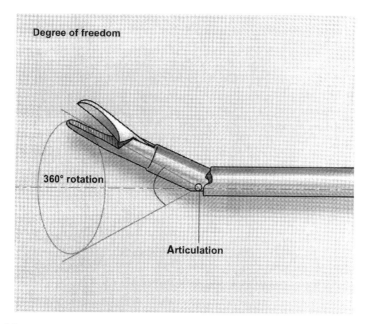

Figure 3.9   Laparoscopic instrument with articulation. *Source*: © 1999–2012 WeBSurg® IRCAD®, All Rights Reserved.

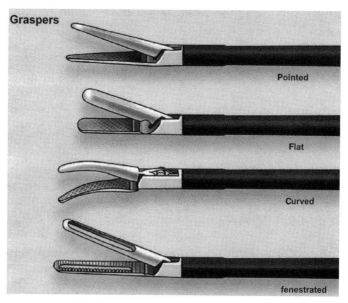

Figure 3.10   Graspers, with tip designs developed for specific surgical tasks. *Source*: © 1999–2012 WeBSurg® IRCAD®, All Rights Reserved.

with "windows" in their centers, are particularly useful, as they provide traction of tissues while providing atraumatic handling of tissue such as the bowel. Fine-tipped dissecting graspers allow for blunt dissection, by dissecting layer by layer in densely adhered tissues and clearing surrounding structures for better visualization. Similarly, curved-tip forceps provide a gentle curve to the dissecting tool that facilitates dissecting around tubular structures, such as blood vessels and bile ducts. Caution must be taken to avoid puncturing these critical structures with the sharp grasper tips.

3.3.2.2 *Endoscopic Scissors* Endoscopic scissors are mainly used to cut avascular tissues during the lysis of adhesions, or taking down of scar tissues. They are also used to cut blood vessels and ducts after they have been tied or clipped. Endoscopic scissors have a variety of different shapes and sizes (Fig. 3.11). For fine precision, microscissors are available, which have a 1–2 mm tip length and are best suited for making a small opening in a vital structure, such as when working with the cystic duct to introduce a cholangiogram catheter (Mutter et al., 2005). Endoscopic scissors may be straight, curved, short blade, medium blade, or long blade, each tailored for a particular function.

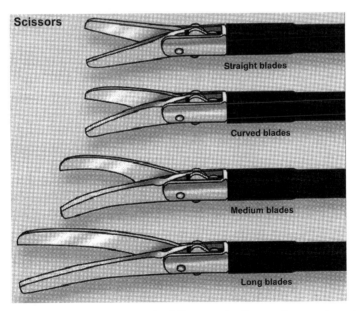

Figure 3.11 Endoscopic scissors, with different embodiments for a variety of cutting tasks. Some endoscopic scissors is also used for monopolar electrosurgery. Electrical current is passed from an electrical grounding pad through the body to the tips of the scissors to cut or coagulate tissues, using diathermic heating rather than mechanical cutting. Endoscopic electrosurgery scissors are electrically insulated. *Source*: © 1999–2012 WeBSurg® IRCAD®, All Rights Reserved.

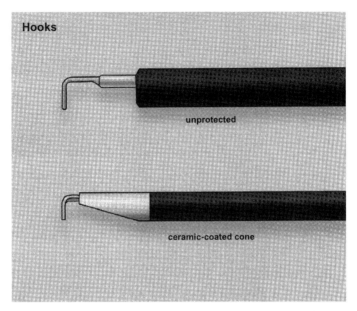

Figure 3.12 Electrosurgery hooks, often used dissection tools during laparoscopic surgery. *Source*: © 1999–2012 WeBSurg® IRCAD®, All Rights Reserved.

3.3.2.3 ***Hooks*** Hooks are primarily used for monopolar electrosurgery and are also insulated along their entire length (Fig. 3.12). Certain manufacturers make hooks with a ceramic cone to protect the distal end. The cone keeps the plastic from being damaged and also keeps current from diffusing at regions adjacent to the tip of the hook. Although these are considered powered devices, they are commonly used as nonpowered devices for dissection of tissues during laparoscopic surgery, as the hooks are convenient for separating tissues.

3.3.2.4 ***Needle Holders*** Needle holders are particularly important, as these instruments facilitate intracorporeal hand suturing (Fig. 3.13). As suturing is a precise technique, the design of these instruments must reflect their task and they must be resilient to allow for a firm grip despite a distance of about 35 cm from the handles to the jaws. Thus, needle holders are typically designed in nondisposable form. Needle control is a function of the length of the instrument; a longer instrument results in transmission of reduced force from the hands to the jaws of the instrument. The goal is to develop instruments with powerful jaws that translate into strong grip (Scott-Conner, 2006).

Most needle holders have a jaw with a relatively flat grasping surface, making it possible to turn the needle in all directions, as in open surgery. Some needle holders have jaws with a dome-shaped notch at the tip, which visually orientates the needle in a perpendicular direction and makes it easier to grasp the needle because of its curvature (Mutter et al., 2005).

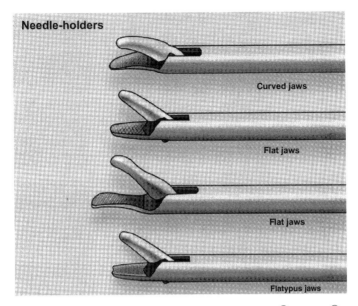

Needle-holders

Curved jaws

Flat jaws

Flat jaws

Flatypus jaws

Figure 3.13   Needle holders. *Source*: © 1999–2012 WeBSurg® IRCAD®, All Rights Reserved.

3.3.2.5 Handles   Laparoscopic instruments have a wide range of handles, with some directed 90° in relation to the working axis and others parallel to the working axis (Fig. 3.14). Others have handles that are positioned between these extremes. Needle holders often have handles that are designed along the axis of the instrument, which makes the handle easy to manipulate as it forms into a cylinder. Ringed handles are made of different shapes and sizes to fit surgeons' fingers. Some have room for one finger and others for several fingers. Some rings are designed with a prong, where the surgeon can rest additional fingers (Mutter et al., 2005).

3.3.2.6 Locking Mechanism   A ratchet or locking mechanism is commonly incorporated into a laparoscopic instrument to allow locking of the jaw. The locking mechanism is usually built into the handle of a laparoscopic instrument, allowing the surgeon to lock, tighten, or release the jaws as needed. Graspers with a locking mechanism are helpful for grasping a structure, such as the gallbladder, and locking the hold in place for the duration of the operation without the tissue slipping from the grasper. Once the jaws of the grasper are opposed against the tissue, the ratchet apparatus prevents slippage, reducing surgeon or assistant fatigue during prolonged retraction. The goal is to utilize the locking grasper to maintain a firm hold of the desired tissue without piercing or tearing the tissue. When smooth, gentle grasping of structures is desired, such as during handling of the bowel, nonlocking graspers are selected to avoid causing trauma to the structures.

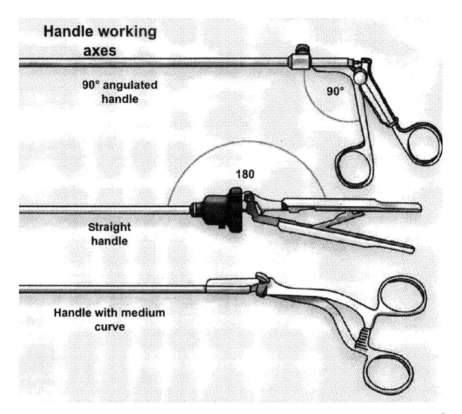

Figure 3.14 Handles of laparoscopic instruments. *Source*: © 1999–2012 WeBSurg® IRCAD®, All Rights Reserved.

3.3.3 Additional Laparoscopic Instruments

3.3.3.1 Clip Appliers Clips are commonly used during surgery to occlude small blood vessels or luminal structures such as lymphatics or the cystic duct. They are U-shaped metal pieces, often titanium, with a size varying between 5 and 10 mm. If the size of the structure exceeds 8 mm, it is likely that a clip would not sufficiency occlude the entire width of that structure, incurring serious implications such as bleeding or a bile leak (Mutter et al., 2005).

Clip appliers differ from the standard laparoscopic instrument described earlier, in that they dispense disposable clips through the shaft of the instrument (Fig. 3.15). As the handle of a clip applier is closed, the ends of the clip come together around a tissue structure, and the clip is released. Clip appliers are available as disposable or reusable units.

3.3.3.2 Staplers Staplers are used to divide hollow organs such as small intestine, stomach, or colon. They are also used to create an anastomosis, which is a connection between hollow or tubular structures such as two loops of intestine. In

Figure 3.15   Clip applier, with metal clip help between the two tips of the instrument (inset).

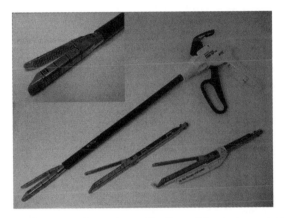

Figure 3.16   Linear laparoscopic stapler, with disposable stapler cartridges of differing lengths shown at bottom and expanded view of the stapler at top right (inset).

a stapled anastomosis, nonabsorbable and nonmagnetic titanium staples are applied by the stapler. The most common type of stapler is the linear GIA (gastrointestinal anastomosis) stapler. It fires three rows of staples on each side with a cutting line in the middle (linear transection). Different staple heights can be selected based on the tissue thickness. They are also designed with or without articulation. The articulation provides additional leverage for maneuvering in difficult areas of the abdomen, such as during pelvic surgery or when the trocar position may be limiting the ideal stapler position (Fig. 3.16).

   While the development of staplers reflects a revolution in surgical technology and operative technique, the adoption of these devices into clinical practice has been a process in evolution. During their early introduction, staplers were thought to be inferior to hand-sewn methods of anastomosis and tissue reconstruction. After many years of research and clinical practice, the safety of mechanical stapling devices was shown to be comparable to standard surgical techniques. The stapling technique is now accepted as being easier and faster compared to the hand-sewn technique, especially in the laparoscopic setting. However, there are some exceptions. Contraindications to stapling include tissue that is too thick to allow properly formed staples such as in obstructed or edematous intestine, and tissue that is too thin, as in dilated bowel, where the staples do not firmly approximate the tissue and increase the risk of bleeding and anastomotic leak (Deitch, 1997).

3.3.4 Specimen Retrieval Bags

In an open surgery, specimens, such as malignant tissues, are easy to take out because of the large size of the incision. Specimen retrieval is somewhat limited during laparoscopic surgery because of the small size of the incisions. The technique and tools used for specimen retrieval are critical in that they must accomplish multiple key objectives. First, it is important to avoid enlarging the incision, as this can lead to incisional hernia in the future. Next, the device and extraction process must provide complete protection of the incision site to avoid contamination with bacteria or cancer cells (Leroy and Marescaux, 2009). Additional important features of the retrieval tool include its strength, size, deployment, and retrieval. A retrieval bag is usually used, and is typically made of transparent polyurethane or nylon material with a polyurethane coating. Nylon bags tend to be more resistant and are the preferred material for large specimens, particularly if they require fragmentation before retrieval (Scott-Conner, 2006).

A specimen retrieval bag is usually folded and stored in a plastic sheath introducer, which typically fits into a 10 mm diameter trocar. The bag is deployed inside the abdomen by advancing a plunger that is attached to the bag through the sheath, with the plunger also unraveling the bag. The bag has a flexible metal ring around its mouth allowing the bag to be held open. The specimen is placed within the wide-mouthed bag using any kind of grasper, and the bag is closed by pulling tightly on a string around the mouth of the bag. The ring is then pulled upward through the sheath to close the bag, and the drawstring is cut. The introducer sheath, bag with the specimen, and the drawstring are all pulled up into the trocar and extracted through the incision by pulling it with hands from outside.

3.3.5 Disposable Instruments

When the minimally invasive revolution began in 1987, most of the trocars were made to be reusable and were made of many disassembled parts that were screwed together. The benefit of reusable trocars is the reduced cost, but this saving is often outweighed by the requirements of cleaning and sterilization. In addition, the labor-intensive disassembly and reassembly required for the cleaning process often results in damage to the mechanical components of the trocar. Surgeons have, over time, preferred disposable equipment because of the cost and effort associated with the maintenance of reusable tools, as well as their diminished efficacy over time. Disposable trocars are available in many more sizes and diameters than reusable trocars, providing additional versatility to accommodate instruments such as circular staplers. Overall though, disposable trocars are more costly than reusable trocars because of their single-use design.

Other frequently disposable tools include gas connection tubing, scissors (the blade blunts rapidly and reuse is suboptimal), clip appliers, and staplers, where many clips and staples need to be applied, and instruments that are rarely used or only for speciality circumstances. Examples of reusable equipment are frequently used high quality instruments, usually in a laparoscopic instrument set, that typically are costly in disposable form. Cautery hooks, graspers, and needle holders are examples of reusable instruments that can be cleaned and sterilized easily in an autoclave (Mutter et al., 2005).

From a global perspective, the decision to use disposable or reusable instruments may depend not on individual surgeon choice but rather on local health policies as determined by limited human and sterilization resources. For example, disposable tools are rarely used in developing nations because their cost is much greater than local labor costs of maintaining reusable tools. In countries in Europe or in the United States, where sterilization methods are intense and labor is costly, disposable equipment is a more favorable option. In most laparoscopic cases, surgeons commonly use a combination of disposables and reusable instruments, depending on optimal function.

3.4 INNOVATIVE APPLICATIONS

Many innovations have been developed, which are transforming the way surgeries are performed. These include not only the development of more advanced laparoscopic tools but also robotics and even a transformation in the design and function of these laparoscopic tools.

Multiple minimally invasive surgical robotic systems have been developed, including the Zeus (Computer Motion) and more recently the da Vinci (Intuitive Surgical), which allow a surgeon to perform laparoscopic surgery with the assistance of a surgical system. The surgeon sits at a "master" control console that allows the surgeon to manipulate robotic arms and view the abdominal cavity. A robotic "slave" apparatus features multiple arms that hold an endoscope and detachable laparoscopic surgical tools. The da Vinci system is now widely accepted, with units actively used at almost 1500 hospitals, and most frequently for urological, gynecological, cardiac, and abdominal procedures (Intuitive Surgical, Inc, 2011). Surgical robotics is described further in Chapter 5.

Additional tools have been evolved for performing true single incision surgery. For example, the SILS (single incision laparoscopic surgery) single flexible port, introduced by Covidien in April 2009, is a multiple instrument access port that is introduced under direct vision through a 2 cm incision at the umbilicus to minimize postoperative pain and to hide the scar. The SILS port has three operating channels, meaning that three 5 mm ports are introduced side by side, allowing for the use of multiple instruments all through the same single incision. A 5 mm laparoscope with 30° angulation is used as the optical system. The remaining two ports are used for straight laparoscopic instruments such as a grasper and hook, if removing the gallbladder. The challenge of this configuration is that the instruments and the laparoscope are within a small distance from each other and are all parallel to each other, making triangulation of tissue exceedingly difficult at times.

To respect the principles and safety of open and laparoscopic surgeries in solving this technical issue in the SILS, specialized articulating forceps have been designed. The articulating feature turns a rigid instrument into one whose axis moves outside of the visual field, allowing for better retraction and freeing the space necessary for the neighboring instrument to perform the necessary task safely (Mutter and Marescaux, 2009). With the turn of a knob, these articulating instruments project an elongated neck (or insert) with an angle that is away from the main shaft axis. In addition, these instruments require a stiff shaft design to compensate for the stress generated by organ manipulation during an operation. These tools represent revolutionary solutions that will allow further advances in the conduct of surgery.

Likewise, maturation of current devices and instrument technology may also lead to the implementation of natural orifice transluminal endoscopic surgery (NOTES) in the clinical arena with an adequate margin of safety and efficacy. This realm has brought a great deal of initial excitement and promise, but to date, the instrumentation is not advanced enough to meet the challenges of the envisioned surgical technique. NOTES is performed by introducing instruments through natural orifices such as mouth, vagina, and rectum and eliminates incisions on the abdomen. Intraabdominal organs such as the gallbladder can be removed through the mouth or vagina. This means that the instruments not only have to be long in order to reach the target organ but flexible endoscopes must also enter the natural orifice and guide the procedure. Many NOTES instruments that have been prototyped thus far, such as the Bard EndoCinch that suctions mucosa and drives a needle through the tissue, the USGI G-Prox that uses baskets to deploy sutures through tissue, the Olympus Eagle Claw that deploys a long curved needle from the end of a flexible endoscope, and the Ethicon EES that has a round needle that it drives through tissue and creates a knot.

While promising, these instruments are generally large and bulky and are difficult to deliver to the operative site of interest. However, development of these prototypes has led to significant technological advances, such as the ability to tie a suture through a flexible endoscope, that may become a springboard for future endolumenal technologies. To further this vision, future designs must provide for long, flexible instruments that perform complex tasks in simple ways, improve endoscopic suturing capabilities, and enable stapling through flexible staplers.

3.5   SUMMARY AND FUTURE APPLICATIONS

Over the past three decades, the field of surgery has seen a major boom with respect to laparoscopy and the development of laparoscopic instruments that enable innovative solutions to surgical problems. While initially introduced by gynecologists, laparoscopy has not only revolutionized the fields of urology and gynecology but also truly changed general surgery.

Minimally invasive surgery gives many benefits to the patient, including less pain, faster recovery, and an improved working environment for the surgeon. At present, there is no doubt that minimally invasive surgery is not only an integral component of general surgery but also the future of surgery.

An increasing number of instruments are being designed for very specific applications. Instruments have become increasingly complex, with greater functionality and freedom of movement. Such instruments mirror the evolution of surgery and technology toward increasing automation of procedures. Automation and incorporation of robotics is expected not only to improve surgical outcomes, but to enable surgeons to operate on patients from remote settings.

CHAPTER 4

Surgical Instruments in Ophthalmology

ALLEN Y. HU, ROBERT M. BEARDSLEY, and JEAN-PIERRE HUBSCHMAN
Jules Stein Eye Institute, University of California, Los Angeles, CA

4.1 INTRODUCTION

Given the small dimensions of the human eye, eye surgery is predicated on the ability to visualize and to precisely manipulate delicate tissues within a confined space. Not only is a thorough understanding of anatomy required for successful surgical outcomes but also knowledge of basic principles of physics and biomechanics.

The anatomy of the eye is broadly defined to include the eyelids, lacrimal gland, canalicular system, conjunctiva, Tenon's fascia, extraocular muscles, and globe, all enclosed by the bony orbit. Within the globe (Fig. 4.1), the anterior segment consists of, from anterior to posterior, the cornea, the anterior chamber, the iris and its aperture the pupil, and the intraocular lens. The cornea and lens are of particular importance to ophthalmologists, as a majority of ophthalmic surgical interventions are performed on these structures, such as corneal transplant, refractive surgery, and cataract extraction. Posteriorly, within the eye and behind the lens are the posterior chamber and the vitreous body. The wall of the eye consists of three layers: the tough fibrous sclera on the outside wall, the vascular uveal tract in the middle, and the delicate neurosensory layer known as the *retina* lining the inner surface of the globe adjacent to the vitreous body.

The origins of ophthalmic surgery date back hundreds of years, as ancient surgeons used needles to attempt to dislodge a cloudy intraocular lenses into the posterior chamber in a technique called *couching*. Patients were strapped down and a long probe was jabbed repeatedly through the cornea into the space behind

Medical Devices: Surgical and Image-Guided Technologies, First Edition.
Edited by Martin Culjat, Rahul Singh, and Hua Lee.
© 2013 John Wiley & Sons, Inc. Published 2013 by John Wiley & Sons, Inc.

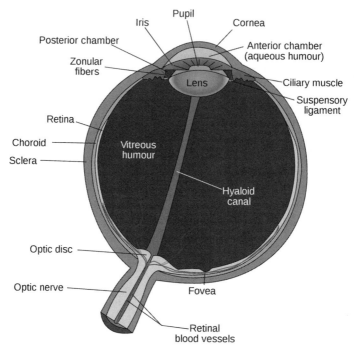

Figure 4.1 The internal eye anatomy. Retrieved from http://en.wikipedia.org/wiki/File: Schematic_diagram_of_the_human_eye_en.svg.

the lens to push the tissue into the vitreous. Tremendous advances is surgery have been made since that time aided principally by the utilization of sterile techniques, anesthesia, microscopy, and synthetic materials. Ophthalmic surgery is roughly divided based on anatomic location into various subspecialties. Oculoplastic surgeons operate on orbital components, including tumors, orbital fractures, and the optic nerve, as well as the exterior adnexal structures, including eyelid restructuring and repair and lacrimal gland tumors. Strabismus surgeons concentrate on the extraocular muscles to correct deviations of alignment leading to diplopia or double vision. Anterior segment surgeons work within the eye on the cornea, iris, and lens, while vitreoretinal surgeons dedicate their practice to posterior segment surgery, including vitrectomy and retinal detachment repair.

The surgical microscope (Fig. 4.2) revolutionized ophthalmic surgery. Initially, little more than binocular telescopes worn by the surgeon in the 1870s, the newest optical systems provide the surgeon and assistant with a well-lit binocular wide-field stereoscopic view that is used while the surgeons are seated. The light intensity is adjustable, and filters protect the patient and surgeon from phototoxic effects. Most microscopes offer a range of magnification from approximately 10–30×, which can be adjusted with foot pedal controls. Moreover, foot pedal control allows the surgeon to focus from the posterior capsule to the cornea or from the mid-vitreous

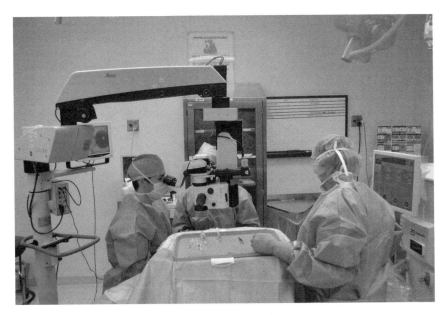

Figure 4.2   The surgical microscope.

to the retina. Motors on the microscope also provide motion control of the optical system in the x–y axis, allowing for the surgeon to compensate for minor patient movements that can change the operating axis (Zabriskie, 2006). This tool has enabled precise visualization inside the eye, which has made advanced ophthalmic surgeries possible. Two such procedures, cataract surgery and vitreoretinal surgery, are described in detail.

4.2   CATARACT SURGERY

4.2.1   Basic Technique

Over 2 million cataract extractions are performed each year in the United States, making the procedure the most common ophthalmic surgery performed, and among the most common surgeries performed in any field. Cataract is a clouding of the eye's naturally clear lens, a structure that is only about 1 cm wide and 6 mm deep. The lens focuses light rays on the retina to produce a sharp image of what we see. When the lens becomes cloudy, light rays cannot pass through easily, and vision is blurred. Cataract development is a normal process of aging, but cataracts also develop from eye injuries, disease states, or medications. Although many methods are used to remove cataracts, the most common technique is phacoemulsification, in which the opacified lens is broken up, or emulsified, into smaller pieces with the assistance of ultrasound energy and removed from the eye through a 3 mm

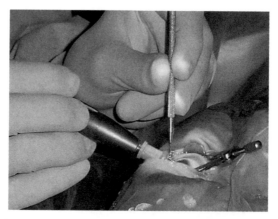

Figure 4.3   Modern cataract surgery, with (left) phacoemulsification probe used to emulsify the lens and (center) a steel chopper inserted to assist the procedure. Retrieved from http://en.wikipedia.org/wiki/File:Cataract_surgery.jpg.

surgical incision through the cornea (Fig. 4.3). This process is performed using a phacoemulsification machine, which includes infusion, aspiration, and ultrasound energy delivery components, and is described in the following text.

Phacoemulsification surgery starts with two small wounds made within the cornea adjacent to its attachment to the white sclera, one for the phacoemulsification handpiece and lens insertion port (Fig. 4.4a), and the second for support and eye manipulation. A steel chopper is often inserted into the second port to assist in breaking the lens nucleus down into smaller pieces. Removal of the thin epithelial layer surrounding the lens is done through a process called a *continuous curvilinear capsulorhexis*, whereby the surgeon gently tears a circular 6 mm diameter opening within the 15 μm thick anterior capsule. With the lens now accessible, it can then be broken up by phacoemulsification (Fig. 4.4b), and the pieces aspirated out with vacuum. The denser nuclear pieces are removed with ultrasound energy, and the less-dense lens cortex is removed using irrigation and aspiration alone. An intraocular lens implant is then inserted with anther instrument through the first port and placed into the remaining lens capsule (Fig. 4.4c).

4.2.2   Principles of Phacoemulsification

The fluid dynamics of phacoemulsification requires constant fluid circulation through an irrigation sleeve to maintain the pressure in the eye, as material is removed by aspiration. Because of the small volume of the anterior and posterior chambers of the eye, the fluid circulation during phacoemulsification surgery is important to ensure efficient removal of the cataract while preventing complications due to tissue collapse. If outflow, or aspiration through the handpiece, is allowed to exceed inflow, or irrigation, even for just a fraction of a second, this can cause chamber instability, collapse of the eye, and unintended damage to the intraocular tissue (Devgan, 2009).

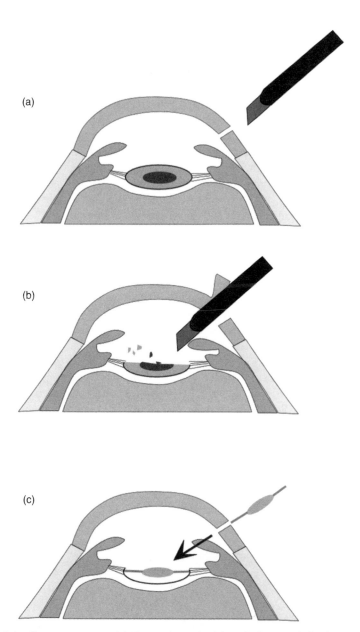

Figure 4.4 Cataract surgery. (a) One or more incisions is first made in the eye and the anterior capsule of the eye is removed, (b) the lens is then broken up through phacoemulsification and aspirated out, and (c) the lens is replaced with an intraocular lens implant. Retrieved from http://en.wikipedia.org/wiki/File:PEA_and_IOL.svg.

The basic formula that governs fluid flow during phacoemulsification surgery is Poiseuille's law: $Q = \Delta P \pi r^4 / 8 \eta L$, where Q is flow rate, ΔP is the pressure gradient, r is the radius of instrument tip, η is the fluid viscosity, and L is the length of the tubing. Given that the viscosity of the fluid and length of the tube are essentially constants, flow is essentially proportional to the change in pressure and radius of the tubing.

In the case of phacoemulsification surgery, ΔP is the pressure gradient between the infusion pressure fluid and the vacuum generated by the phacoemulsification machine. The source of fluid inflow is a bottle of balanced salt solution hanging on the machine, and the height of the bottle is adjusted relative to the patient's eye: the higher the bottle, the higher the infusion pressure.

Fluid outflow consists of the fluid that is removed via the phacoemulsification probe as a result of the vacuum level generated by a fluid pump.

4.2.3 Phacoemulsification Instruments

The primary tool of phacoemulsification is the handpiece (Fig. 4.5) that is placed with its tip inserted into the eye through the corneal wound. This tool delivers the ultrasonic energy to break up the opacified lens, aspirates fluid while creating a vacuum for holding material to the aspiration port, and provides inflow of fluid to the anterior chamber. The handpiece is connected to a machine that provides the aspiration pressure, provides the irrigation fluid, and controls the level and intensity of the ultrasonic energy delivered to the handpiece. The design of handpiece probe tips varies with regard to angles and size of the lumen depending on the application desired. Steeper tip bevels (less angled) are better for cutting the more dense lens nuclear material, whereas a tip with a flatter bevel has an oval-shaped port with a larger surface area, which allows for greater surface area to generate greater adherence, or holding force, of lens nuclear material (American Academy of Ophthalmology, 2006).

Current phacoemulsification machines have in common foot pedal controls with a least three positions. The first position allows entry of fluid into the handpiece

Figure 4.5 Phacoemulsification handpiece (top) and the irrigation/aspiration handpiece (bottom). The phacoemulsification handpiece has three lines: irrigation infusion, aspiration/vacuum, and phacoemulsification, while the irrigation/aspiration handpiece lacks the ultrasound line and is used for the less dense lens cortex removal.

through the irrigation port that provides inflow (controlled by adjusting the height of the infusion bottle), cools the handpiece, and keeps the anterior chamber formed. The second position engages the aspiration mode at a constant or variable rate, depending on the settings selected by the surgeon (Seibel, 2005). In the third position, ultrasound energy is delivered. Many other microsurgical instruments are utilized in anterior segment surgery, from microforceps to scissors to cannulae, each designed to be operated through the small peripheral incisions.

4.2.4 Phacoemulsification Systems

While there are a variety of phacoemulsification machine manufacturers and configurations, they all generate the required vacuum and aspiration based on one of the two pump types: Venturi and peristaltic.

A peristaltic system utilizes a series of rollers to displace fluid to produce flow in flexible, compliant tubing that is wound tightly around a rotating wheel. As the wheel rotates, a segment of fluid trapped between two rollers is moved, creating a vacuum behind, which is relieved by more fluid coming up to the tubing. Therefore, the flow rate is directly proportional to the speed of the rotary mechanism. Significant vacuum is generated only when the tip is occluded. The surgeon sets a desired flow rate and vacuum limit. The flow rate determines the rise time, a measure of how rapidly the vacuum builds up when the flow is restricted.

A Venturi pump uses compressed gas (nitrogen or air) to create inverse pressure. The Venturi effect is the reduction in fluid pressure that results when a fluid flows through a constricted section of pipe. As there is an intrinsic vacuum preload at the pump level, the vacuum rise time is almost instantaneous on pressing the foot pedal. Vacuum and flow rates are proportionally linked and cannot be independently adjusted, as they can on a peristaltic pump. However, in contrast to the peristaltic pump, the flow rate and vacuum cannot be set independently with a Venturi system.

When the lens material occludes the tip, the vacuum builds until the material is evacuated. The vacuum (measured in mm Hg) determines how well a particular material that has occluded the handpiece tip will be held to it. The surgeon then adds ultrasonic power at a variable or fixed level. Ultrasonic power is generated by applying a time-varying electric field across a piezoelectric crystal in the phacoemulsification headpiece. The oscillating crystal deforms in response to the electric field, resulting in a conversion of electrical to mechanical energy. This mechanical energy is produced by oscillation of the tip at a frequency preset for each machine, which varies from 27 to 60 kHz (American Academy of Ophthalmology, 2006). The amplitude of the movement, or stroke length (ranges from 50 to 100 μm), is varied when the power is changed. As the tip moves forward, compression of gas atoms in solution occurs; as the tip moves backward, expansion of gas atoms occurs and bubbles of gas are formed. When the bubbles implode, they release heat and shock waves that contribute at the tip to activity, which disassembles the nucleus. Cavitation can be enhanced by changes in the needle shape. For example, the distal bend in an angled handpiece tip adds a nonaxial vibration to the primary oscillation. This nonaxial vibration augments the axial aspiration and produces at the cutting tip an elliptical motion that increases the mechanical breakdown of the nuclear material.

Phacoemulsification power can also be modulated by varying the duty cycle, which is the period when the phacoemulsification power is being delivered. In an attempt to deliver ultrasonic power more efficiently, modes such as pulse and burst were developed. In pulsed mode, the number of pulses per second can be set and the level of power with each pulse dependent on the foot pedal excursion. By contrast, in the burst mode, phacoemulsification involves the delivery of a present power in single bursts whose frequency increases as the foot pedal is depressed further (Devgan, 2009).

4.2.5 Future Directions

In the near future, there may be a new application of an existing technology that may completely alter the way the cataract surgery is performed. The femtosecond laser is a device that emits coherent optical pulses with duration on the order of 10^{-15}, or femtosecond, has been used extensively in ophthalmology because of its ability to alter delicate tissue in a precise, predictable way. Tissues can be cut very precisely at an exact depth with practically no heat development. Clinical trials utilizing femtosecond laser to perform the incisional steps of cataract surgery, including cornea wound creation, capsulorrhexis creation, and disassembly of the lens, are currently ongoing (Nagy et al., 2009). It may be that in the future while the lens will still have to be removed by aspiration manually, the phacoemulsification step and its inherent risks will be replaced by a simple laser procedure to reduce the hard cataract to a soft, easily removed pulp.

4.3 VITREORETINAL SURGERY

4.3.1 Basic Techniques

Vitreoretinal surgery is a minimally invasive technique, similar to laparoscopic surgery, whereby rather than large incisions to access the inner tissues, small entry ports are created, and a variety of instruments are introduced to remove or replace intraocular structures. Surgical manipulators, lights, lasers, and infusing and aspirating devices are all used in both types of surgery. However, rather than by a relative freedom to place ports in any location, as one can generally do within the abdomen, the eye is much more confined. The only reliably safe anatomic entry point to the posterior segment is located at a narrow band around the eye 3–4 mm posterior from the corneoscleral junction known as the *pars plana*, the space in the eye wall posterior to the ciliary body but anterior to the retina. Placing entry ports too far in either direction can lead to significant complications and surgical failure.

The vitreous body, a clear gel that occupies the posterior compartment of the eye, located between the crystalline lens and the retina occupies about 80% of the volume of the human eye. It is mainly composed of water (99%), collagen fibrils, and hyaluronic acid. The vitreous gel is integral in the pathogenesis of many posterior segment diseases. For example, in diabetic retinopathy, bleeding into the vitreous cavity can severely reduce visual acuity. Only with the removal of vitreous gel and blood, vision can be restored. However, given the gellike nature of vitreous gel and its strong attachments to the retina, vitrectomy (removal of the vitreous gel) has to be performed in a precise, controlled manner, as excessive traction from the adherent vitreous gel exerted on the retina may result in complications such as retinal tears and detachment. Interestingly enough, removal of the vitreous gel has no untoward effects on the eye and its overall health. The gel is replaced initially by a saline solution and eventually by the body as a plasma exudate and never reforms once removed. The eye does not collapse once the vitreous is removed, as the structural integrity is maintained by the fibrous sclera, not from pressure within the globe itself.

4.3.2   Principles of Vitrectomy

Surgery within the posterior segment of the eye is bound by the same principles as anterior segment surgery. The volume of the eye must be maintained as the material is removed. However, adding complexity to posterior segment surgery is the lack of ambient light, the increased density of the material (the vitreous gel) through which the instruments move, and a limited view of the entire retina. The first vitrectomy was performed in 1971 using a 17-gauge (1.42 mm diameter) one-port system with a combined infusion cannula and vitreous cutter (Machemer et al., 1972). Over the past 30 years, numerous innovations have improved outcomes. Current pars plana vitrectomy, while maintaining the principles developed earlier, such as entry at the pars plana and a closed system technique, utilizes three 20-gauge (0.91 mm of diameter) ports, placed 3–4 mm posterior to the corneoscleral junction. One port allows the intraocular infusion of the balanced saline solution to maintain a level of intraocular pressure as desired by the surgeon. The remaining ports are used to introduce various instruments such as a light pipe, vitreous cutter, intraocular forceps and scissors, or laser probe into the vitreous cavity to illuminate the posterior segment and manipulate tissues (Fig. 4.6).

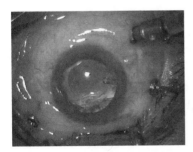 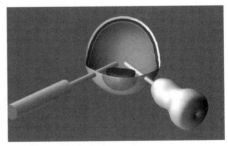

Figure 4.6   (a) Pars plana vitrectomy, retrieved from http://en.wikipedia.org/wiki/File: Vitrectomy-23g.jpg and (b) placement of two intraocular instruments into the eye, with light pipe (left), vitreous cutter (right), and lens (center) retrieved from http://en.wikipedia. org/wiki/File:Vitrectomy.jpg.

Vitrectomy is performed using the operating microscope in conjunction with either a direct contact lens or an indirect noncontact lens. Both lenses are used to magnify the surgeon's view into the eye; the direct contact lens can be held by the assistant or sewn onto the eye at the corneoscleral junction; the indirect noncontact lens is fixed above the eye. Direct contact visualization systems allow for greater magnification and enhanced three-dimensional perception at the expense of a smaller field of view. On the other hand, indirect noncontact viewing systems provide a wider viewing angle and better visualization through media opacities, small pupils, and gas-filled eyes.

4.3.3 Vitrectomy Instruments

The basic tools of vitrectomy are an illuminating device, a vitreous cutter, an endolaser probe, various small graspers, peelers, forceps for delicate intraocular work, and an infusion cannula to maintain the ocular pressure. A light source is required during vitrectomy to visualize the tissues being removed. Most vitrectomy machines are equipped with a built-in light source employing either yellow or white light from a halogen or metal-halide light source. The light pipe, a fiber-optic cable encased in a plastic handpiece, is connected to the light source and can function as a separate instrument or can be attached to the infusion cannula. A recent innovation has been the introduction of a xenon light source, which can provide bright illumination through a narrow light pipe for small-gauge surgery, while eliminating light wavelengths below 400 nm that are known to be *phototoxic* (Williams, 2008).

The workhorse of vitrectomy is the vitreous cutter, whose basic function is to remove vitreous gel (Fig. 4.7). Most modern high speed (600–5000 cuts/min) vitrectomy cutters are pneumatically driven with a side-cutting guillotine port near the tip (Fig. 4.8). Two separate tubes connect the handpiece to the cutter on the vitrectomy machine and to the vacuum to allow for aspiration during surgery.

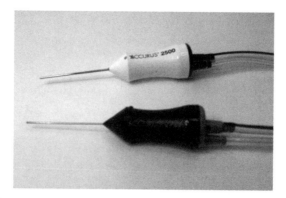

Figure 4.7 Vitrectomy handpieces. A guillotine at the tip of each needle slides open to engage vitreous gel entering the port. On closure of the port, the vitreous strand held by the needle is cut and removed.

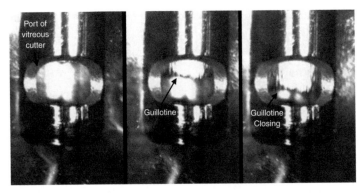

Figure 4.8 Needle tip of a vitreous cutter in a vitreous gel, with needle facing downward. The guillotine is visible inside the port using a high speed camera, shown closing/cutting from left to right.

Regardless of the design, the vitreous cutter is required to have a cutting rate sufficient to prevent traction, and a suction rate that allows removal of the vitreous gel while minimizing traction on the underlying surfaces.

Vitrectomy instruments are also used during macular surgery. The macula is the center of the retina and is responsible for a majority of vision that is used in reading, driving, and fine, detailed work, as it holds the highest concentration of photoreceptors within the retina. It is also affected by a variety of disease states such as macular holes, epiretinal membranes, and age-related macular degeneration. During macular surgery, epiretinal tissue removal from the retinal surface is essential. Various sizes and types of microforceps and scissors (horizontal or vertical cutting) are commercially available for this purpose. The conventional method to remove membrane-type epiretinal tissue is to use a barbed-top disposable needle or a microvitreoretinal blade with a bent tip to create a new or engage an existing membrane edge, lift it up using the angled tip, and grasp it with membrane forceps to strip the tissue from the retinal surface with gentle and smooth traction force. If the membrane is too fragile or thin, a diamond-dusted membrane elevator is used to gently touch the retinal surface in order to give firm purchase on the tissue. In some cases, scissors can be used to dissect and separate epiretinal tissue from underlying retina, segment and isolate blocs of tissue, or cut areas where epiretinal tissue is adhered to the underlying retina.

The endolaser is a relatively new tool that has significantly expanded the applications for vitreous surgery by allowing the minimally invasive ablation of tissues. Similar to the light probe, the endolaser probe is also a fiber-optic cable encased in a plastic handpiece, which is passed into the eye through the vitrectomy ports, to treat diseased retina millimeters from the pathology. The intensity of the laser burn depends on the duration and power settings of the laser, as well as the working distance of the endolaser tip from the retina. The endolaser is used in the setting of retinal vascular diseases, such as diabetic retinopathy, to ablate and induce tissue atrophy in ischemic areas of the retina. In the setting of a retinal tear, the endolaser can be used to create a fibrous scar between the retina and the underlying layers around the tear to prevent fluid from propagating underneath the retina and causing a retinal detachment.

4.3.4   Vitrectomy Systems

Vitrectomy machines, similar to phacoemulsification machines, are highly complex devices that provide irrigation at an adjustable level to control intraocular pressure, aspiration for removal of material, a light source, and energy for the vitreous cutter. In addition, the functions must be operated by a foot pedal, similar to the phacoemulsification machine controls.

Two major types of drive systems are used for vitrectomy handpieces: guillotine electric and guillotine pneumatic. The first type is an electric drive with a sinusoidal transmission, which translates the rotary motion of an electric motor shaft to the linear guillotine motion of the inner cutter tip. The profile of the motion remains constant (although faster) as the cut rate is increased. Thus, the duty cycle (ratio of time the cutter port is open or closed) remains constant as the cut rate is varied.

The second type of handpiece, the pneumatically driven cutter, uses air pulses provided by the surgical system to extend the cutter tip. One type of pneumatic cutter is a guillotine-type cutter where the tip is attached to a diaphragm. The surgical system provides an air pulse, extends the diaphragm, and the cutter tip closes, completing a cut. When the air is released, a spring returns the tip to the open position. This closing and opening have a characteristic constant time. As the cut rate is increased, the duration of open time per cut decreases, while the closed time remains constant (decreasing the duty cycle). The second type of pneumatic drive mechanism utilizes a dual drive line system. The surgical system supplies intermittent air pulses that alternate between each line. The first pulse serves to close the port while the second pulse opens the port. Although the dual driveline system can improve the duty cycle compared to a traditional pneumatic cutter utilizing a spring, the duty cycle still decreases at a high cut rate with this system.

The pressure within the eye is regulated by simple gravitational flow of fluid within a confined, essentially nonexpandable space. By increasing or decreasing the height above the eye of a bottle of saline solution, the surgeon can control the pressure within the eye to meet the surgical needs, whether a tamponade for bleeding, facilitation of subretinal fluid removal, or assisting in vitreous gel removal.

4.3.5   Future Directions

A recent advance in vitreous gel surgery has been the development of less invasive, 23- and 25-gauge transconjunctival vitrectomy systems (Charles, 2006; Williams, 2008). A standard 20-gauge (0.9 mm diameter) vitrectomy system requires incisions in the sclera to obtain access to the vitreous gel, as the instruments are relatively large, and these wounds must be closed with sutures. However, 23-gauge (0.6 mm diameter) or 25-gauge (0.5 mm diameter) entry wounds are thought to be self-sealing and do not require suture closure. The trocar/cannula assembly consists of microcannulae preloaded on the 23- or 25-gauge needle trocars. The microcannulae

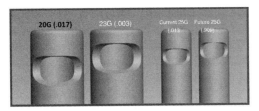

Figure 4.9   Vitrectomy port size comparison. *Source*: Courtesy of Alcon.

serve to maintain alignment between the entry holes in the conjunctiva and sclera, as well as to provide unobstructed instrument access, which is especially important because of the small size of the 23- or 25-gauge wounds. The microcannula consists of two components: a polyimide cannula and a polymer cannula hub. The polyimide cannula is mounted within the polymer hub or collar. The cannula forms the entry port and utilizes a distal bevel to facilitate insertion through the tissue. The polyimide cannula material provides both strength and flexibility, allowing the cannula wall to be thin and to avoid collapsing or buckling. The hub maintains the position of the cannula and prevents it from sliding too far into the vitreous chamber and also allows surgeons to grasp the cannula.

Many surgeons believe the potential advantages of smaller-gauge sutureless techniques include shortened operative time and faster patient recovery. However, the reduction in port diameter translates into reduced flow rates compared to 20-gauge cutters (Fig. 4.9). Slower flow rates translate into more time required to remove the same amount of vitreous. Other concerns are about increased risk of postoperative hypotony and infection, as the sclerotomies are allowed to seal on their own and not sutured closed. A third drawback for the use of smaller-gauge instruments is that the increased flexibility of these tools limits their usage, especially during procedures with high stress at the sclerotomy sites or when grasping of taut epiretinal tissue with smaller-gauge forceps (Williams, 2008).

4.4   OTHER OPHTHALMIC SURGICAL PROCEDURES

This chapter focuses primarily on the tools and techniques used in phacoemulsification and vitreoretinal surgery. However, there are many other common surgical procedures in ophthalmology, which benefit from recent technological advances, including corneal transplantation, laser refractive surgery, and trabeculectomy. Each is performed by subspecialists in anterior segment surgery, and also by multitudes of generalist ophthalmologists.

Corneal transplantation consists of removing the native, diseased cornea and replacing it with a healthy donor cornea from a recently deceased individual. Most often, the host cornea is removed secondary to scarring from infection or chronic inflammation, trauma, or damage from congenital or acquired conditions. The surgeon uses either a circular blade called a *trephine* or a femtosecond laser to create an 8 mm diameter corneal button and removes the host cornea, opening the anterior chamber in the process. A donor cornea is cut to be slightly larger and placed within the empty space and sutured into place using extremely fine stitches. Unlike solid organ transplants, a cornea rarely requires long-term immunosuppression and is rarely rejected. In addition, some surgeons will remove a cataract, perform iris surgery, or even vitreoretinal surgery, while the cornea is temporarily off, in a procedure called *open sky*. Ongoing studies are currently looking at synthetic corneas for transplantation and other novel techniques for replacing only the diseased layer of the cornea, rather than the entire tissue.

Laser refractive surgery is an increasingly common set of procedures that include LASIK (laser *in situ* keratomileusis) and PRK (photorefractive keratectomy). Essentially, a femtosecond laser is used to reshape the cornea by photodestruction of minute pieces of tissue to produce either a steeper curvature to correct nearsightedness or a flatter curvature to correct farsightedness. There is excellent recovery after the procedure and the procedures are at very low risk. Increasingly common are procedures that forgo any corneal restructuring and simply place a corrective lens within the eye's anterior chamber, thus precluding the need for either a laser or glasses.

Finally, trabeculectomy is a common procedure performed for elevated eye pressure with optic nerve damage, or glaucoma. The goal of the surgery is to reduce the intraocular pressure and to maintain it at a low level by assisting in the outflow of the fluid within the eye. This is performed by removing a piece of tissue where the cornea meets the iris and shunting this fluid track to create a bubble of intraocular fluid underneath the conjunctiva. Interestingly, the flap that controls the rate of outflow is held under tension by numerous sutures, which are visible through the conjunctiva. If the surgeon feels that the outflow needs to be increased, he or she can shoot a small laser at the sutures to break one or more to release tension on the flap and allow more fluid to exit the eye and further reduce the pressure.

4.5 CONCLUSION

Ophthalmic surgeons have applied advances in medical device technology to develop minimally invasive devices and instrumentation for the treatment of eye diseases. These technologies are based off of principles from a number of fields, including pneumatics, optics, lasers, ultrasonics, and biomaterials. Given the aging population and the current and future burden of eye disease, it is expected that a great deal of resources will be applied to advance the surgical techniques, devices, and materials to as of yet undiscovered heights.

CHAPTER 5

Surgical Robotics

JACOB ROSEN

Department of Computer Engineering, University of California, Santa Cruz, CA

5.1 INTRODUCTION

The recent introduction of surgical robotics into the operating room offers a significant breakthrough in the way surgery is conducted. It combines technological and clinical breakthroughs in developing new robotic systems and surgical techniques to improve the quality and outcome of surgery. These breakthroughs are based on more than a decade of innovation in the field of robotics in both academia and industry.

The scope of this chapter covers the fundamental concepts and approaches utilized in surgical robotics. It is acknowledged that the majority of the commercially available surgical robotic systems are based on scientific foundations and innovations that emerged out of the research community, and many references and pointers are provided to previous publications of these academic efforts. In this chapter, however, the detailed discussions are limited to FDA- and CE-approved surgical robotic systems. The reader may refer to additional recently published reviews (Elhawary et al., 2008; Fichtinger et al., 2008; Hager et al., 2008; Karas and Chiocca, 2007; Kazanzides et al., 2008; Gomes, 2011). Recent attempts to define the state of the art were compiled in a book entitled: Surgical Robotics—Systems Applications and Visions (Rosen et al., 2011).

5.2 BACKGROUND AND LEADING CONCEPTS

Surgery may be performed primarily through three main modalities (Fig. 5.1): (i) a surgical procedure in which the surgeon interacts with the tissue directly

Medical Devices: Surgical and Image-Guided Technologies, First Edition.
Edited by Martin Culjat, Rahul Singh, and Hua Lee.
© 2013 John Wiley & Sons, Inc. Published 2013 by John Wiley & Sons, Inc.

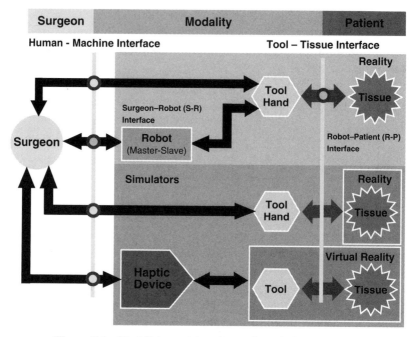

Figure 5.1   Modalities and interfaces of surgical procedures.

with his/her fingers and through manipulation of surgical tools using an open or minimally invasive approach, (ii) a surgical procedure in which the interaction with the tissue is mediated by a surgical robotic system in conjunction with cameras and imaging modalities that provide visual information prior (preoperative) and during (intraoperative) the operation, and (iii) a surgical procedure conducted in a simulation environment in which the operational medium can be either real or simulated tissue. These three modalities have a common human–machine interface in which information is shared between the surgeon and the operation modality. This information-rich layer may be analyzed in order to monitor the surgical process at high levels and to assess the surgeon's operational skills.

The promise of surgical robotics is to deliver high levels of dexterity and vision to anatomical structures that cannot be approached by the surgeon's fingers and viewed directly through the surgeon's eyes, while simultaneously minimizing the impact and trauma to the tissue surrounding the surgical site. Making this technology available to surgeons has led to the development of new surgical techniques, which would otherwise be impossible. The surgical robot and the various imaging modalities act as mediators between the surgeon's hands and eyes and the surgical site, respectively; however, these two elements are part of a larger information system that will continue to evolve and affect every aspect of surgery and healthcare, in general. It is likely that the clinical knowledge accumulated through the use of these new systems, and an understanding of their potential capabilities, will lead to the development of new and more capable surgical robotic systems in the future.

5.2.1 Human–Machine Interfaces: System Approach

Two human–machine interfaces are established with the introduction of a surgical robotic system: the surgeon–robot interface (S–R) and the patient–robot interface (R–P). Each has a unique set of requirements that dictates its design capabilities and functions. These two interfaces may be used to classify the various surgical robotic systems as depicted in Figure 5.2.

5.2.1.1 Surgeon–Robot (S–R) Interface: System Architecture The S–R interface is defined by a wide spectrum of control levels provided to the surgeon over the surgical robotic system (Fig. 5.1). Assuming that a certain level of control is required to complete a task, this control level can be distributed between the human operator and the robotic system at different ratios. The distribution of the control level between the surgeon and the robotic system defines the level of automation allocated for the task. Figure 5.3 depicts how the level of automation affects time delay, the need for imaging modality, accuracy, and the approach to hard and soft tissues.

The level of automation is bounded by two extreme scenarios. The right-hand side in Figures 5.2 and 5.3 (horizontal axis) describes a scenario in which the surgical robotic system is fully autonomous. In this mode of operation, the surgical robot executes a predefined trajectory, maintaining full control over the execution of a plan that was predefined by the surgeon. This process requires a careful registration that fixes the organ in a specific position and orientation in space while registering it with respect to the base of the surgical robotic system. The surgeon initiates the execution of the process and monitors its progress. Other than to terminate the procedure in case of emergency, the surgeon will not be able to change the preoperative planning during its execution. This surgical approach and control level are suitable for hard tissues such as bone that can be scanned by various imaging modalities, positioned, oriented, and registered in space with respect to the surgical robotic system or to soft tissues such as the brain, which is mechanically constrained by the skull. This mode of operation is commonly used in industry assembly lines, in which robots are incorporated to perform preplanned tasks. It is therefore natural that Robodoc, one of the first robotic systems introduced to surgery, followed the same approach. As such, Robodoc was used to mill the femur bone in preparation for a stem implant in a total hip arthroplasty (i.e., total hip joint replacement) (Spence, 1996; Bargar et al., 1998; Nishihara et al., 2004; Schulz et al., 2007).

The left-hand side of Figures 5.2 and 5.3 (horizontal axis) describes a scenario in which any movement of the surgical robotic system is in direct response to a

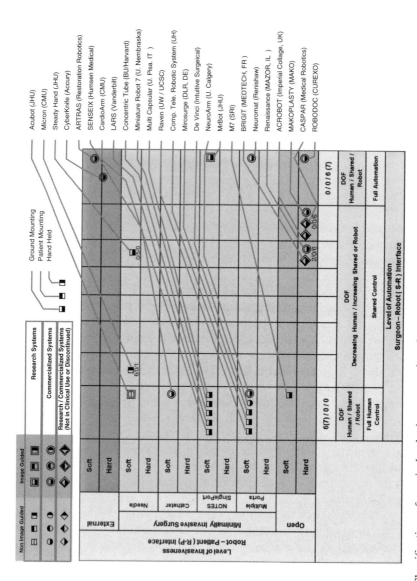

Figure 5.2 Classification of surgical robotic systems based on a surgeon–robot (S–R) interface (horizontal axis) defining the level of automation and a robot–patient (R–P) interface dictating the level of invasiveness.

real-time position command input provided by the surgeon. The system architecture used to enable this approach is *teleoperation*, utilizing a master/slave configuration. The master is defined as the surgical console and the slave serves as the surgical robot itself, interacting with the patient's tissue through the surgical tools. Teleoperation in a master/slave configuration is another technology that emerged in the 1950s as a safe method for handling radioactive materials. This approach was introduced to surgical robotics via research systems such as SRI's M7 (Green et al., 1995), MIT's Black Falcon (Madhani et al., 1998), and commercial systems, including Zeus®, by Computer Motion and Intuitive Surgical's da Vinci® (DiMaio et al., 2011). These systems were primarily designed to operate on soft tissues. Owing to its mechanical properties, soft tissue, unlike hard tissue, changes its geometry during surgical procedures (Section 2.3). As a result, preoperative scanning and preplanning along with autonomous operation is not the preferred mode of operation. Instead, the teleoperation architecture brings the surgeon back into the surgical scene to control and execute every motion of the surgical robot.

The level of automation incorporated into a robotic surgical procedure varies widely and is defined by how the surgical task is shared between the robotic system and the surgeon. At one end of the spectrum, the surgical procedure may be broken down into subtasks, and selected subtasks can be automated. At the other end of the spectrum, an effort is made to develop a control strategy in which both the surgeon and the robot hold a set of surgical tools simultaneously and collaborate during the surgical procedure.

5.2.1.2 Robot–Patient (R–P) Interface: Surgical Approach and Levels of Invasiveness   The robot–patient (R–P) interface determines the level of invasiveness (vertical axis in Figs. 5.2 and 5.3). The level of invasiveness spectrum spans across a range of surgical approaches, including (i) the invasive open-procedure approach, which requires a large incision to expose the targeted anatomy, (ii) variations of minimally invasive surgical approaches with a gradual reduction of invasiveness, such as multiple tools inserted through ports, NOTES (natural orifice transluminal endoscopic surgery, defined later), catheters and needles, and (iii) a noninvasive approach in which energy (radiation) is provided by an external source to a localized space to provide a localized therapy. All of these surgical approaches may, to some extent, be applied to both soft and hard tissues. As the level of invasiveness decreases, the level of manipulation also decreases and, as a result, the surgeon has fewer degrees of freedom to mechanically manipulate the tissues. Figure 5.3 shows how other factors are affected by a reduction in the level of invasiveness from the patient's perspective: factors such as potential infection, scar tissue, and recovery time, and from the surgeon's perspective: factors related to manipulability, vision, pre- and intraoperative planning, and tissue damage to the surrounding tissues.

An open procedure, in which a large incision is made to fully expose the anatomical structure(s), is still the common practice for many surgical procedures. A minimally invasive surgery (MIS), also known as *minimal access surgery* (MAS), aims to minimize the impact on tissues surrounding the surgical site. In this approach,

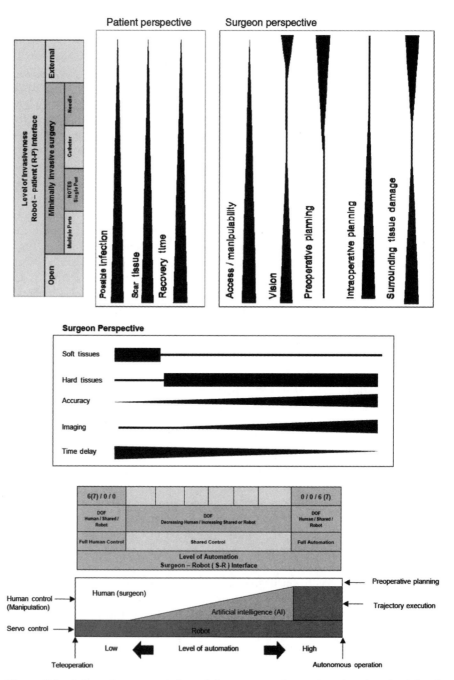

Figure 5.3 Schematic representation of the surgeon-robot-patient domains: (top) level of invasiveness in the robot–patient (R–P) interface, (bottom) level of automation in the surgeon–robot (S–R) interface, and their impact on various operational parameters from the patient and surgeon perspectives.

holes of 5 mm or smaller are made in the skin and three or more ports—including one camera and two endoscopic tools—are inserted. Usually, additional holes are created to accommodate additional tools used for tissue retraction and to provide alternate approaches to the surgical site. The cavity under the skin (typically in the abdomen) is inflated with CO_2 in order to create a space for the surgeon to interact with the tissue under the skin. If the incision diameter is reduced to 2 mm, the skin will tend to seal by itself without stitches and the scar tissue will be kept to a minimum. During this type of procedure, an endoscopic camera provides a view of the internal anatomy, which is projected onto a screen. It was the MIS approach that enabled the important and relatively widespread introduction of robotic systems into the operating room.

The MIS, as it is practiced at present, represents an important first step in the effort to deliver effective tools to the surgical site, while minimizing trauma to the surrounding tissue. There are three evolving approaches that further minimize the impact on surrounding tissues: NOTES, needles, and catheters. Radiosurgery is a noninvasive approach.

NOTES: NOTES is a relatively new approach to surgery, still in an experimental phase, without fully developed surgical tools and robotic systems. As part of this surgical approach, an assembly of tools and one or more endoscopic camera(s) are incorporated into a flexible snakelike tool that is inserted into the body through natural orifices (e.g., the mouth, urethra, anus, eye socket, or nose) and provides access to internal organs while avoiding an external incision and potentially scar tissue.

Needles: Needles are commonly inserted into the body to either inject medicine, to deposit radioactive seeds, or collect a biopsy. Steering a long and narrow needle through a nonhomogeneous tissue is a challenging task, previously explored via an image-guided robotic device (Simone and Okamura, 2002). On the basis of the principles of soft tissue biomechanics and structural beam theory, it is possible to steer the tip of a needle by manipulating its base (Cowan et al., 2011). Spinning the needle may provide another mode of stabilizing the needle, thus improving its steering capability (Engh et al., 2010).

Catheters: Catheters are typically introduced into the body on a guide wire through the vascular system. Catheters are capable of carrying a variety of end effectors, such as balloons, for mechanically widening narrowed or obstructed blood vessels. They may also carry stents that are deposited as part of an angioplasty procedure. As another example, catheters with steerable tips are used to treat atrial fibrillation (AF), the most common cardiac arrhythmia. Tissue ablation is conducted by the settable catheter-controlled teleoperation system so that the surgeon is not exposed to the X-ray radiation associated with fluoroscopy used for imaging (Saliba et al., 2006; Saliba et al., 2008; Reddy et al., 2007).

Radiosurgery: Radiosurgery is a medical procedure that allows noninvasive treatment of tumors. As part of this surgical technique, ionizing radiation is used to ablate the tumor via radiation generated by an external source. The CyberKnife, reviewed in this chapter, is a commercial system that utilizes this approach (Kilby et al., 2010).

5.2.2 Tissue Biomechanics

Tissues are the target medium of surgery, and their biomechanical properties play an important role in both the preoperative planning and the execution of the surgical procedure itself. Tissues may be classified into two categories: (i) hard tissue, which is primarily bone, and (ii) soft tissue, such as tendons, muscles, nerves, and blood vessels, which accounts for all of the remaining tissues in the human body. During surgery, hard tissue does not experience large deformations, unlike soft tissue, which undergoes large deformations in response to internal and external loads. Soft tissues are nonhomogeneous, nonisotropic, nonlinear, and viscoelastic materials—properties that make them difficult to model and that make their response to loads or displacements are difficult to be predictable. Furthermore, soft tissues are attached to each other in ways that generate internal stresses, which are again difficult to assess or predict. Once the tissue is cut, or the connective tissues are dissected, the internal stresses are removed and the soft tissue may change its geometry significantly.

There are several aspects of experimental tissue biomechanics methodologies that are unique to surgery and to surgical robotic applications, in particular. First, during surgery, tissues are exposed to loads resulting from tool–tissue interactions. These interactions generate loads that are significantly different from normal physiological loads. For example, internal organs are subjected to localized compression, tension and shear loads applied by endoscopic tools, and loads that they would never experience otherwise under normal circumstances. Figure 5.4 shows the stress–strain relationship of various internal organs. These data were acquired by utilizing an endoscopic tool that applied loads similar to those applied during surgery (Rosen et al., 2008). Second, biomechanical properties change significantly, depending on the conditions under which they were collected: *in vivo*, *in vitro*, or *ex corpus*. To the extent possible, *in vivo* data is preferred, as they can provide the most accurate tissue characterization. Several experimental robotic devices with *in vivo* data collection capabilities have been developed. Among these are the Motorized Endoscopic Grasper (MEG) (De et al., 2007), ROSA, and TeMPeST 1-D (Ottensmeyer and Salisbury, 2001; Ottensmeyer et al., 2004). Third, *preconditioning* is a process used for testing soft tissue biomechanics, in which the tissue is subjected to multiple loading cycles before the data acquisition cycle. This process "stabilizes" the mechanical properties of the tissue and is known as *tissue conditioning*. Despite positive effects on the consistency of the data collected following tissue conditioning, this approach cannot be applied in the context of surgery. The surgeon who palpates the tissue may experience different stress–strain relationships for each palpitation, and these cycle- and time-dependent changes must be accounted for during data collection in order to fully characterize the tissue. Figure 5.4a and b show the difference in the stress–strain relationship between the first and the

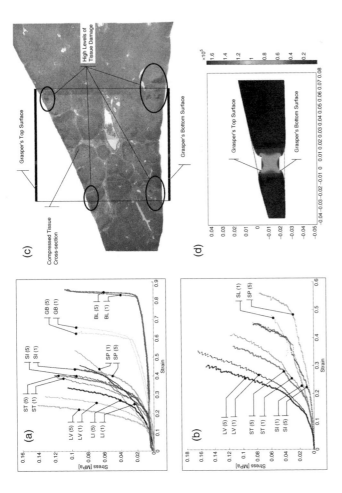

Figure 5.4 Biomechanics properties of soft tissue (swine internal organs) and tissue damage. Examples of stress–strain curves for all organs under study, as measured with the motorized endoscopic grasper at 5.4 mm/s loading velocity. First and fifth cycles show (a) *in vivo* and (b) *ex corpus* (organs legends: BL, bladder; GB, gallbladder; LI, large intestine; LV, liver; SI, small intestine; SP, spleen; and ST, stomach). The loading cycle number (1 or 5) is defined in the brackets. Liver response to compression loads of 40% strain. (c) A cross-section of a liver generated as an assembly of multiple tissue slices using standard pathological techniques following an application of compression strain by a Babcock grasper attached to the MEG. Vascular tissue damage is indicated by dark gray areas across the tissue slices. The horizontal arrow indicates the approximate span of the grasper jaws. (d) Von Mises stress distribution and the displaced cross-section of liver as predicted by a linear FEM. The geometrical dimensions are expressed in meters and stresses are expressed in Pascal (Rosen et al., 2008; Ottensmeyer and Salisbury, 2001).

71

fifth palpations. Fourth is the issue of tissue damage generated as a result of loads or energy transmission by the surgical tools. *Dissection* is a form of controlled tissue damage that results from the application of mechanical shear stresses or the application of electrical or other energy sources to generate a cut in the tissue while controlling potential bleeding. In this case, the damage is intentional, a derivative of the surgical requirements. However, unintentional tissue damage may also occur as a result of the mechanical interaction between the tissue and the tool during tissue manipulation or retraction. This unintentional tissue damage may have short-term effects that lead to recoverable tissue function with or without scar tissue. Or it may result in uncontrolled bleeding that must be resolved through the surgical procedure itself. In the worst case, tissue damage may have long-term effects that lead to necrosis and tissue death. The extent of tissue damage caused during surgery as a result of loads applied by the surgical tools depends in part on the distribution of stress, which itself is based on the design of the contact surfaces, the level of the applied loads, and the time duration that these loads are applied to the tissue. Sensor-based surgical tools—along with knowledge of biomechanics of tissue damage—can be used to monitor these parameters and mitigate tissue damage. Figure 5.4c and d show the correlation between histological analysis of tissue damage (identified by marking dead cells) and the stress distributions predicted by a finite element analysis.

5.2.3 Teleoperation

Surgical robotic systems that heavily rely on the surgeon's control of the system are based on classical master/slave teleoperation architecture. This architecture consists of two modules: the surgeon console (master) and the robot (slave). The surgeon console includes a set of input devices for the hands and feet, a display system, and, in some cases, voice command components. The device that interacts with the surgeon's hands and fingers acts as an input device that generates position commands to the surgical robot. It also serves as a haptic device that, along with its embedded actuators, can render forces and torques that are reflected back to the surgeon, providing information about the interacting forces between the surgical robot tool tip and tissues.

The robotic system interacting with the patient (slave) includes a minimum of three robotic arms: two are used to manipulate the surgical instruments and a third is used to control the endoscopic camera. Additional arms may include other surgical tools for operation or tissue retraction. The surgeon controls the position of the robotic arms by manipulating the two input devices at the console. The endoscopic camera arm is controlled by one of the input devices or by voice commands from the surgeon, and the view of the internal anatomy acquired by the endoscopic camera is transmitted back to the surgeon's console. If two endoscopic cameras are embedded into the endoscope, a three-dimensional view of the anatomy can be displayed to the surgeon.

A surgical robotic system using a teleoperation architecture enables two modes of operation: a bilateral control mode and a unilateral control mode (Fig. 5.5a).

(a)

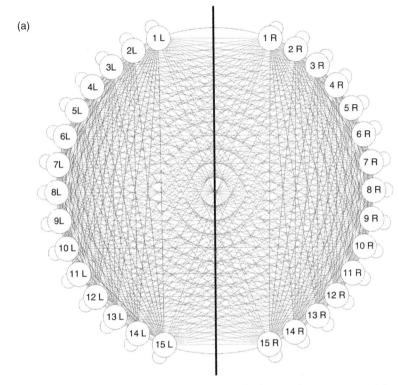

Figure 5.5   A block diagram of a typical bilateral teleoperation system used in surgical robotic systems. (a) A simplified block diagram of the teleoperation scheme. Note that the actuators and controllers on the master console are eliminated if force feedback is not incorporated into the system. (b) A detailed overview of the system architecture, including the surgical console (master) and surgical robot (slave) connected through a communication layer with three options. A wired communication (option 2) is the common practice and FDA approved for clinical use. Other alternative communication layers (options 1 and 3) were studied as part of the experimental evolution research systems (Lum et al., 2007; Brett et al., 2008). The surgeon initiates the movement of the robot by moving the stylus of a haptic master input device. The position of the stylus is sensed by position sensors embedded in the master joints and acquired by the A/D converter that is connected to the master PC via USB. Using a UDP protocol, the position command is transmitted through the network layer to the remote site and received by the slave PC. Using inverse kinematics, the position command is translated into joint command and sent via the D/A to the servo controllers. The servo controllers generate voltage commands to the DC actuators of the surgical robot, which in turn move the robot to the commanded position. A video stream of the surgical site is first compressed in the remote site by either software or hardware and then streamed through the network layer to the surgical console. In the surgical console, the video following its decompression is presented to the surgeon on a monitor. A foot paddle controlled by the surgeon allows him or her to engage and disengage the master and the slave by activating the brakes.

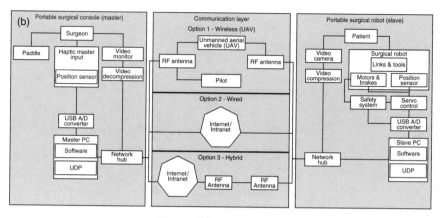

Figure 5.5 (*Continued*)

In both modes of operation, in the feed-forward flow of information, the surgeon generates position commands to the robot by moving the input devices located at the surgeon's console. The position commands are transferred through a controller to the surgical robotic arms (slave), and the actuators move the arms and the surgical tools to the proper positions. This flow of information is common to both the bilateral and the unilateral control modes. Force feedback is the flow of information that is included in the bilateral control mode and eliminated in the unilateral control mode.

There are two primary methods of generating source signals for the force feedback in bilateral master/slave teleoperation architecture, thus allowing the system to reflect haptic sensation to the surgeon as the tool tip interacts with the tissue. The primary method used in surgical telerobotic systems and approved for clinical use is based on the difference between the position command generated by the surgeon using the input devices at the console (master) and the actual position achieved by the robot. This error is usually scaled by a constant and reflected as a force rendered by the master's actuators. As the difference between the position command and the actual position increases, the force feedback to the operator increases proportionally and vice versa. In spite of the fact that the bilateral mode of operation does not require additional sensors for generating force feedback, the high level of friction caused by non-direct-drive actuators and the high inertia due to large robotic arms may degrade the quality of the force feedback signal using this algorithm. An alternative approach for incorporating force feedback requires the use of force/torque sensors located as close as possible to the end effector in order to diminish the mechanical and dynamic interferences. Force and torque acquired by the sensors are sent back to the surgical console to be rendered by the haptic device and delivered to the surgeon's hands.

Given the harsh environments associated with tool sterilization (high temperature steam) and operation, attaching force/torque sensor wires and connectors to the tool and protecting them from this environment remains a technological challenge, but research efforts have shown promising results in this area (Fig. 5.6)

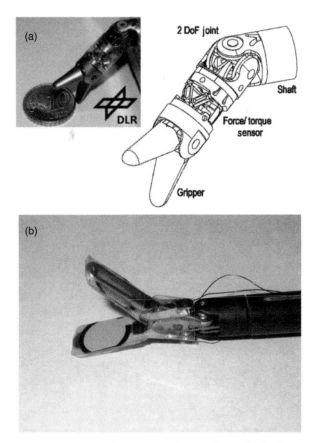

Figure 5.6   Force sensor mounted on the proximal end of a surgical tool (a) DLR endo-scopic tool instrumented with a six axis force–torque sensor and (b) UCLA Tactile sensor mounted on the faces of a robotic endoscopic grasper.

(Franco et al., 2009; Thielmann et al., 2010). Alternative sterilization methods such as gas sterilization may relax these requirements. For MIS tools, in particular, plac-ing force sensors at the distal end of the tool is further limited by the 5–10 mm diameter of the port.

5.2.3.1 Haptics   Haptics (from the Greek word for "touch"), in the context of surgery, refers to the perception of a surgeon and the technology associated with conveying this perception. Surgeons rely heavily on haptic perception to assess soft tissue. This assessment is conducted in part by palpating the tissue with the fingers in an open surgical procedure. The stiffness of a tissue is either increased or decreased as a result of damage or disease. Variation in stiffness of a specific organ may also help to target a localized tumor. As surgical approaches become less invasive, the surgeon is gradually removed from the surgical site, and the interaction with the tissue is facilitated by mediating surgical tools and surgical robotic systems.

Surgeons have regained some of the haptic capabilities by visually assessing the deformation of the tissues in response to the interaction with the surgical tools. Although this technique may be useful for soft tissue stiffness assessment, it cannot be used to assess suture tension, given the relatively high stiffness of sutures. As a result, a suture may break during knot tying due to lack of haptic sensation.

Surgical robotic systems have the capability of regaining the haptic sensation through a force feedback control algorithm embedded in the surgeon's console. Experimental results using the hand, a regular MIS grasper, and a robotic device with force feedback for ranking the stiffness of materials with similar stress–strain characteristics, as soft tissue of internal organs indicated that the performance with the robotic device was closer to the performance of the human hand in rating material stiffness than to the performance obtained by MIS grasper (Rosen et al., 1999). Even in the hand-in-glove conditions, the test operators were able to rank the material stiffness correctly in all cases. This fact emphasizes the need for advanced instruments for increasing the haptic sensation beyond the capability of an unaided hand.

5.2.3.2 Time Delay

During actual teleoperation, physical distance and a network separate the patient site from the surgeon sites with time varying delays. When a surgeon makes a gesture using the master device, motion information is sent through the network to the patient site with a network time delay (T_n). The manipulator moves, and the audio/video device observes the motion. Digital A/V is compressed (T_c), sent from the patient site to the surgeon site through the network (T_n), then decompressed (T_d) and observed by the surgeon. The surgeon has experienced a total delay of $T = 2T_n + T_c + T_d$, from the time the gesture was made to the time the action was observed.

Lab experiments showed that the completion time of the task and the length of the tool tip trajectory significantly increased in correlation with the time delay. For teleoperation with a time delay of 0.25 and 0.5 s, the task completion time increased by a factor of 1.45 and 2.04, and the length of the tools' trajectory increased by a factor of 1.28 and 1.53. There were no statistical differences in the number of errors or in the completion time and tool-tip path length between experienced surgeons and nonsurgeons (Fig. 5.7) (Lum et al., 2009).

5.2.3.3 Indexing, Motion Compensation, and Scaling

Indexing   Indexing is the process by which the surgeon disengages the master from the slave, repositions the input devices, and reengages the master and slave to continue the operation. Indexing is enabled by brakes mounted on motors of the robot, which fix the position and orientation of the robot in space, while the robot is disengaged from the surgical console. Indexing allows the surgeon to keep the robot's hands and arms within the optimal workspace and to maximize manipulability and personal comfort. Indexing is limited to positioning only, and not to orientation. As a result, the orientation of both the master and the slave must be locked during position indexing. If locking the orientation is not possible for

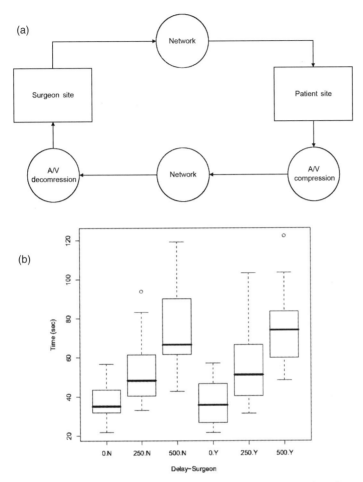

Figure 5.7 Time delay and its effects on surgical performance in telesurgery. (a) Schematic block diagram representing all time delay sources. (b) Completion time of an FLS block transfer as a function of time delay (0, 0.25, 0.5 s) for surgeons (Y) and nonsurgeons (N) using the Raven surgical robotics system in a teleportation mode.

some reason, for example, during a tool change, the master orientation may have to adjust itself to match the orientation of the slave before the reengagement.

Motion Compensation A tremor is an involuntary muscle contraction and relaxation generating movements of one or more body parts that may occur at rest or while the body is in motion. For a surgeon holding a surgical tool, a tremor may affect his or her ability to effectively interact with the tissue. It is particularly critical in microsurgery and ophthalmology, where the accuracy and repeatability required to perform the surgical procedure may exceed human performance capabilities. Surgeons are trained to lock body parts in order to reduce tremors.

For example, in open and MIS surgeries, surgeons will hold the upper arm close to their body, essentially eliminating movements of the shoulder and upper arm and allowing only elbow, wrist, and finger movements. In microsurgery, the palms are usually at rest against a stationary surface so that movements of the entire arm are eliminated and only hand and finger movements are enabled. Surgical robotics may provide two forms of motion compensation depending on the system architecture.

In master/slave teleoperation architecture, the surgeon controlling the master is physically removed from the surgical tool. Because of this physical separation between the surgeon hand/fingers and the surgical tool, motion compensation is sometimes introduced in the control algorithm to eliminate the tremor. There are two types of architectures in which the surgical tool is held by the robot and the surgeon simultaneously. When the base of the robot is grounded—such as in the Freehand/JHU (Simone and Okamura, 2002) or the Rio®/Mako® (Hagag et al., 2011)—the control system may attenuate a specific bandwidth associated with the human tremor. If the entire system is handheld and the base is not grounded, as in Micro CMU (Ang et al., 2001), the actuators connected between the base of the tool and the portion held by the hand are actuated in such a way that the tool tip remains stationary, and thus the tremor from the human hand is reduced.

Scaling   In a master/slave teleoperation architecture, the movements provided to the system by the surgeon through the master input device can be amplified (*scaled up*) by a scale factor greater than 1 or attenuated (*scaled down*) by a scale factor less than 1. In actual surgery scenarios, scaling down is more often utilized than scaling up, as scaling down increases precision (and also increases duration). Scale factors are task specific.

5.2.4 Image-Guided Surgery

In image-guided surgery, imaging modalities track surgical tools, using images acquired before or during the operation in order to guide the intervention. This surgical approach, sometimes called *computer-assisted surgery* (CAS), includes the following critical steps:

1. image acquisition, which can be accomplished by a variety of imaging modalities, including X-ray, PET, CT, MRI, ultrasound, and tomography,
2. image analysis,
3. diagnostics,
4. preoperative planning with or without surgical simulation,
5. surgical procedure, which includes registration and navigation, and
6. postoperative verification.

These steps, with some variation, are described in Section 5.3 for a number of robotic systems that rely on imaging modality to perform surgery. In the context of surgery, the various imaging modalities can be classified into two categories: (i) online or real-time imaging systems that provide immediate visual feedback to the surgeon during the surgery (e.g., ultrasound, fluoroscopy) and (ii) off-line imaging systems that require image-acquisition time and/or postprocessing time, and therefore cannot be used intraoperatively (e.g., CT). Imaging modalities such as MRI require significant postprocessing calculations in order to produce the image; however, this process occurs quickly enough to provide feedback to the surgeon intraoperatively. As a result, several robotic systems were developed with MRI compatibility with nonmagnetic materials, which allow the surgeon to conduct the surgery within the MRI bore (Elhawary et al., 2008).

5.2.5 Objective Assessment of Skill

Conducting a surgical procedure involves high level cognitive decision making in conjunction with low level manual control of the surgical tools. Basic and advanced surgical training must produce surgeons who can be trusted to conduct unsupervised surgical intervention in a clinical setting.

Methods for evaluating surgical proficiency remain mostly subjective. While surgical simulators and surgical robotic systems can capture the physical parameters associated with surgery, capturing the cognitive parameters is more challenging. Data on physical parameters such as surgical tool type, tool kinematics—which includes position and orientation of the tools in space, as well as their forces and torques—and the camera view of the surgical site (Misra et al., 2008; Satava, 2008; Gallagher and O'Sullivan, 2011) can serve multiple purposes: it can facilitate objective assessments of technical skills, can be used for mentoring during and after the operation, and can help form a clinical record of the surgical procedure.

On the other hand, capturing the high level decision making processes that occur during surgery is difficult. One technique is to simply ask the surgeon to verbalize the mental decision making processes as they occur. Given the difficulty in capturing the many cognitive processes associated with surgery, there are no quantitative data that documents how the surgeon's mental load is distributed between high level decision making and low level tool manipulation, both of which are needed to complete the surgery. It is assumed that the decision making load is higher than the motor-skill load for a proficient surgeon and that the attention required for these two tasks may vary based on the level of training.

The decision making and motor skills of a surgeon are assessed through his or her training period and during the professional certification exam. This assessment is fundamentally subjective, and as the medical profession faces greater demands for accountability and patient safety, there is a critical need for the development of consistent and reliable methods for objective evaluation of clinician performance during procedures. The methodology for assessing surgical skill as a subset of surgical ability is gradually shifting from subjective scoring by an expert—which may constitute a biased opinion based on vague criteria—toward a more objective, quantitative analysis.

Developing an objective analysis of surgical skill based on task deconstruction or decomposition is an essential component of a rigorous objective skills-assessment methodology. A broader understanding of procedures is achieved by exposing and analyzing the internal hierarchy of tasks while providing objective means for quantifying training and skills acquisition (Rosen et al., 2006; Reiley et al., 2010). There are three primary approaches or models for task decomposition and its associated skills assessment and training applications: (i) black box, (ii) gray box, and (iii) white box. In the *black box approach*, the models and their states are abstract and do not correlate with specific events in reality; for example, surgical suturing is represented by a single model with abstract states and model architecture. In a *white box model*, every state of the model represents a specific and well-defined event in reality; for example, each tool-tissue interaction (grasping pushing, etc.) is represented by a unique state with a specific signature of forces, torques, and velocities (Rosen et al., 2006). An intermediate approach decomposes a step of an operation into more fundamental tasks, and each task is represented by a single black box model. The level of granularity in this so-called *gray box approach* is higher than the black box approach but lower than of the white box approach.

A useful analogy that may explain the white box approach for decomposing the surgical task is the human spoken language. On the basis of this analogy, the basic states, which are made up of tool/tissue interactions, are equivalent to "words" of the MIS "language," and the states form the MIS "dictionary" or set of all available words. In the same way that a single word can be pronounced differently by different people, the same tool/tissue or tool/object interaction can be performed differently by different surgeons. Differences in force/torques (F/T) magnitudes account for this different "pronunciation," yet different pronunciations of a "word" have the same meaning, or outcome, as in the realm of surgery.

A cluster analysis was used to identify the typical F/T and velocities associated with each tool/tissue and tool/object interaction in a surgery's "dictionary" or, using the language analogy, to characterize different pronunciations of a "word." Utilizing the "dictionary" of surgery, the Markov model (MM) was then used to define the process of each task or step of the surgical procedure, thus "dictating chapters" of the surgical "story" (Fig. 5.8)

5.3   COMMERCIAL SYSTEMS

5.3.1   ROBODOC® (Curexo Technology Corporation)

5.3.1.1   Clinical Procedure — Problem and Needs   Hip joint replacement is a relatively common orthopedic procedure normally conducted to relieve chronic arthritis pain, or in cases where the joint has been fractured or otherwise severely damaged as a result of trauma (Fig. 5.9). In a total hip replacement procedure, or "total hip arthroplasty" (THA), both the head of the femur and the acetabulum are replaced. A hemiarthroplasty replaces only half of the anatomical joint, typically the femur head. In both cases, the anatomical joint is replaced by a metal or ceramic

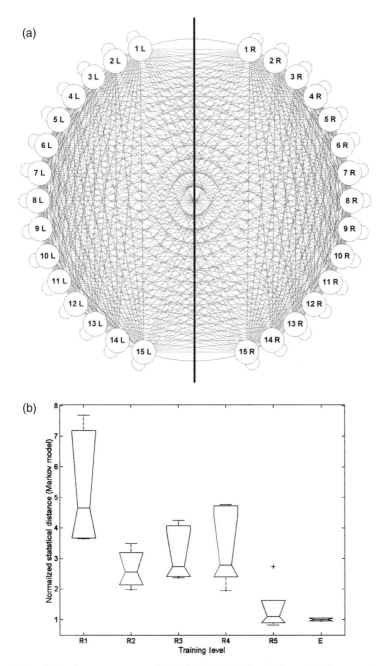

Figure 5.8 Objective assessment of skill in MIS. (a) A multistate Markov model representing a generic MIS procedure conducted with two endoscopic tools. (b) The learning curve of minimally invasive suturing. Normalized statistical distances between surgical residences, R. (R1, first year; R5, fifth year) and experienced surgeons, E. The statistical distance between surgeons in training compared to experienced surgeons decreases, as the surgeons progress through their 5 years of training (Rosen et al., 2006).

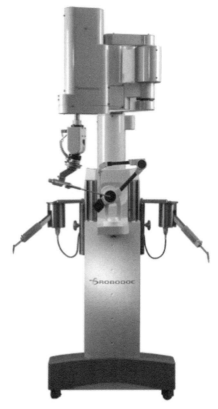

Figure 5.9 ROBODOC Surgical Assistant System by Curexo Technology Corporation.

prosthetic implant. The femoral component consists of a stem and head and is inserted into the femur. The acetabular cup is implanted into the hip socket of the pelvis. During the procedure, cartilage and bone are removed, and the bone is reshaped to accept the prosthesis. The acetabular cup is screwed into the pelvis, and the stem of the femoral component is either cemented or, more recently, pressed fit (cementless) into the femur.

5.3.1.2 System Architecture The ROBODOC® system by Curexo Technology Corporation includes two major subsystems: ORTHODOC® Preoperative Planning Workstation and ROBODOC® Surgical Assistant. ORTHODOC converts the CT scan of the patient's joint into a three-dimensional bone image, which can be manipulated by the surgeon to view bone and joint characteristics, thus allowing for optimal prosthetic selection and accurate alignment. Using ORTHODOC's digital library of prosthetic images, the surgeon selects the best size, type (anatomical or straight stem), and brand of femoral stem prosthesis. This virtual surgery creates a precise preoperative plan customized for each patient. The preoperative plan is then transferred to ROBODOC, which executes the plan by milling the bone with submillimeter accuracy, thus preparing the bone to receive the prosthetic implant with a precise fit. ROBODOC is also capable of removing bone cement for revision surgeries (Spence, 1996; Bargar et al., 1998; Nishihara et al., 2004; Schulz et al., 2007).

5.3.2 daVinci (Intuitive Surgical)

5.3.2.1 *Clinical Procedure — Problems and Needs* In a classical MIS, surgical tools are inserted through ports, along with an endoscopic camera, into the human body. A cavity in the human body is inflated with CO_2, and the operation is conducted under the patient's skin. This revolutionary surgical technique minimizes the trauma and maximizes the recovery of the patient. However, it requires tools that reduce the quality and number of degrees of freedom (DOF) for tissue manipulation: from seven DOF down to five DOF. The tools also reduce haptic sensation, as long endoscopic instruments do not convey sensation in the same way that fingers do. The technique also limits the view of the surgical site, as endoscopic cameras display the site in two dimensions, whereas the human eye can view the site in three dimensions. Despite these drawbacks and difficulties, the MIS is becoming a common practice for an ever increasing number of procedures, because of patient benefits and positive clinical outcomes. The challenge is therefore to retain the benefits of MIS while reducing as possible the deficiencies of this approach with respect to open surgery.

5.3.2.2 *System Architecture* The da Vinci system by Intuitive Surgical Inc. is a surgical robotic system that utilizes an MIS approach. The system follows the classical master/slave teleoperation architecture, which includes a surgical console (master) that controls the patient-side subsystem (slave). The system was originally based on technologies developed by SRI, MIT, and IBM (Green et al., 1995; Madhani et al., 1998) and evolved through two prototypes (Lenny and Mona) into three FDA-approved versions of the same product named da Vinci (1999), da Vinci S (2006), and da Vinci Si (2009).

The most recent version of the system, da Vinci Si, includes a patient-side subsystem with four arms mounted to a post next to the operating room table. Three of the four arms carry MIS surgical tools (5–8 mm in diameter) and the fourth arm is equipped with two endoscopic cameras with a single shaft (12 mm in diameter) that can reproduce a three-dimensional view of the surgical scene. There are approximately 50 different tools that can be mounted on the robotic arms through a surgical sterile barrier. Similar to the rest of the system, the tools are cable driven. The four DOF of the tool along with the three DOF of the slave, together form a seven-DOF mechanism. The tools may be divided into two groups, including double-wrist-joint 8 mm tools and snakelike 5 mm tools. The double-joint internal wrist of the MIS tools sets the da Vinci system apart from the manually controlled MIS tools and provides a superior manipulability. Although snakelike manual tools do exist, they are more difficult to manipulate manually through their proximal end outside of the patient.

The patient- and master-side arms are structured as extended four-link parallelograms with a pivotal point (also known as *remote center of rotation*) located outside of the mechanism. At the patient side, this center is located at the port where the tool is inserted into the patient through the skin, and at the surgeon side, this point is the center of the gimbal mechanism attached to the surgeon's hand. By locating the remote center of the mechanism at the port, the system eliminates site-to-side translation of the tool, which can damage the skin, but still maintains permissible in/out translation and rotation along any direction. Both of these mechanisms are actuated through mechanical cables, and in some DOF through single-stage gears connecting each joint to the corresponding actuator located at the moving base of mechanism. The actuators also serve as a counterbalance for the entire arm. On the patient side, the four robotic arms are connected to a single post through passive linkages. Counterbalance mechanisms along with electromechanical clutches allow positioning and orientation of the base of each arm with respect to the patient.

The surgical console includes two robotic arms that are used as the primary input devices to the system. By manipulating these two devices, the surgeon provides position and orientation commands to the arms on the patient side. The console also includes two screens that are fed by the two endoscopic cameras to recreate a three-dimensional view of the surgical scene. A series of foot peddles allows the surgeon to index the system and to control other surgical functions. The surgeon supports his/her arm on a horizontal bar and uses the index finger and thumb inserted through finger loops to interact with the input devices. A surgeon looking down toward his or her hands through an eyepiece can view the 3D display positioned between the surgeon's eyes and hands. Using this unique setup along with a precise mapping from the surgeon's wrist joint to the surgical tool's wrist joint, the surgeon perceives the surgical instruments as a natural extension of his or her own hands. This console configuration along with the system technical capabilities is the major contributor to the "intuitive" sensation of operating the system.

Two surgical consoles are electronically linked, allowing two surgeons to control the tools and endoscopic cameras. Although the system is based on a teleoperation architecture, in which the master and slave can be separated by a large distance, the approved mode of operation is limited to a scenario in which the patient-side and the surgeon-side are colocated, keeping the surgeon and the patient in the same room (DiMaio et al., 2011) (Fig. 5.10).

5.3.3 Sensei® X (Hansen Medical)

5.3.3.1 Clinical Procedure — Problems and Needs Cardiac arrhythmia (abnormal heart rhythm) is used to describe a large number of conditions associated with abnormal electrical activity in the heart (Fig. 5.11). The effects of cardiac arrhythmia may vary between non-life-threatening abnormal heart beat to life-threatening predisposition to stroke, embolism, and cardiac arrest, leading to sudden death. AF is the most common cardiac arrhythmia and involves the two upper

Figure 5.10 The da Vinci Si by Intuitive Surgical Inc. (a) Surgeon-side two surgical console and (b) Patient-side subsystem. *Source:* © 2012 Intuitve Surgical, Inc.

(a)   (b)   (c)

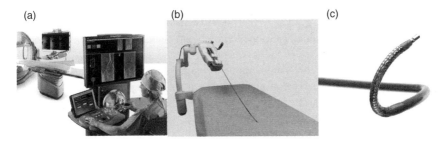

Figure 5.11 Sensei® X by Hansen Medical: (a) surgical console, (b) surgical robotic arm, and (c) Artisan Catheter. *Source:* Courtesy of Hansen Medical, Inc.

chambers (atria) of the heart. Its name comes from the fibrillating (i.e., quivering) of the heart muscles of the atria, rather than a normal coordinated contraction.

If rhythm control cannot be maintained by medication or cardioversion, then catheter ablation may be used. Catheter ablation is an invasive procedure that involves a flexible catheter inserted into the heart through the veins. The catheter delivers high frequency electrical impulses that ablate the heart tissue responsible for the abnormal conduction of the electrical signal pathways.

5.3.3.2 *System Architecture* Sensei® X by Hansen Medical is configured as a master/slave robotic teleoperator. The surgical console includes a single parallel robotic haptic device, along with an array of switches and knobs operated by the left hand. With these two input devices, the surgeon steers and navigates the catheter. The display includes visualization of fluoroscopy, intracardiac ultrasound images, three-dimensional mapping system images, and real-time electrograms.

The surgical robotic arm (slave) is attached to the surgical table. It can be manipulated with respect to the patient but remains fixed during the operation. It carries a set of actuators that manipulate the various degrees of freedom of the two sheaths and catheter. The Artisan Catheter includes two steerable elements, an outer sheath and an inner sheath. The outer sheath allows deflection in a single plane, and the inner sheath is steerable and maneuverable in all directions. An ablation catheter is placed within the lumen of the Artisan Catheter, allowing the surgeon to ablate the tissue once the tip of the catheter is navigated to the targeted anatomical structure (Saliba et al., 2006; Saliba et al., 2008; Reddy et al., 2007).

5.3.4   RIO® MAKOplasty (MAKO Surgical Corporation)

5.3.4.1 Clinical Procedure — Problems and Needs   Total knee arthroplasty (THA) or "knee replacement" and unicompartmental arthroplasty (UKA) or "partial knee replacement" are surgical procedures aimed to replace the entire or part of the knee joint to treat disability or relieve pain arising from either trauma or various joint disorders due to infection or age, usually involving arthritis or other inflammatory condition. The knee is generally divided into three elements: the inside (medial), the outside (lateral), and the joint between the kneecap and the femur (patellofemoral). Between 10% and 30% of patients experience wear limited to a single element, typically the medial element, making them candidates for the UKA.

Modern total knee replacement implants include a femoral head, a tibial plate, and a patellar plate (usually not introduced with a robotic procedure). The diseased or damaged weight-bearing joint surfaces of the knee are replaced with either a cementless or cemented implants, including metal and plastic components shaped to maintain the kinematics of the knee. The soft tissues are removed and the bones are cut by a *reamer* (a handheld drill) and *broaches* (serrated cutting tools) to create specific planes and cavities in the cortical bones that accommodate the implant. The bone preparation for this procedure is challenging and is the primary motivation for a robotically assisted solution. Large gaps created between the bone and the implant may result when large bone elements are removed by the broach. These gaps may generate a suboptimal stress distribution and stress transfer between the implant and the bone. Undersizing or oversizing the implant may cause a variety of problems, including unstable joint fractures and pain. Accurate fit, placement, and prosthesis selection are facilitated by the abilities of robotic and robotic-assisted systems to execute a preoperative plan (Kazanzides, 2007).

5.3.4.2 System Architecture   The MAKO Robotic Arm Interactive Orthopedic System (RIO®) comprises three major hardware components (Fig. 5.12). The robotic arm supports the cutting system that allows the surgeon to create the desired resections of bone. The camera stand supports both the computer monitor used by the surgeon to view the bone resections and the localizing camera system for tracking the patient anatomy through the use of tracking arrays mounted on the bone. The guidance module is used by the physician's assistant or a surgical technician to assist the surgeon navigating through the implant planning and surgical application. The surgical tool is held by the surgeon and the robot simultaneously. The surgeon guides the tool using virtual fixtures so that the robotic arm is passive

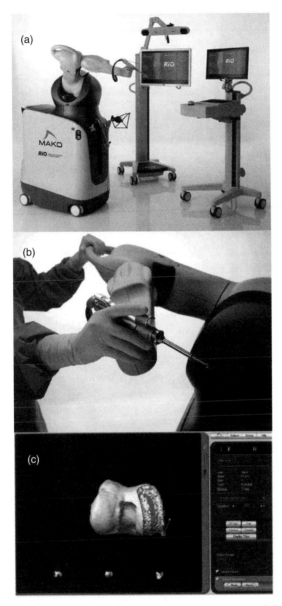

Figure 5.12 Robotic Arm Interactive Orthopedic System (RIO®) by MAKO: (a) Overview of the entire system, (b) shared control of the surgical tool—the surgical robotic arm and the surgeon holding the surgical tool together, and (c) screenshot of the RIO system software showing the model of the patient's femur bone with the planned resection volume. *Source*: Courtesy of MAKO Surgical Corp.

as long as the motions are within the boundaries of the preplanned space. The robotic arm applies haptic force feedback only if the surgeon attempts to move the cutting burr outside the predefined surgical plan. The robotic arm is a six-DOF serial manipulator attached to a mobile cart that is fixed during the operation. The total range of motion of each joint is designed to accommodate both right-handed and left-handed surgeons, as well as to provide sufficient workspace to perform the worst case surgery envisioned. The accuracy of positioning the tip of the tool is less than 1 mm.

The cutting burr spins at up to 80,000 rpm, removing the bone volume to be replaced by the implant. A variety of cutting burrs can be used and exchanged in the system during a surgery. Irrigation is provided to the area of the resection using irrigation tubing attached to the end effector assembly (not shown) to cool the bone during cutting, preventing thermal necrosis that can lead to loosening of the prosthetic implant over time.

Figure 5.12c is a screenshot of the RIO system software, showing the model of the patient's femur bone with the planned resection volume in gray as well as a portion of the bone already removed. This is the interface the surgeon uses to guide the bone resections. The surgeon is expected to cut the gray colored bone away up to the planned boundaries shown as white. If the surgeon attempts to move the cutting burr outside of the planned gray resection areas, the RIO robotic arm applies a force on the surgeon's hand preventing any cutting outside the planned boundaries. The ability to passively move with the predefined space with minimal resistance due to friction, backlash, and internal effects is enabled by the back drivability of the system. This operational requirement is met using a cable-driven transmission and a tungsten wire rope, implemented in each DOF (Hagag et al., 2011).

The RIO system is currently used for implantation of medial and lateral UKA components, as well as patellofemoral arthroplasty. These procedures follow the following four steps:

1. *Preoperative Imaging*: Preoperative CT scans are obtained, consisting of 1 mm slices for the knee joint and 5 mm slices for the hip and ankle. The scans are then reconstructed to obtain a three-dimensional view of the anatomy. Initial preoperative planning is conducted using 3D CAD models of the implants.

2. *Preoperative Planning*: The preoperative plan is based on four main parameters: metrics of component alignment, 3D virtual visualization of implant position, intraoperative gap kinematics, and dynamic lower limb alignment assessment. Preoperative planning based on CT scans is limited, as the CT scan is not capable of imaging soft tissues. As a result, the plan must be modified intraoperatively to achieve precise gap balancing and long-leg alignment. Bone resection volumes are defined automatically by the system, and boundaries for the cutting instrument are set to prevent inadvertent surgery to areas outside these predefined zones.

3. *Operation and Intraoperative Soft-Tissue Balancing*: Following the setup and initialization of the robotic system, a standard orthopedic leg holder is used to

restrain the leg. Anatomical surface landmarks are registered before the skin is incised. After skin incision, small articular accuracy checkpoint pins are inserted into the tibia and femur, and the two bone surfaces are registered at these points to match them to the CT models. Virtual kinematic modeling of the knee and intraoperative tracking allow real-time adjustments to be made to obtain correct knee kinematics and soft-tissue balancing. The surgeon moves the arm by guiding its tip within the predefined boundaries. The robot gives the surgeon active tactile, visual and auditory feedback during burring. Following the preparation of the bone, the implant is attached.

4. *Postoperation Follow-Up*: In 24-h overnight hospital stay for pain control, antibiotics and anticoagulation are often used. The patient is mobilized the same day with physical therapy and a continuous passive motion (CPM) system that flexes and extends the knee overnight to begin motion and determine comfort level.

5.3.5 CyberKnife (Accuray)

5.3.5.1 Clinical Procedure — Problems and Needs   Radiosurgery allows noninvasive treatment of both benign and malignant tumors (Fig. 5.13). It is also known as *stereotactic radiotherapy* (SRT) when used to target lesions in the brain, and *stereotactic body radiotherapy* (SBRT) when used to target lesions in the body. Radiosurgery operates by directing highly focused beams of ionizing radiation. The ionizing radiation is used to ablate, by means of a precise dosage of radiation, tumors and other lesions that could be otherwise inaccessible or inadequate for open surgery because of potential damage to nearby anatomical structures such as arteries, nerves, and other vital organs. As part of the selective ionizing radiation,

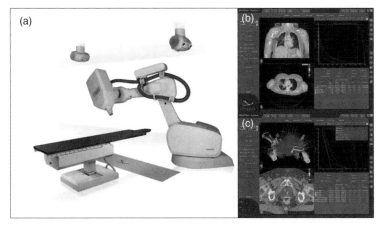

Figure 5.13   CyberKnife system: (a) along with the treatment delivery graphical user interface (GUI), (b) lung treatment plan image, and (c) prostate treatment plan image. Images used with permission from Accuray Incorporated.

ions and free radicals are formed from the water in the cell and the surrounding tissue, which produce damage to DNA, proteins, and lipids, resulting in the cell's death. The technological and clinical challenges are to deliver the correct dose of radiation to a specific location in space is order to ablate the target tissues while minimizing the damage to the surrounding tissue under dynamic conditions, such as breathing or unexpected patient movements.

The CyberKnife by Accuary can delivers a therapy anywhere in the body, where the radiosurgery is clinically indicated (with FDA 510(k) regulatory clearance). Common treatment sites include intracranial, head and neck, spine and paraspinal, lung, prostate, liver, and pancreas.

5.3.5.2 System Architecture Procedures that use the CyberKnife system include the following steps, which are described as follows: scanning, planning, treatment, and follow-up.

1. *Scanning*: Before treatment with the CyberKnife system, the patient undergoes imaging procedures using CT, MRI, angiography, or PET to determine the size, shape, and location of the tumor.
2. *Planning*: The image data is then digitally transferred to the CyberKnife system's treatment planning workstation, where the treating physician identifies the exact size, shape, and location of the tumor to be targeted and the surrounding vital structures to be avoided. A physician then uses the CyberKnife software to generate a treatment plan to provide the desired radiation dose to the identified tumor location while avoiding damage to the surrounding healthy tissue. As part of the treatment plan, the CyberKnife system's planning software automatically determines the number, duration, and angle of delivery of the radiation beams.
3. *Treatment*: During the CyberKnife procedure, a patient lies on the treatment table, which automatically positions the patient. Anesthesia is not required, as the procedure is painless and noninvasive. The treatment, which generally lasts between 30 and 90 min, typically involves the administration of between 100 and 200 radiation beams delivered from different directions, each lasting from 10 to 15 s. Before the delivery of each beam of radiation, the CyberKnife system simultaneously takes a pair of X-ray images and compares them with the original CT scan. The radiation is generated by 1000 MU/min 6MV X-band linear accelerator that is carried by a robotic arm. During treatment, the six-DOF robot, a KR240-2 Kuka, with manufacturer specification for position repeatability of better than 0.12 mm, moves in sequence through the nodes selected during treatment planning. An optimized path traversal algorithm allows the manipulator to travel only between nodes at which one or more treatment beams are to be delivered, or through the minimum number of additional zero-dose nodes required to prevent the robot trajectory intersecting fixed room obstacles or a "safety zone" surrounding the couch and patient. At each node, the manipulator is used to reorient the linear accelerators such that each beam originating at the node can be

delivered. Using an image-guided approach along with three stereo CCD cameras mounted on a boom that is attached to the ceiling continually tracks, detects, and corrects for any movement of the patient and tumor throughout the treatment to ensure precise targeting without the clinician's intervention. The patient typically leaves the facility immediately on completion of the procedure.

4. *Follow-Up*: Follow-up imaging, generally with either CT or MRI, is usually performed in the weeks and months following the treatment to confirm the destruction and eventual elimination of the treated tumor.

In 2010, Accuray released a new product called *CyberKnife VSI*. The basic concept remains unchanged, but significant improvements and additions to the system technology implemented in the past decade have made the early technical publications obsolete. For a recent review, see Kilby et al. (2010).

5.3.6 Renaissance™ (Mazor Robotics)

5.3.6.1 Clinical Procedure – Problems and Needs Spinal fusion, that is, fixing the relative motion between two or more adjacent vertebrae by joining them through an implant made out of screws, rod, plates, and cages, as well as bone graft is an orthopedic surgical procedure that is performed in cases of fracture of vertebral body, degenerative disc disease (disc herniation, instability of facet joint, compressive radiculopathy), spine tumors, and scoliosis (Fig. 5.14). As part

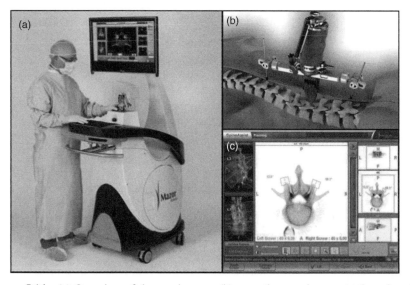

Figure 5.14 (a) Overview of the renaissance, (b) mounting renaissance to the spine, (c) 3D planning of pedicle screws to be introduced into L3 vertebra. *Source*: Courtesy of Mazor Robotics.

of the procedure, two screws per fused vertebra are inserted on the left and right pedicles, which are the segments that connect the body to the arch of the vertebra. A rod is fixed to the head of the pedicle screws of two vertebrae to achieve the fusion. The introduction of the implant's screw into the pedicle is conducted under fluoroscopic imaging and requires exposing the pedicles, drilling a pilot hole for each screw, and inserting the screw. The introduction of the pedicle screw is complicated because the fluoroscopic imaging provides anterior–posterior and lateral images of the anatomy that is not aligned with screw insertion plane. As a result, the screw is misplaced in 10–40% of cases by more than 2 mm from its ideal position, and in about 3% of cases, the screw misalignment reaches 5 mm, resulting in damage to the nerves. The difficulty and the risk of nerve damage increase as the procedure is performed in the thoracic and cervical spine as the size of the vertebrae decreases in these regions.

5.3.6.2 System Architecture   Renaissance™ (formally SpineAssist®) by Mazor is a miniature bone-mounted robotic system. The base of parallel architecture robotic device is directly attached to the spine and its end effector includes a metal tube guide for surgical instruments such as a needle or drill that can be positioned and oriented to a desired location near the mounting site of the base (Shoham et al., 2003; Shoham et al., 2007; Kantelhardt et al., 2011).

The surgical procedure incorporating the robot includes the following steps:

1. *Preoperative Planning*: The surgeon plans the desired orientation, entry point, and depth of one or more drill or needle procedures based on computer tomography (CT), or magnetic resonance imaging (MRI) images;
2. *Intraoperative Robot Attachment*: The sterilized robot with the targeting guide is rigidly attached with a minimally invasive attachment jig to the bony structure close to the surgical site;
3. *Robot Registration*: A precise geometric relation between the coordinate systems of the robot, the target anatomy, and the plan is established;
4. *Robot Positioning*: The robot controller moves the targeting guide to its planned position and locks the robot in place; and
5. *Manual Execution*: The surgeon executes drilling or needle insertion through the positioned guide.

Steps (4) and (5) are repeated for each planned location.

5.3.7   ARTAS® System (Restoration Robotics, Inc.)

5.3.7.1 Clinical Procedure — Problems and Needs   The total number of hair follicles for an adult human is estimated at 5 million, with 1 million on the head of which 100,000 alone cover the scalp (Fig. 5.15). Most cases of hair loss are because of androgenic alopecia (AGA). Fifty percent of men by age 50 and 40% of women by menopause have some degree of AGA. The treatment options

(a)   (b)

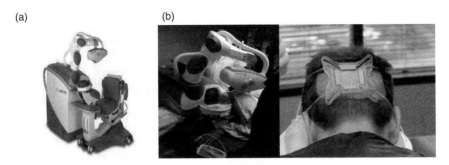

Figure 5.15   The ARTAS® System. *Source*: Courtesy of Restoration Robotics, Inc.

are either medical or surgical. Hair transplantation is one among several surgical procedures and is considered as a permanent solution to baldness. Restoration is possible because the hair follicles on the sides and back of the scalp are insensitive to the hormones that cause androgenic alopecia, so there is less chance of fallout. During surgical hair transplantation, hair follicles are redistributed in bald areas, where they grow hairs for the rest of the individual's life.

5.3.7.2   System Architecture   ARTAS™ by Restoration Robotics is an inter-active, computer-assisted system utilizing image-guided robotics to enhance the quality of hair follicle harvesting. The ARTAS system includes an interactive, image-guided robotic arm, special imaging technologies, small dermal punches, and a computer interface. The system is positioned over the patient's donor area of the scalp. The robotic arm is equipped with two cameras that serve as a stereo-vision sensor that can identify and detect follicular units on the patient's scalp. The physician chooses which follicular units are to be harvested and inputs this patient-specific information into the system. The system is capable of adjusting itself and compensating for the patient's head movements using visual servoing, as well as calculating the angle, ordination, and position of each follicular unit on the scalp surface. The type and number of follicular units, as well as the pattern of harvesting can be selected. Using this information, the imaging system semiau-tonomously guides the robotic arm and its tool to extract the follicular units one at the time. Each follicular extraction made by the robotic system tool involves 1 mm incision that does not require sutures or other wound-closure treatments that are needed following other harvesting techniques. The follicular units are stored until they are implanted into the patient's recipient area using current manual techniques (Petrou, 2009).

5.4   TRENDS AND FUTURE DIRECTIONS

Progress in science and engineering typically follows one of the two paths: a common step-by-step evolutionary process or a rare leap-forward revolutionary

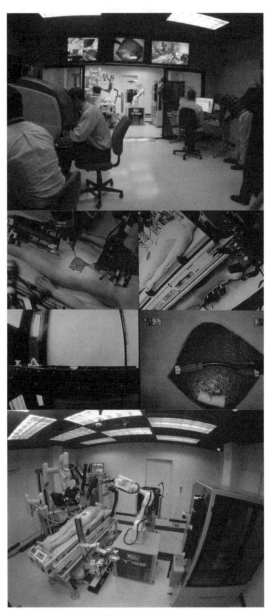

Figure 5.16 Operating room of the future. Trauma pod—Phase 1, a fully automated operating room.

Figure 5.17 Raven II—A research platform for studying surgical robotics. The image depicts a teleportation experiment in which two surgeons located at the University of Washington collaboratively teleoperated Raven II located at the University of California, Santa Cruz, such that each surgeon controls one pair of robotic arms.

process due to a breakthrough idea concept or principle. Predicting the first pattern may be based on extrapolation of existing trends, whereas the second pattern is almost impossible to predict. In the context of surgery, events such as the introduction of MIS, NOTES, and robots into the operating room may be considered as breakthroughs, which changed the common practice in surgery. In addition to the evolutionary/revolutionary patterns, there are also synergetic and symbiotic relationships between innovation in surgical practice and the technology and sciences associated with it. In the case of surgical robotics, the technology was made available in many cases ahead of its time followed by an evolutionary process of developing new surgical approaches to utilize it to improve the outcome of surgery. In contrast, the surgical concept of NOTES was ahead of the technology that would enable it to fully explore its limits.

We can identify three current evolutionary processes that may at some point also experience a revolutionary breakthrough. The first relates to the level of invasiveness of the surgical procedure. There is an effort to reduce the level of invasiveness, which leads to minimizing the impact and trauma to the surrounding tissue, reducing the risk for infection, quicker recovery, and shorter hospitalization periods. This trend faces the challenge that a reduction in the level of invasiveness is associated with smaller tools with fewer degrees of freedom and therefore limited manipulability. Moreover, in the vast majority of the surgical robotic systems, the actuators are left outside of the human body and the actuation is transmitted through cables and rods. The ability to reduce the tool tip and still maintain an external actuation source is limited by the material and mechanical properties of the transmitted mechanisms. In many respects, current designs have exhausted these capabilities.

The alternative approach is to transmit energy to an external source and hence prevent the need to invade the body, or to introduce the entire robotic system into the body without any electrical or mechanical physical connections. Packing the energy source into a small form factor capsule along with actuation, sensing, manipulation capability, and computational power remains a significant challenge (Menciassi et al., 2011). From the actuation perspective, it is possible to use external electromagnetic fields to guide and navigate an internal robotic system through the body.

The second trend is associated with an ongoing effort to improve visualization capabilities. Endoscopic cameras along with imaging modalities provide a view and representation of the anatomical structures. However, the physiology and function are not visually represented in conjunction with the anatomy. For example, if neural activity and blood flow can be merged with the anatomical representations of the brain, heart, and prostate, the outcome of procedures such as brain surgery for treating epilepsy, cardiac procedures to treat cardiac arrhythmia, and prostatectomy along with many other procedures in which both nerves and blood vessels must be spared would be significantly improved.

The third trend is related to the level of automation and control of the surgeon over the execution of the surgical procedure. Automation can be addressed at two interfaces: (i) the interface between the surgeon and the peripheral activity in the operating room (i.e., sterile and circulation nurses) and (ii) the interface between the surgeon the surgical site. *Trauma Pod*, a research program funded by the Defense Advanced Research Projects Agency (DARPA), demonstrated in 2007 that the entire operating room can be fully automated without the need for human presence (Garcia et al., 2009). The functions of the sterile nurse were replaced by a tools changer, an equipment dispenser, along with a robotic arm that replaces tools for the surgical robot and provides disposable equipment to the surgical site. The surgeon who teleoperated the robotic system, in this case the da Vinci system, issued verbal commands that triggered a fully automated tool changing and equipment dispensing. The functions of a circulating nurse were replaced by an IT system that tracked the tools and supplies throughout the procedure. Automating the surgical procedure itself using a surgical robot is currently demonstrated for hard tissues, whereas operating on soft tissue is conducted under full human control. It is anticipated that the wide spectrum between these two extremes will continue to be explored by automating subtasks of the operations while developing operational modes in which the surgeon and the robot share surgical robotic tools (Fig. 5.16).

Surgical robotic systems are primarily closed architecture systems. This approach to system design prevents any change or modification to the system by any entity other than the company that developed that system and in that way avoids any liability issues. However, these circumstances present a major difficulty to the research community for using such a system as a research platform in which change is the order of the day. In order to accommodate the needs of the research community, a surgical robotic platform named *Raven* was developed in the past decade at the University of Washington and the University of California, Santa Cruz. As a research platform, Raven is a completely open architecture system from both the software and the hardware perspectives. Its two generations were extensively tested in different modes of operation, and the final version provides a technologically mature platform for research purposes (Fig. 5.17) (Lum et al., 2007; Lum et al., 2009; Rosen et al., 2006; Rosen et al., 2011).

The revolutionary process involved with the introduction of surgical robotics system into the operating room is still in its infancy. It is anticipated that the number of operations conducted with surgical robotics will continue to grow and their use will eventually become a common practice. As innovation in surgical science and technology will continue to evolve in its unique and unexpected manner, it is likely that surgical robotic systems will have significant impacts on surgical outcomes and human health.

CHAPTER 6

Catheters in Vascular Therapy

AXEL BOESE

INKA — Intelligente Katheter, Otto-von-Guericke University Magdeburg, Magdeburg

6.1 INTRODUCTION

The average human life expectancy is rising worldwide because of improvements in standard of living and access to medical care. This increased life expectancy has led to an increase in age-conditioned and civilization-dependent illnesses such as cardiac infarction or apoplexy. These usually acutely arising diseases can be treated carefully by minimally invasive interventions. An intervention in medicine is an active treatment of a disease, and the term *minimally invasive* describes a method causing only slight injury or trauma to the patient. In minimally invasive vascular therapies, a small hole through the skin can be used to allow access into the vascular system to a location deep within the vasculature that is the cause of the illness. Catheters are used to perform these local therapies deep within the body but are controlled externally by an interventionist (Fig. 6.1).

A catheter is a tubular instrument made of glass, rubber, or plastic for the purposes of injection, instillation, draining and discharging body fluids, or as a carrier for measurement probes (Zetkin and Schaldach, 2005). Catheter interventions are also often performed with the assistance of a guide wire, which is a small wire that can be placed inside a catheter to allow steering of the catheter through the vasculature. Common applications for catheters in the human body are therapy of vascular or heart diseases, urethral use for emptying of the bladder, administration of intravenous fluids or drugs, or for the delivery of local anesthesia in the spine (epidural anesthesia).

Vascular imaging is often performed to aid the navigation of catheters and guide wires into the desired location during minimally invasive interventions. The visualization of blood vessels with the help of X-ray systems or magnetic resonant imaging (MRI) is called *angiography*. In X-ray angiography, a liquid-contrast agent

Medical Devices: Surgical and Image-Guided Technologies, First Edition.
Edited by Martin Culjat, Rahul Singh, and Hua Lee.
© 2013 John Wiley & Sons, Inc. Published 2013 by John Wiley & Sons, Inc.

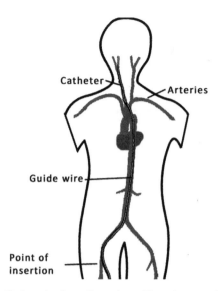

Figure 6.1 Catheterization of arteries with catheter and guide wire.

containing X-ray dense materials such as iodine is injected into the desired vessels during angiography. Because the contrast agent is radiopaque, this technique allows the structure of the blood vessels to be represented visually for a short time following injection (Fig. 6.2). Catheters used for interventions under the control of X-ray systems are typically made of polyethylene, polyurethane, or silicone and are supported by an inlay of radiopaque materials, such as a metal, which allows the catheter itself to be visualized. For angiography under MRI control, a contrast agent is not necessary; however, catheters must be MR compatible.

Together, the small puncture site and the ability to visualize and access regions deep within the vasculature using image guidance enable catheter interventions to be performed with high accuracy and minimal trauma to the patient. In addition, these advantages result in relatively short operation times and low subsequent treatment costs. However, the application of harmful ionizing radiation and contrast agents is unfavorable, perforations of the vessel wall and thrombosis can occur during these procedures, and infections caused by contaminated catheters have been reported in the literature. Despite these challenges, minimally invasive catheter interventions are therapy methods with high outcomes and overwhelming benefits for many patients.

6.2 HISTORIC OVERVIEW

The first documented use of a vascular catheter was by Sir Christopher Wren in England in the seventeenth century, who used a sharpened quill from a bird's feather for intravenous infusion experiments on animals (NC3Rs) In the nineteenth century, hypodermic needles and syringes were invented, which allowed

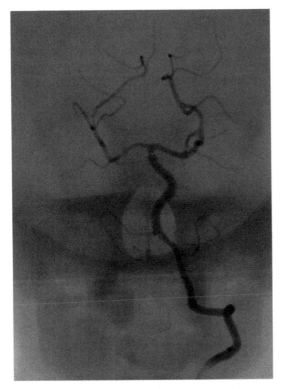

Figure 6.2   X-ray angiography of arteries using contrast agent.

more reliable intravenous administration. The next major step in catheterization and angiocardiography was achieved by Forsmann in 1929, who advanced a flexible tube through a vein in his own arm into the right atrium of his heart using a latex urethral catheter (Wilms and Baert, 1995). Clinicians Moniz, Reboul, Rousthoi, and others contributed to the development of this technique between 1930 and 1940. In 1941, Cournand and Richards first demonstrated the safety of cardiac catheterization in humans by performing intravascular blood pressure and flow measurements within the right atrium of the heart. In these procedures, cardiac catheterization was found to have no side effects or complications for the patient, and this technique became routine in hemodynamic evaluation (Peters and Cleary, 2008). Seldinger published his famous method for percutaneous catheterization in 1953, significantly improving the safety of interventional procedures (Kalso, 1985). This technique is still the gold standard for angiographic interventions at present.

   With the improved safety and subsequent expansion of catheter interventions, new tools and applications were developed. Balloon angioplasty catheters were invented in 1977 by Gruentzig, and featured a balloon integrated onto the catheter tips that allowed clogged vessels to be expanded and their blood flow improve Stents were subsequently developed to serve as metal scaffolds to keep vessels o after balloon dilation, and in 1989, were first implanted by Palmaz into an artery of the chest using a balloon angioplasty catheter. Balloon expandable stents are still in use. Other recent advances with catheters have focused on the improvement of catheter materials, catheter construction, and biocompatible coatings.

6.3   CATHETER INTERVENTIONS

Catheters are primarily used for the therapy of vascular illnesses in the heart, brain, and peripheral vessels, both in the arteries and veins. The type, shape, and specifications of a catheter are dependent on the region of the body and the planned therapy. The size of the catheter primarily depends on the diameter of the vessel; catheter diameter should generally be less than two-thirds of the vessel diameter to maintain adequate blood flow. Diameters of vessels frequently used for catheter interventions are listed in Table 6.1.

The first step of an intervention using a vascular catheter is the insertion into the vasculature and is most commonly performed using the Seldinger technique. Various access points can be used throughout the body, such as the basilar artery in the arm or the femoral artery in the leg. After the application of a local anaesthetic, the skin and the underlying vessel are punctured with a hollow needle, or cannula (Fig. 6.3, step a). The correct position of the cannula within a vessel is verified by inducing bleeding (Fig. 6.3, step b). After removing the interior part of the cannula, the mandrine, a guide wire can be inserted into the vessel (Fig. 6.3, step c). The cannula is then extracted while holding the guide wire in place with a finger (Fig. 6.3, step d and e). The puncture location must then be widened using

Table 6.1   Typical Diameter of Blood Vessels in the Human Body

Vessel	Region of Body	Diameter, mm
Femoral artery	Leg	7–11
Abdominal aorta	Thorax	15–25
Transverse aorta	Thorax	20–30
Brachiocephalic artery	Thorax	12
Brachial artery	Arm	5
Subclavian artery	Arm	7
Common carotid artery	Neck	7–8
Vertebral artery	Neck/head	0.9–4.1
Internal carotid artery	Head	3.3–5.4
Middle cerebral artery	Head	2.4–4.6
Basilar artery	Head	2.7–4.3
⁻ior inferior cerebellar artery	Head	0.3–1.5
n femoral vein	Leg	12–13
ena cava	Body	15–30
gular vein	Neck	5–6
vein	Arm	7

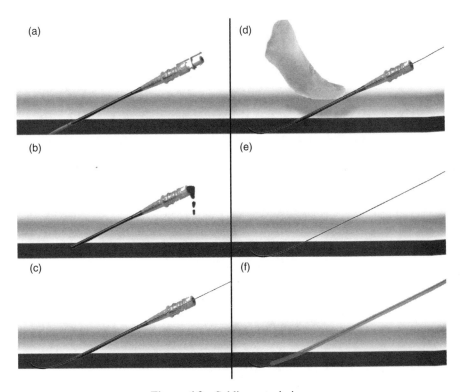

Figure 6.3   Seldinger technique.

a dilator or scalpel to allow the catheter to pass into the vessel over the guide wire (Figs. 6.3–step f).

After insertion into the vessel lumen using the Seldinger technique, the guide wire and catheter are alternately advanced forward into the vasculature. The size, stiffness, and shape of both the guide wire and catheter must be considered, as both influence steering through the vascular system, especially through bifurcations and complex vessel structures. Steering of the catheter can be simplified using catheter with different tip configurations or preformed or ductile guide wires. Guide wir can often be shaped manually before insertion to facilitate advancement of the v into complex structures (Fig. 6.4). The catheter can then subsequently be adva along the guide wire into the desired vessel region.

Two common techniques are used to bend the guide wire and catheter t and to steer them into the desired vessel. The first technique features a p catheter, such as one with a soft, angled tip, and a stiff guide wire. By a stiff guide wire into the internal lumen of the catheter, the angled ç straightens, and the catheter tip is advanced into the vessel. When th into close to a vessel bifurcation, the stiff guide wire is removed, and the th the catheter is bent into the branching vessel. A softer guide wire is the the catheter lumen and the catheter is advanced further into the

(a)   (b)

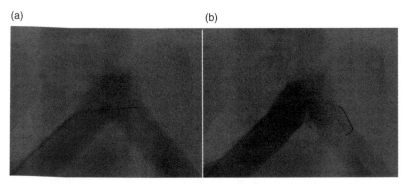

Figure 6.4   Steering of a catheter and guide wire in the external iliac artery (cross-over technique) under X-ray guidance.

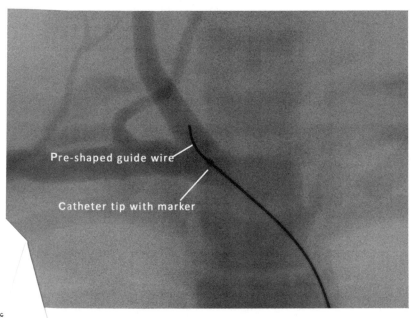

Pre-shaped guide wire

Catheter tip with marker

6.5   Catheter and preshaped guide wire in the brachiocephalic artery.

rved shape influencing the shape of the soft guide wire. The second
ring technique features a straight catheter and a preshaped guide wire.
stra catheter is placed near a vessel branch, and a preshaped guide
exits into the lumen of the catheter. The preshaped guide wire becomes
into th ide the catheter lumen, but the guide wire acts as a spring, as it
location tip and regains its original shape. This bent tip can then be fed
guide wire g vessel and advanced forward until the wire is at the desired
ion. The straight catheter can then be advanced forward over the
ng the guide wire's path into the vessel branch.

After reaching the desired target, local therapies are performed, such as administration of medication, breakdown of blood clots with drugs (thrombolysis), expansion of narrowed vessels (balloon angioplasty), delivery and placement of a stent, or mechanical removal of vascular occlusions. Catheters and guide wires must be cleaned and rinsed several times during the intervention to prevent thrombosis. Therefore, heparine and saline solution are flushed through the catheter or the catheter and wire are removed and wiped several times throughout the course of the interventional procedure. When the procedure is complete, both the guide wire and catheter are removed.

6.4 CATHETER AND GUIDE WIRE SHAPES AND CONFIGURATIONS

6.4.1 Catheters

6.4.1.1 Specifications When designing a catheter for interventional use, it is important to follow guidance documents and standards provided by the US Food and Drug Administration (FDA) and the International Organization of Standards (ISO). The international standard norm ISO 10555-(1–7) (ISO, 2010a) specifies general standards for sterile single use intravascular catheters and is the standard most relevant for use outside of the United States. The following information is required to be supplied by the manufacturer: description of the product, the outside diameter, effective length, name or trade name and address of the manufacturer, lot designation, expiration date or use before date, any special storage or handling instructions, indication of sterility, method of sterilization, indication for single use, any known chemical and/or physical incompatibilities with substances likely to be used with the catheter, and instructions for use and warnings, as appropriate. Much of this information is listed on the catheter package (Fig. 6.6).

The outside diameter of an intravascular catheter is most often indicated in either millimeters or French units (F), where 0.33 mm is equal to 1 F. The minimum inside diameter and/or the suitable maximum guide wire size is also important, as well as the lengths of each section of the catheter (each with different rigidity) and the positions of radiopaque markers within the catheter shaft. Some interventional catheters also have preformed tips or standard ports for injection of contrast agents or medications that must be specified.

6.4.1.2 Construction and Materials Design and construction of vascular catheters have to fulfill several requirements, with the primary requirement to optimize function of the catheter. The tip of a catheter must be rounded to prevent perforation of the vessel. The diameter of the catheter must be appropriate for the application, and the stiffness must be sufficient for steering, but soft enough to prevent damage to the vessel walls. For this reason, the rigidity often varies along the length of a catheter, with the distal end softer than the proximal end. Radiopaque structures such as bismuth, molybdenum, lead, or barium sulfate may be placed on or within the catheter in order to visualize important interfaces of a catheter (Peters

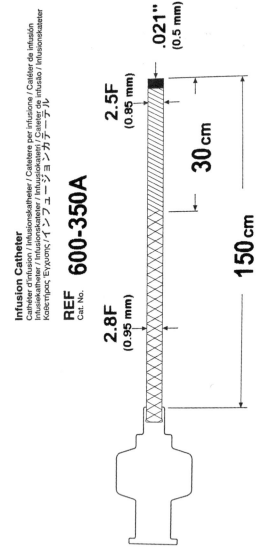

Figure 6.6 Package information of a vascular catheter.

Figure 6.7   A traumatic tip and X-ray marker of a balloon catheter.

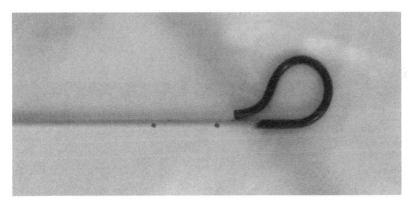

Figure 6.8   Pig tail or flush catheter with side-holes.

and Cleary, 2008), such as the proximal and distal ends of a balloon (Fig. 6.7) or of a flexible tip. These structures are often integrated into the catheter as external rings, as inlets of metal mesh inside the catheter wall or as small metal particles in the wall material itself.

Many catheters have holes toward the tip for improved mixing of injected contrast agents with flowing blood (Fig. 6.8). Better mixing improves visualization of the blood vessels during X-ray angiography by revealing the full vessel structure instead of diluted streaming of the contrast agent in the blood. This is especially critical in areas with high blood flow and volume, such as the aorta.

Few materials are suited to the design of vascular catheters. Polyethylene, polyurethane, or silicone are most commonly used for vascular catheters, because they are biocompatible, easy to manufacture, and flexible. A mixture or combination of different materials is often used to optimize the performance of a catheter (Fig. 6.9). Biocompatibility is critical in catheter design, as vascular catheters come into contact with blood, the vascular endothelium, and other tissues such as skin and connective tissues. During intravascular procedures, catheters can be coated with biofilms, and can act as substrates for thrombosis and microbial colonization. These risks are reduced by coating the catheter surface with antithrombotic coatings such as Teflon (PTFE) or silicone. However, catheters must also be compatible with chemical compounds such as contrast agents or medications.

Figure 6.9   Two catheters of different size and material.

Figure 6.10   X-ray image, showing internal support structure of catheters from Figure 6.9.

The rotational stiffness of the catheter is particularly important in order to transfer the required torsion from the proximal to the distal end of the catheter and to accurately steer the catheter through the vasculature. A rigid inner structural material, such as a mesh of nylon, dacron, or steel, is often embedded into the soft outer catheter wall to support the wall and to improve rotational steering (Fig. 6.10). Clearly, vascular catheter design requires careful consideration of the physical, biological, and chemical properties of the materials, coatings, and structures used in catheters. Some common design requirements for vascular catheters are listed in Table 6.2.

6.4.1.3   *Catheter Tips and Applications*   There are several types of blood vessel bifurcations inside the human body, and the geometry and even the structure of the vessel tree changes from patient to patient. A variety of catheter types and shapes can be used to accommodate this range of geometries and structures. In some cases, the tip of a catheter can be customized by the interventionist by forming the tip to the desired shape using hot steam. Catheters with preformed tips are also available for common interventions. A standard multipurpose 6 F catheter with a preformed 45° tip, used for diagnosis of vascular disease, is shown in Figure 6.11. An overview of preformed catheters and their applications is given in Figure 6.12, with the catheters defined by either inventor, function, or shape.

Catheters can have one or more lumens, or inner channels through the body of the catheter. These lumens can be arranged in parallel through the catheter, in

Table 6.2   Design Requirements for Vascular Catheters[a]

Biological	Physical	Chemical
Nonirritant—provokes minimal inflammatory response	High tensile strength, rotational stiffness	Absence of leachable additives (e.g., catalysts and plasticizers)
Noncarcinogenic—low tendency to cause neoplasia (cancer)	Resists compression—maintains patent lumen	Stable during storage
Nonthrombogenic—low tendency to cause blood clotting	Optimum flexibility	Stable on chemical sterilization
Nontoxic	Low friction coefficient	Stable on implantation (nonbiodegradable)
Resists microbial adhesion	Dimensional stability	Permits adhesives in fabrication (possibility of bonding dissimilar materials)
Resists biofilm deposition	Tolerates physical sterilization methods (e.g., heat, steam, irradiation)	Accepts surface coatings (e.g., hydrogel, antithrombotic, antibacterial)
Low friction	Ease of fabrication (e.g., heat forming or welding) Nonpermeable (water, gases, solvents) Radiopacity—ability to image catheter with X-rays MRI (magnetic resonance imaging) compatible	Compatibility with chemical compounds and solvents (absence of absorption and chemical reaction)

[a](Catheter Design, 2010).

which one lumen is used for guide wire placement and the others for measurement of internal blood pressure or temperature or administration of medication. Some examples of multiple lumen catheters are central venous catheters, which are placed near the heart, or pulmonary artery (Swan–Ganz) catheters, which are fed into the pulmonary artery (Fig. 6.13a).

Thrombectomy catheters are used for the mechanical removal of blood clots and can utilize a variety of operating principles. The vortex principle can be used to rotate the catheter within the vessel and fragment the thrombus. The Bernoulli and the Venturi principles can also be applied using a water jet that causes a negative pressure inside the catheter and sucks the thrombus inside (Lichtenberg, 2010). In Figure 6.13b and c, a triple lumen thrombectomy catheter that operates based on the Venturi principle is shown. Other devices for the mechanical removal of thrombus function such as a corkscrew, eliminating blood clots by retrieving them and pulling them into the catheter.

Another important application for vascular catheters is the opening of vessel occlusions. These occlusions, called *stenoses*, are caused by calcification or plaque

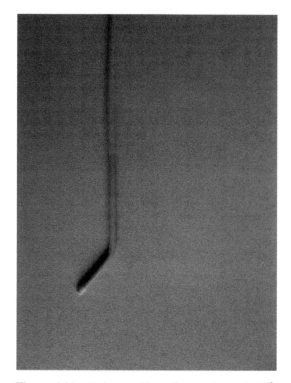

Figure 6.11   Catheter with preformed tip angle 45°.

Shape	Name	Example of use
	Flush catheter	Flush aortography
	Headhunter	Selective catheterization carotis
	Sidewinder	Cross-over, cerebral angiography
	Newton	Cerebral angiography
	Bentson	Cerebral angiography
	Mani	Cerebral angiography
	Multipurpose	Multipurpose
	Cobra	Cross-over
	Sheperd hook	Selective catheterization kidney

Figure 6.12   Different catheter types and examples of use.

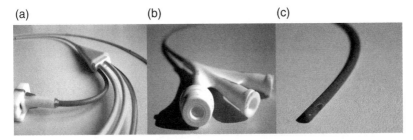

Figure 6.13 Multilumen catheters: (a) pulmonary artery (Swan–Ganz) catheter, (b,c) thrombectomy catheter.

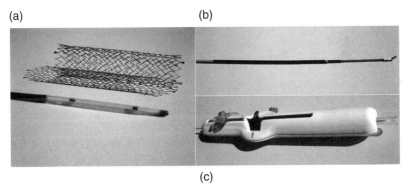

Figure 6.14 (a) Balloon catheter and nitinol stent, (b) mounted stent on a balloon catheter, and (c) inflation pump for angioplasty.

on the inner vessel wall or blood clots in areas of low blood flow. Vessel structures beyond the stenosis can suffer from reduced blood flow, affecting the brain's function when the stenosis occurs in the head or neck. Stenosis in the peripheral vessels in the legs or arms can lead to pain and loss of power of muscles, and occlusions in the coronary vessels of the heart can lead to dysfunction or decreased heart output.

For the therapy of these occlusions, balloon angioplasty is performed, in which a special balloon catheter is placed within the stenosis, and the vessel is expanded by inflating the balloon. To keep the vessel open, a stent, or grid of metal or plastic (Fig. 6.14a), can be placed around the balloon tip of the catheter (Fig. 6.14b) and delivered to the location of the stenosis (Baim and Grossman, 2006). The balloon is inflated using an inflation pump (Fig. 6.14c) to widen the stent and to secure it within the occluded vessel so that the vessel remains open.

Balloons integrated onto catheter tips may also be used to block the circulation of blood to provoke occlusion in vessels during angioplasty and to keep removed calcifications from traveling further down the blood stream. Balloons may also serve as tools for mechanical removal of calcifications and plaques. Smaller balloons can anchor the tip of an interventional device at a fixed position within a vessel, as the replacement of a guide wire or insertion of different tool often results in movement of the catheter tip. Balloon catheters typically feature two concentric lumens, with

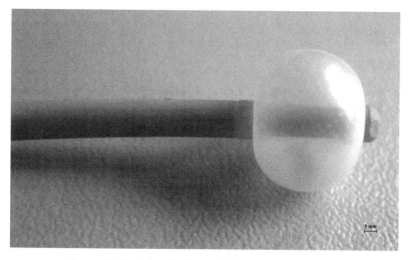

Figure 6.15 Inflated silicone balloon on a catheter tip.

an inner lumen used for guide wire insertion and contrast agent injection, and an outer lumen for inflation of the balloon. The balloons themselves are made out of thin plastics or a flexible material such as rubber (Fig. 6.15).

Atherectomy catheters are also used to remove calcifications, especially in the extremities. Small blades mounted at the catheter tip or on the balloon of an atherectomy catheter can aid in mechanical scraping of calcified deposits from a vessel wall. The separated material is then aspirated trough the inner lumen of the catheter, such that it does not travel further into the vascular system.

Visualization of the inner wall structure of a vessel can be performed internally using intravascular ultrasound (IVUS) catheters. These catheters feature a miniaturized ultrasound transducer integrated within the tip of an IVUS catheter (Fig. 6.16). A thin section or slice of the vessel can be scanned, by emitting and receiving high frequency ultrasound signals from the transducer. By slowly pulling the catheter tip out of the vessel, a number of image slices can be combined, and the vessel wall can be scanned over a portion of its length. The translation of the catheter is often motorized and carefully controlled to maintain a constant pull-back speed ($\sim$0.5 mm/s) that can be accurately related to the image sampling rate to provide high quality 3D ultrasound imagery.

Catheters have also been developed specifically for cardiac electrophysiology in order to treat heart arrhythmias, or irregular heartbeats, that can lead to a loss of heart capacity. These devices feature integrated electrodes that can measure the electrical behavior of the myocardium to detect arrhythmia (Fig. 6.17). Once located, the arrhythmia is treated by ablating the responsible cardiac muscle tissue with high frequency electrical pulses (Mario et al., 2010). Steerable cardiac electrophysiology catheters use a cable-driven steering mechanism in the catheter lumen that allows the catheter tip to be bent, such that the electrode can provide measurements of most inner surfaces of the heart.

Figure 6.16   Ultrasound transducer incorporated into an IVUS ultrasound catheter.

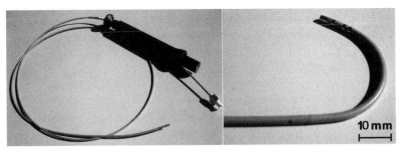

Figure 6.17   Steerable catheter for cardiac electrophysiology. (a) Whole catheter with steering mechanism inside the grip and (b) electrode at the bent tip with three ring markers for visualization under X-ray angiography.

6.4.2 Guide Wires

6.4.2.1 Specifications   The international standard for guide wires, ISO 11070 (ISO, 2010b), describes the requirements, test conditions, and restrictions for their manufacturing and usage. According to the standard, the nominal size of a guide wire is designated by the maximum outside diameter. The length of the guide wire must be given to an accuracy of ±5%. The following information must be supplied by the manufacturer: description of the device, name or trade name and address of manufacturer, lot designation, expiration date or use-by date, any special storage and handling instructions, indication of sterility, method of sterilization, indication for single use, any known incompatibilities with substances likely to be used with the device, and instructions for use and warnings, as appropriate.

The diameter of a guide wire depends on the catheter that is used. In practice, the diameter of a guide wire is specified in inches and in millimeters, and many procedures are performed using 0.035 in. (0.89 mm) wires. Smaller vessels may be accessed using 0.018 in. (0.46 mm) and 0.014 in. (0.36 mm) wires.

6.4.2.2 Construction and Materials The structure of most thick guide wires (e.g., 0.035 in.) is in principle the same, having a thin flexible wire wrapped around an inner, thicker wire core (Fig. 6.18). This enables an accurate transfer of torsion and rotation from the proximal end in the user's hand to the distal end inside the patient. These guide wires have rounded tips to prevent penetration of the vessel wall, and an additional safety wire (Fig. 6.21) inside the outer wire is provided to prevent the rounded tip from getting lost in the patient during the intervention. A conical rather than cylindrical inner core can be used to vary the stiffness over the length of the guide wire (Fig. 6.18).

Thinner wires (e.g., 0.018 in.) are fabricated with a single wire construction. These guide wires often feature a springlike tip that is attached to the distal end to protect from penetration of the vessel wall during insertion (Fig. 6.19).

A specific guide wire may be selected for a desired application, or several guide wires may be used throughout the course of an intervention. A guide wire with a stiff end is better suited to pass through a stenosis or calcification or to straighten a catheter, while a guide wire with a soft end is better for steering into a vessel bifurcation. During the beginning of an intervention, a guide wire with a flexible tip is often selected to reduce the risk of perforation of the vessel (Schneider, 2008). Stiffer wires are often used to pass through a stenosis or to straighten a catheter. Softer guide wires are more desirable in small vessels, such as those in the brain, which have a high risk of rupture. To reduce the cost of interventions, a guide wire with a variable stiffness can be selected, such as one that has both a soft end

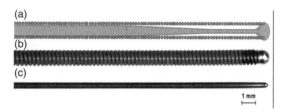

Figure 6.18 Guide wire construction: (a) Illustration and (b) photo of thicker (0.035 in.) guide wire with wire wrapped around a conical core, and photo of thinner (0.018 in.) guide wire with (c) single wire construction.

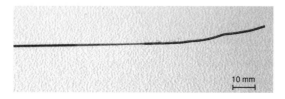

Figure 6.19 Guide wire with 0.018 in. diameter and a springlike tip.

and a rigid end. Both ends can then be used for the intervention, according to the requirements.

A coating of silicone, PTFE (Teflon), or polyurethane is applied to the guide wire surface to reduce the coefficient of friction (Schneider, 2008). Guide wires must be biocompatible and, in general, must adhere to the same biological, chemical, and physical requirements as catheters (Table 6.2). As with catheters, guide wires are used during X-ray angiography and, therefore, can also include radiopaque markers or materials that improve visualization of the guide wire tip or of the entire guide wire.

6.4.2.3 Guide Wire Tips and Applications   As with catheters, guide wires can be ordered to have preshaped tips or tips that can be manually shaped by the interventionist (Fig. 6.19). An angled or curved wire is used for selective catheterization to provide better access into a vessel branch (Fig. 6.20). The stiffness and the length of the flexible tip can be varied in special guide wires with extractable cores (Fig. 6.21). In this case, the stiffness is reduced during a procedure by pulling

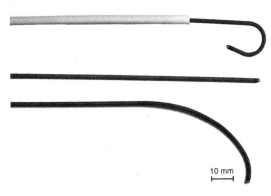

Figure 6.20   Different shapes of guide wires: J-type wire inside 5F catheter, straight wire, and manually shaped wire.

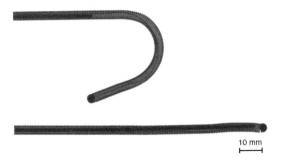

Figure 6.21   Guide wire with extractable core and safety wire inside.

an internal core wire back inside the guide wire assembly so that only the floppy spring structure of the outer wire remains as the tip. This technique can also be used for bending in order to steer the guide wire into a branching vessel.

6.5 CONCLUSION

Catheter interventions make it possible to access internal structures deep within the circulatory system. However, these interventional procedures are limited by the size of the vessels, with smaller vessels requiring a smaller diameter catheter to maintain blood flow. Unfortunately, smaller catheters are limited by a lower rigidity and are more difficult to steer through the vasculature because of the influence of tube diameter on the transfer of torsion. Small vessels are also more sensitive to perforation, requiring catheters and guide wires accessing these vessels to have softer and more flexible tips. However, this flexibility in bending also results in a reduction of steering control.

Reaching the smallest vessels of the brain is not only difficult but also time consuming. However, in emergency situations, such as treatment of a stroke or a vessel rupture, time is critical. The acute therapy of ischemic stroke improves patient outcomes only within the first 5–6 h following the first symptoms. Two hours following stroke, the affected tissue damage is typically irreversible. Six hours after the occurrence of the first symptoms, there is no significant advantage to the acute therapy, and the risk of intracerebral bleeding or other complications is higher than the potential for saving brain functions (NINDS) (Nedeltchev, 2003; ESO, 2008). The anatomic structure of vessels can also add significant complication and time to a therapy, especially in the case of vessel structures with many bends or with bending angles of more than 90°.

Despite these challenges, the development of new tools, technologies, and techniques are expected to lead to further improvements in vascular access, both in terms of speed and accuracy. New coatings and materials can further reduce friction and thrombogenesis. Robotics and steering mechanisms are being developed by companies such as Hansen (Sensei) and Siemens (magnetic navigation), and may provide significant improvements, especially in cardiac interventions. The Hansen Sensei system uses internal wires to steer the tip of the catheter through the vasculature by varying the tension of these wires. The Siemens system features large external magnets beside the patient's table that can bend a catheter with a small magnet on its tip. Both systems are designed for electrophysiology, where accurate mapping of the internal heart is important and the diameter of the catheter can be relatively large. The development of microelectromechanical systems (MEMS)-based sensors and electronics will enable local internal measurements, such as temperature, oxygen saturation, blood pressure, and blood flow velocity to be performed simultaneously at the catheter tip. Local blood flow inside aneurysms is of particular interest to physicians who make therapeutic decisions for the treatment of aneurysms.

High quality imaging systems with new functions and better image resolution may also improve patient outcomes by allowing smaller structures or lesions to be detected. For brain interventions, hybrid interventional suites are becoming increasingly available, and allow for the fusion of X-ray and MR information in order to provide a more complete set of visual data for the physician. Development of MR compatible materials for catheters and guide wires may also increase the number of interventions that can be performed under real-time MRI guidance, allowing for improved contrast and visualization of soft tissue. Together, improvements in catheter and guide wire design, control, and visualization are expected to enable more effective use of catheters and to bring about an expansion of minimally invasive interventional procedures.

PART III

ENERGY DELIVERY DEVICES AND SYSTEMS

CHAPTER 7

Energy-Based Hemostatic Surgical Devices

AMIT P. MULGAONKAR, WARREN GRUNDFEST, and RAHUL SINGH

Center for Advanced Surgical and Interventional Technology (CASIT), University of California, Los Angeles, CA

7.1 INTRODUCTION

Hemostasis, from the Greek root heme (blood) and stasis (halt) can be defined in the clinical context as control of bleeding without the occurrence of pathologic thrombotic events (Levy et al., 2010). More simply, the most basic definition of hemostasis is the cessation of bleeding. Management of bleeding resulting from injury or surgery remains a major challenge in medicine. In normal conditions, hemostasis naturally occurs through the actions of the coagulation cascade, in which a symphony of multiple systems and mechanisms act together to arrest hemorrhage, or blood loss. However, severe trauma and the normal anatomic insults inherent in surgical procedures can overwhelm the body's natural clotting response.

Current strategies for surgical hemostasis can roughly be divided into four major categories: (i) treatment using drug agents or blood products, (ii) treatment using mechanical means, such as clamps, clips, ligatures, patches, or glues, (iii) treatment using biologically active adherent materials, and (iv) energy-based surgical methods that coagulate as part of their action.

The use of energy is a fundamental part of the modern surgical practice. Energy-based surgical methods include a wide variety of tools and technologies, including electrocautery, electrosurgery (Fig. 7.1), ultrasonics, and laser-based systems. While each of these techniques relies on different methods of operation, they all achieve their hemostatic functionality through the same basic set of physical and thermal mechanisms. Electrocautery relies on direct application of thermal energy to the surrounding tissues by means of a heated instrument. Electrosurgery acts through means of a high frequency electric current applied to the desired site of

Medical Devices: Surgical and Image-Guided Technologies, First Edition.
Edited by Martin Culjat, Rahul Singh, and Hua Lee.

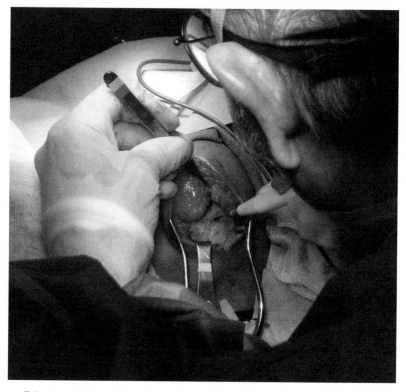

Figure 7.1   Surgical excision of a tumor using electrosurgical coagulation. Retrieved from http://en.wikipedia.org/wiki/File:Electrosurgery.jpg.

action. Ultrasonic coagulator systems make use of the mechanical action of a high frequency piezoelectrically driven element directly on tissues. Laser-based systems rely on the emission of energetic photons to transfer thermal energy to the target area.

This chapter aims to provide a basic introduction to the history and principles of operation behind these four energy-based surgical methods. Owing to its continued prevalence in modern surgical practice as well as the rich history and tradition of the technique, this chapter specially focuses on the use of electrosurgical techniques. For more information on ultrasonics, please see Chapter 14, and on lasers, Chapter 9.

7.2 HISTORY OF ENERGY-BASED HEMOSTASIS

The use of thermal cautery for hemostasis has been shown as far back as the medical practices of the ancient Egyptians around the thirtieth century BCE (Breasted, 1930; Vender et al., 2005). Hippocrates described the use of cautery for a variety of conditions in the fourth century BCE. Rods specifically designed for cautery were also found in the ruins of a house in the Roman city of Pompeii buried by volcanic ash from Mt. Vesuvius in 70 CE (Wall et al., 2008).

Boiled oil was occasionally used to achieve wide-scale hemostasis during surgical procedures, cauterizing tissues to which it was applied (Malis, 1996). However, this method resulted in frequent charring of adjacent living tissues due to the large amount of thermal energy released by the oil and the inherent imprecision of using a liquid thermal medium (Bulsara et al., 2006). Thermal cautery, which generally resulted in a mass that eventually sloughed (shed) away, was one of the primary hemostatic tools of the ancient surgeon.

The use of thermal cautery eventually waned with Ambroise Paré, the French surgeon credited as one of the fathers of surgery, demonstrating in the mid-sixteenth century the superiority of ligatures and other hemostatic methods to that of indiscriminate cautery (Wies, 1929; Vender et al., 2005). It was not until 200 years later that the advent of electrically heated elements allowed for thermal cautery to be used with the precision necessary for modern surgical procedures.

Early observations of the effect of electricity on tissues were generally the result of accidental self-inflicted electrical-discharge injuries of the early experimenters working with the as then not yet fully understood phenomenon of electricity. One of the first such experiments was by Cusel, who in 1847 used electricity to destroy an epithelial neoplasm (Fiandesio and Valobra, 1968; Valobra and Fiandesio, 1968). In 1881, William Morton discovered that painful electrical shock could be avoided by passing a high frequency alternating current (AC) of 100 MHz through biological tissues. Jacques-Arsene D'Arsonval showed similar results with a lower frequency (10 kHz) in 1891 (d'Arsonval, 1891). He additionally noted that the current influenced carbon dioxide elimination and oxygen absorption, while also causing a rise in tissue temperature. This phenomenon of tissue heating was termed *diathermy* by Franz Nagelschmidt in 1897 (Kobak, 1925). By focusing the current at the end of a fine electrode and varying the separation of the electrode to the tissue, the heat produced was capable of being modulated. The resulting d'Arsonval's device was capable of heating tissues by concentrating current at the end of an electrode and was used to treat skin lesions as early as 1900. The techniques and apparatus were subsequently improved, with a grounding pad added in 1907. In 1910, Dr. William Clark developed an electrosurgical spark-generator capable of producing a much "hotter" spark, subsequently allowing for deeper and more efficient diathermic tissue penetration. The first endoscopic medical procedure involving fulguration, or impingement of an electrical spark on tissue by separating the electrode from the tissue surface, was reported in 1914, but this method had not yet come into widespread use (Hall, 1976; Malis, 1996; Bulsara et al., 2006).

Much of the impetuous for the development of advanced methods of surgical hemostasis in the first decade of the twentieth century came from the field of neurosurgery. Clinically accepted practices and techniques, such as suturing or clamping, were not suitable for use with delicate brain tissues (Cushing, 1911) or were too cumbersome for the limited operative fields encountered in neurosurgery (Voorhees et al., 2005). The need to manually stem blood loss at each step drastically lengthened procedures, increasing the risk of life-threatening hemorrhages developing elsewhere. In the mid-1920s, Dr. Harvey Cushing, a prominent neuro-surgeon, started a collaboration with Dr. William Bovie, a physicist working on developing a more efficient electrosurgery system (O'Connor et al., 1996). This system, which consisted of a foot-operated electrosurgical loop capable of shaving away cancerous tissues, was first used to successfully remove a previously inoperable brain tumor in 1926. The system, now known as a *Bovie knife*, built upon the similar works of several contemporaries (Voorhees et al., 2005), and forms the basis for modern electrosurgical systems. For an in depth look at the individual advances that led to the Bovie and its attendant revolution of surgical hemostasis, the reader is directed to Ward's contemporary summary of the development of electricity in medicine and electrosurgery (Ward, 1932).

The original electrosurgical devices based on the Bovie are what we now refer to as monopolar electrosurgical systems, in that the current passes from an electrode with a single terminal to a distantly placed dispersive grounding pad. In 1940, Dr. James Greenwood developed the "two point" coagulator that forms the basis for the modern bipolar electrosurgical systems. His observations of the damage caused by conventional monopolar systems to surrounding tissues due to the strong currents required for inducing hemostasis in brain tissue led him to replace the conventional monopolar system with a pair of fine-tipped forceps attached to a current-generating power source (Greenwood, 1940). This allowed him to pick up and coagulate individual vessels, as only tissues between the forceps tips experienced current flow, and therefore heating. In general, monopolar systems tend to be more effective for hemostasis because of the deeper heating of tissues, while bipolar electrosurgical systems result in less tissue damage because of the lower power required for the closely spaced electrodes (O'Connor et al., 1996) Dr. Leonard Malis subsequently designed and built the first commercial bipolar coagulator in 1955 (Malis, 1996).

Following the discovery of the piezoelectric effect by the Curie brothers in 1880 and the development of high powered transducer systems by Wood and Loomis in 1927, ultrasonic energy was discovered to have a variety of effects on tissues, including the transmission and generation of heat and the destruction of tissues through cavitation (Wood and Loomis, 1927; Mason, 1981). This was applied in 1968 to the removal of cataracts through a technology called *phacoemulsification* (Kelman, 1974; Kelman, 1998). A variety of modern devices now make use of ultrasonics to both cut and coagulate tissues (Fig. 7.2).

In the early 1960s, CO_2 lasers emitting light in the infrared region were shown to effectively coagulate blood. But the expense of these lasers and the lack of an ergonomically effective delivery system restricted their use. Subsequent laser developments in the early 1980s led to the use of Neodymium:YAG (Yittrium Aluminum Garnet), Holmium:YAG, and Erbium:YAG lasers for the coagulation and cutting of tissues, but with various limitations in performance. For example, while Neodymium:YAG lasers transmitted through a quartz fiber-optic probe and operating at 1.06 μm were shown to be capable and effective at hemostatically sealing large vascular organs, they were ineffective with soft tissues because of the increased penetration depth of the laser into tissue compared to blood (Goldman

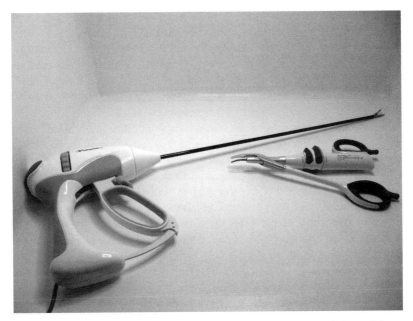

Figure 7.2 Two types of ultrasonic shears, including a ligature (left) and harmonic scalpel (right). Both perform hemostatic functions using ultrasonic vibrations.

et al., 1973; Glover et al., 1978). More recently, a series of diode lasers that are significantly cheaper, smaller, and more compact have been used in a variety of surgical procedures. While effective as coagulation tools, the combination of expense and the need to wear protective eye covering has limited the application of lasers in surgery. While some surgeons were initially enthusiastic about the use of lasers for the coagulation of tissue, there are few quantifiable benefits over standard electrocautery for most applications. Certain lasers have found unique roles in the treatment of dermatologic conditions and in the case of the blue–green argon laser for retinal photocoagulation, but for general surgery, the laser remains focused on a range of narrow niche applications.

7.3 ENERGY-BASED SURGICAL METHODS AND THEIR EFFECTS ON TISSUES

All the hemostatic methods discussed in this work transform energy to heat to achieve hemostasis. The first and most basic form of a hemostatic device uses the application of direct heat to cut or coagulate tissue. Electrocautery uses heat generated from an electrical current flow in the instrument to locally burn, coagulate, or destroy tissue to prevent hemorrhage. In electrocautery, electricity is used to heat an object, and the object is then used in a manner similar to classic "hot iron" cautery (Zinder, 2000).

Electrosurgery is the cutting, coagulating, desiccating, or fulgurating of biological tissues by means of directly applying a high frequency (>100 kHz) electrical current to the site of desired action. This differs from electrocautery in that an electric current passes directly through tissues in the body and induces local heating at the desired site of hemostasis. In this context, coagulating is the act of clotting a wound, desiccating is the drying of tissues, and fulgurating is the destruction of tissues using electrical discharges.

A variety of tools and techniques also exist that harness ultrasonic energy to coagulate tissues and perform surgical hemostasis. At the most basic, these techniques use high frequency mechanical vibrations (above 30 kHz) from piezoelectric transducers to generate localized vibration-induced heating in tissues. This controlled heating denatures proteins, forming a local hemostatic coagulum that can be used to seal both large and small vessels, and coapt (join) tissues. The effect observed is a function of both the shape of the tool in contact with the tissue, as well as the energy level and duration of application of the ultrasonic energy. For example, using a scissor-blade style harmonic scalpel, one can cut tissue at the center of the instrument while simultaneously coagulating the edges of the cut zone through vibration-induced heating.

Several different laser-based systems for surgical hemostatic control have also been proposed and developed for clinical use. These systems all rely on photothermal heating effects for ablation and coagulation, which occur in the $1-10^6$ W/cm^2 power level range(Wall et al., 2008).

7.3.1 Disambiguation

In both the published literature and the common parlance of the practitioner, a number of terms have historically been used interchangeably, and in many cases incorrectly, to refer to these techniques and their specific outcomes (Zinder, 2000). The terms electrosurgery, electrocautery, radiosurgery, endothermy, diathermy, surgical diathermy, electrodessication, and radiofrequency (RF) heating are commonly used and confused, in part due to lack of specific definitions when the terms were being coined, and changing meanings and uses of the terms as the use of energy in medicine has advanced and matured.

The modern definition of diathermy, as specifically defined by the FDA, is the "controlled production of 'deep heating' beneath the skin in the subcutaneous tissues, deep muscles, and joints for therapeutic purposes" (FDA, 2009). However, historically, diathermy has been used more generally to refer to any electrical heating of tissues. Surgical diathermy is often used synonymously with electrosurgery.

While electrocautery and electrosurgery are often used interchangeably, electrocautery is strictly defined as the use of an electrical current to heat the cautery probe, while electrosurgery uses an applied RF field to internally heat the tissue. Simply, electrocautery uses a DC current to heat the cautery probe to externally apply heat (endothermy), as in soldering, while electrosurgery uses an AC current field to internally heat the tissues, as in a microwave.

7.3.2 Thermal Effects on Tissues

All the energy-based surgical methods discussed fundamentally function using thermal energy to achieve the desired hemostatic results. Electrocautery directly imparts the thermal energy, electrosurgery through ionic motion, lasers use light, and ultrasonic coagulation uses direct mechanical action.

Heat is transferred in tissues in the absence of blood flow through conduction. The tissues themselves absorb energy in two distinct ways: by heating due to increased translational motion of the particles in the tissue (Ohmic heating) and through increased vibrational or rotational motion (dielectric heating) (Zinder, 2000). In general, Ohmic heating is dominant below 500 MHz, with dielectric heating dominating above that threshold. Both electrosurgery and ultrasonic diathermic systems rely on Ohmic heating, while laser-based systems rely on dielectric heating.

The Ohmic heating effect is the generation of heat when the electric current flows through a conductor, with the heat generation due to the resistance of the conductor. In medical diathermy, electrical energy is converted to thermal energy (heat), as it passes through tissues because all tissues have a natural electrical resistance. Tissue resistivity, or resistance per unit length, varies widely among tissues in the body (Table 7.1). Muscle and skin are good conductors and hence have low electrical resistance, while fat and bone have higher resistance and conduct electricity poorly.

The end effect of producing diathermic heating within the tissues is a function of the intensity and duration of heat generated. If tissue is quickly heated above 100 °C, intracellular water within the tissue rapidly turns to steam, violently bursting the tissue at the cellular level, and causing tissue vaporization. This process is known as *fulguration* and will result in a charred black-colored coagulum. On the other hand, if tissues are heated slowly, the tissue proteins coagulate into a white mass, similar to that produced by heat on egg albumin (Mitchell and Lumb, 1962). This is known as *desiccation*, or the destruction of tissue, short of carbonization, by dehydration (O'Connor et al., 1996). While fulguration is faster than desiccation at achieving coagulation, it causes far more destruction of the surrounding tissues for an equivalent hemostatic outcome (Glover et al., 1978).

Table 7.1 Biological Tissues Conductivity

Tissue	Resistivity
Blood	30–50 Ω cm
Liver	$\sim$150 Ω cm
Muscle	200–500 Ω cm
Fat	$\sim$1 kΩ cm
Bone	>1 kΩ cm

7.4 ELECTROSURGERY

7.4.1 Electrosurgical Theory

By applying an electrical current to a tissue site, electrosurgery tools are able to cut, coagulate, desiccate, and/or fulgurate. The electrical current produces Ohmic (also known as *Joule*, or resistive) heating at the site, and the amount of heat is directly proportional to the current density. Current density, j, is the electrical current per unit area of cross section, and the generalized form of Ohm's law is given as

$$E = \sigma \times j \qquad (7.1)$$

where E is the electric field and σ is the conductivity of the material. Ohm's law states that the current density is proportional to the electric field applied to the material and the resistivity of the material. The above generalized form of Ohm's law is equivalent of the more common version of relating current to voltage (i.e., $V = I \times R$). For a given current density, j, and material resistivity, g, the power density, P_D, is given from Ohm's law as

$$P_D = j^2 \times g \qquad (7.2)$$

Power density is the ratio of the overall power to the volume it is being applied. The rate at which the local temperature rises as an object is heated is given by

$$\frac{dT}{dt} = \frac{P_D}{(c \times \rho)} \qquad (7.3)$$

where P_D is the power density, c is the specific heat, and ρ is the density of the material. Specific heat is the amount of heat per unit mass required to raise the temperature by 1 °C. As can be seen from Equation 7.3, the greater the power density, the higher the local rise in temperature. For the same overall power, the local induced temperature rise is inversely related to the size of the electrode or area of application. Therefore, with electrodes of unequal size, a larger power density, and consequently higher local temperature increase will result adjacent to the smaller electrode. In monopolar electrosurgery, a tool with a fine tip is used in concert with a larger grounding pad, with the location adjacent to the tip being the targeted surgical region. This is illustrated in Figure 7.3, with the lines schematically depicting the paths on which the current will flow.

While electrical current can be used to induce diathermy, electrical current also stimulates muscle and nerve cells. This effect is due to voltage-gated ion channels that are found in the cell membranes of these tissues. This stimulation may cause pain, muscle spasms, and even cardiac arrest. However, the sensitivity to electrical stimulation decreases and the excitability threshold increases with increasing frequency (>kHz). Nerve and muscle cell stimulation is minimal at 100 kHz (d'Arsonval, 1891) and above; therefore, electrosurgery can be performed safely at these frequencies. The basic circuit for an electrosurgical device is composed of a

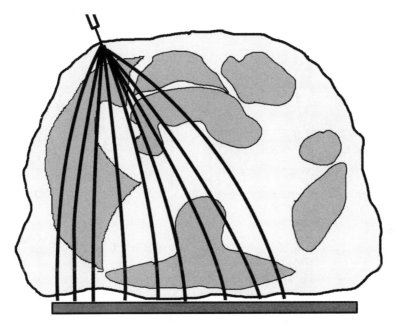

Figure 7.3   Schematic representation of the relative current densities in a cross section of a patient's body during monopolar electrosurgery. A high current density is experienced at the tool tip, with a lower density adjacent to the dispersive grounding pad at the bottom.

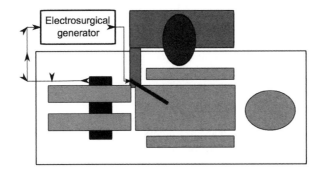

Figure 7.4   Basic circuit for monopolar electrosurgical devices.

generator, an active electrode, the patient, and return electrode (Fig. 7.4). RF energy in the range of 400–600 kHz is most commonly used, as diathermy is still possible while nerve and muscle stimulation is significantly reduced. Generators deliver at least 100 W to the patient, and the applied voltage ranges from 100 V to 5 kV.

Electrical current being passed through biological tissue will induce a rise in heat along the path between the patient electrode and return electrode. The RF current flows from one electrode through the tissue to the second electrode, with

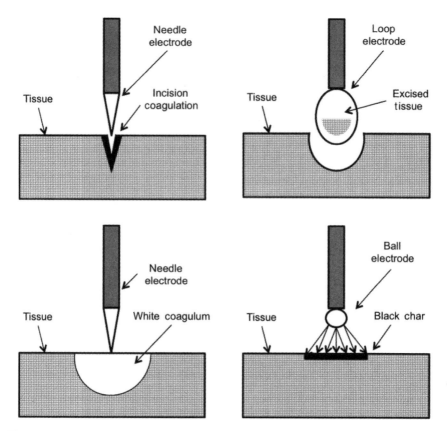

Figure 7.5   Common monopolar tool tips and their effects on tissues: (a) incising tissue, (b) excising tissue, (c) coagulative desiccation, and (d) fulguration. *Source*: Adapted from Bussiere (1997).

the oscillating signal causing agitation of ions within the interstitial fluid of the tissues. The agitation drives water from cells, causing desiccation, and at the same time heats the cells, causing coagulation at a temperature of 48–60 °C (Prutchi and Norris, 2005). Ionic frictional heating causes further temperature increases, resulting in cutting and fulguration. At 200 °C, solid cell components are reduced to carbon. The total power or heat, as well as the method in which the RF energy is delivered, determines the specific clinical tissue response, and whether cutting or coagulation is performed (Fig. 7.5) (Bussiere, 1997).

7.4.2   Cutting and Coagulation Techniques

Pure electrosurgical cutting is the cutting of tissue, both incising (making a mark) and excising (cutting out) (Fig. 7.5c), with minimal lateral thermal damage. To achieve cutting, a high voltage current (up to 180 W) continuous-wave (CW) sinusoidal waveform (0.5–3 MHz) is applied to a small surface area electrode (Fig. 7.5a), such as a needle or electrode loop formed from a thin wire (Fig. 7.5b) (Tektran, 1997). The sinusoidal waveform induces rapid tissue heating to temperatures greater than 100 °C, which leads to vaporization of the interstitial fluid, and thus precise tissue separation. The remaining heat is absorbed by water released from these cells, minimizing thermal damage without coagulation. For tissue cutting, direct contact between the electrode and tissue is not required but increases the precision of the cutting tool.

Coagulation is performed using either desiccation or fulguration. As described earlier, fulguration (Fig. 7.5d) is a noncontact technique often used to control diffuse bleeding, and desiccation is a contact technique often used for coagulation in the case of local bleeding. The typical power is around 50 W for fulguration and 45–90 W (25–50% duty cycles) for desiccation (Tektran, 1997). With both techniques, the average power is reduced below the threshold for cutting using pulsed-wave, rather than continuous-wave (CW), operation. This chain of short RF energy packets produces less heat than would be observed with CW operation in cutting. The intermittent application of energy allows tissue to cool off during energy application, resulting in lower peak temperatures, and leads to coagulation rather than tissue vaporization.

A blended cutting and coagulation mode can also be used, by varying the duty cycle of the pulsed-wave input. A blended current is not a mixture of both cutting and coagulation currents, but rather a modulation of the duty cycle, with a lower duty cycle producing less heat. A cutting current is applied, but with some coagulation effects. The shorter the bursts of RF energy, the greater the relative level of coagulation and less cutting is achieved (Megadyne, 2005).

7.4.3 Equipment

From the time of Bovie to the present day, a basic monopolar electrosurgical system has consisted of three primary components: (i) an electrosurgical generator, (ii) an active electrode probe, and (iii) a dispersive grounding pad (Fig. 7.6). The electrosurgical generator provides power and transforms the mains electricity into high frequency energy that is delivered through the electrode probe to the surgical site. A grounding pad, or return pad, returns current from the patient to the unit.

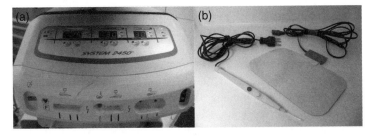

Figure 7.6   (a) Electrosurgical unit and (b) monopolar electrode and grounding pad.

Electrosurgical generators have historically been ground referenced; however, current systems make use of an isolated generator (Massarweh et al., 2006). Ground-referenced generators were originally developed under the presumption that the current entering the patient's body would return to the generator via the grounding pad. However, electricity can also travel through various conducting objects touching the patient, such as monitor leads, the operating room table, or stirrups, leading to ground if they conduct better than the pad itself and potentially resulting in site burns on the patient. In addition, if there is a defective grounding pad or one is forgotten, the current will find other paths to ground, which may again result in alternate site burns. Ground-referenced generators are now considered outdated but may still be occasionally found in operating rooms (Bussiere, 1997).

In 1968, isolated generator technology revolutionized the electrosurgery arena. These generators isolate the current from ground by referencing it within the generator's circuitry, and therefore, eliminating unintended shorts to ground as with a ground-referenced generator. A disadvantage is that if conductivity of the grounding pad is disrupted partially or completely, pad site burns may occur. In response to this problem, return electrode contact quality monitoring (RECQM) systems have been developed. These systems measure the impedance between the patient's skin and the grounding pad and may alert the operator or prevent activation of current if grounding is lost.

The grounding pad provides a means to couple to the patient and complete the electrical circuit back to the generator. Its sole task is to remove the current from the patient safely. Most grounding pads are flexible sticky pads that have conductive polymer glue covering a conductive foil. The pad must be large enough to keep the current density low as the current exits the patient's body; otherwise, a burn may result under the pad. Surface area impedance can be compromised if the application site or the application of the pad is suboptimal. Excessive hair, scars, adipose tissue, bony prominence, fluid invasion, and prostheses are some examples that can lead to increased impedance at the contact between the patient and grounding pad.

In monopolar electrosurgery, current flows from the generator to the active electrode in the surgeon's hand at the surgical site, through the patient's tissue, to a dispersive ground pad, and back to the generator. Heating is very focused and localized proximal to the electrode tip, as the rate of heating is proportional to the square of the current density. Monopolar electrosurgery is the most commonly used electrosurgery modality in surgery and is utilized by a majority of surgeons during laparoscopic procedures (Fig. 7.7).

Bipolar electrosurgery incorporates both electrodes (positive and negative) at the site of the surgery (Fig. 7.8). Only the tissue between these electrodes is affected, and there is no need for a grounding pad with this modality. A more focused target area is affected, thus lower voltage is utilized and less char is formed. Damage to the adjacent tissue is significantly diminished when compared to monopolar electrosurgery.

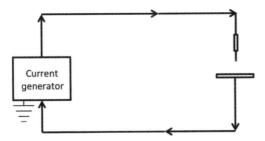

Figure 7.7   Current path in monopolar electrosurgery. Current flows from the tool tip and is returned to the generator/ground through a dissipative pad.

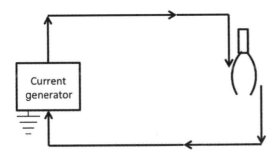

Figure 7.8   Current path in bipolar electrosurgery. Current flows from one tip of the tool to another tip connected the generator ground.

7.4.4   Considerations and Complications

While risks to both the patient and physician have been reduced over the years through advances in technology and techniques, several hazards and potential complications still exist, chief among them being electrothermal injuries, burns, and fires (Hutchisson et al., 1998). Potentially serious burns to both patients and/or operating personnel can occur through a variety of mechanisms.

Electrothermal injuries are primarily a result of inadvertent direct active contact, insulation failure, direct coupling, or capacitive coupling (Wu et al., 2000). Burns can occur when the electrosurgical probe tip is accidentally activated, or if it makes inadvertent direct contact with stray tissues. This is a particular risk in laparoscopic electrosurgery, as the tool's tip may be outside of the endoscopic field of view, and therefore difficult to see. Burns or injuries can also result because of faulty insulation on the active electrode, allowing for the formation of alternate pathways for the current flow. High voltage currents and frequent sterilization of the active electrode may increase the risk of insulation failure and alternate current pathway formation.

Direct coupling occurs when the electrosurgical tool induces current flow in alternate pathways, such as if a surgeon touches the electrode to a metal clamp. This is a particular risk in laparoscopic electrosurgery, as unrecognized current transfer along the laparoscope, electrode, or cannula can occur with electric energy passing through the abdominal wall, causing visceral damage out of the view of the laparoscope or the surgeon (Tucker and Voyles, 1995).

Capacitive coupling can result due to the flow of the electrical current from the active electrode through intact insulation to adjacent conductive objects, such as tissues or trocars. An electrostatic field is thus created between the conductors and may potentially allow the current to flow from one conductor to the other once the net charge exceeds the insulator's capacity. Use of fully conductive metal trocars can minimize potential tissue injury by allowing energy to dissipate over greater surface areas, thus decreasing the localized current density (Massarweh et al., 2006).

Patients can also be burned at the site of the return electrode because of poor contact between the grounding pad and the patient's skin. Poor contact with the grounding pad causes an area of high current density, resulting in a burn. Such burns can also occur if a patient comes into contact with other potential grounding sources (Hutchisson et al., 1998).

Of secondary safety concern to surgeons is the risk of fire involving properly maintained electrosurgical equipment, with an activated cautery tip coming into contact with flammable substances. These substances can range from surgical oxygen and alcohol-based prep solutions to surgical drapes and bedding.

7.5 FUTURE OF ELECTROSURGERY

In recent years, a number of technologies have been developed, which may have significant impacts on the future of electrosurgery. These technologies include floating-ball RF dissectors and bipolar vessel and tissue sealing devices. The TissueLink^{TM} floating-ball RF dissector (Tissuelink Medical, Dover, NH) is a monopolar probe for dissection and coagulation. The device is saline-cooled and is able to keep the surface temperature below 100 °C. A continuous flow of electrically conductive saline between the tissue and the electrode is used to couple the RF energy with the tissue, where it is converted into thermal energy. By maintaining a relatively low surface temperature during operation, a greater depth of tissue coagulation and hemostasis is achievable, compared with traditional electrosurgery (Pearson et al., 2002).

Advanced bipolar vessel sealing devices have been achieved by incorporating tissue impedance sensing capabilities. Measuring both the electrical and mechanical impedance of the tissues in real time allows for the application of specifically tailored electrical currents and mechanical sealing forces. When the applied currents and mechanical forces onto the tissues are optimized, the instruments are capable of achieving hemostasis by sealing and fusing vessel lumen walls. Vessels up to 7 mm in diameter can be effectively sealed, and the strength of the seal is comparable to that of vascular clips, staplers, and sutures. Vessel sealing is capable of withstanding three times the normal systolic blood pressure (Kennedy etal., 1998).

Another vessel sealing device has been introduced and is increasing in the popularity of use. The EnSeal® Laparoscopic Vessel Fusion System (SurgRx, Inc.) uses both increased mechanical pressure and advanced polymers to enhance controlled energy delivery and to maintain tissue temperatures to around 100 °C. The proprietary polymer is temperature sensitive and is embedded with nanometer-sized conductive particles, so as to maintain low tissue temperature and improve the control of the energy being delivered. The two blades in the bipolar device are able to compress tissues with up to 7800 psi (Newcomb et al., 2009) while sealing them with electrical energy (Advincula, 2005). The combination of compression and fine regulation of electrical current yields vessel seals with high burst pressures and minimal thermal spread.

7.6 CONCLUSION

Several energy-based hemostatic techniques have been refined over the past century and are commonly used during surgery at present. Of these, electrosurgery is most widely practiced in the operating room. Electrosurgery is now considered relatively safe and has become an invaluable tool for fine cutting and coagulation of a wide variety of tissues in a range of procedures. Several new technologies have recently been developed, which are expected to make further inroads into advancing the surgical suite.

CHAPTER 8

Tissue Ablation Systems

MICHAEL DOUEK, JUSTIN McWILLIAMS, and DAVID LU

Department of Radiological Sciences, University of California, Los Angeles, CA

8.1 INTRODUCTION

Fifty years ago, the notion that a physician would be able to destroy a tumor deep in a patient's body, utilizing a small needle placed through the skin and guided to the tumor by imaging, was the thing of imagination. At present, this process, termed *tissue ablation*, has become a widely used technique in medicine for a host of different therapies.

In contemporary medical terminology, "ablation" means broadly the localized destruction of target tissues by thermal or chemical means. A growing number of ablative techniques exist. In addition, clinical uses for ablative technologies are widely varied and rapidly expanding. These techniques are utilized by physicians in diverse medical specialties, for a wide range of clinical applications.

It is beyond the scope of this chapter to exhaustively cover each ablative technique and its clinical functions. Rather, this chapter gives the reader a taste of this emerging field by studying some basic principles common to all forms of ablation technologies. Radiofrequency ablation (RFA) is covered in some depth, as this is the most widely used ablative technology at this time; other ablative technologies are discussed more briefly. This discussion centers on one very important clinical application, tumor ablation, though tissue ablation has been utilized in a number of medical fields outside of oncology. Finally, before concluding, some emerging techniques and future directions in tissue ablation systems are described.

Medical Devices: Surgical and Image-Guided Technologies, First Edition.
Edited by Martin Culjat, Rahul Singh, and Hua Lee.
© 2013 John Wiley & Sons, Inc. Published 2013 by John Wiley & Sons, Inc.

8.2   EVOLVING PARADIGMS IN CANCER THERAPY

From the beginnings of recorded history, cancer has afflicted mankind. Evidence of cancer has been found in fossilized human bones and in ancient Egyptian mummies; ancient documents describe this disease entity at length (American Cancer Society, 2009). Early philosophers and scientists speculated on the causes of cancer. The ancients attributed the development of cancer to the Gods. Hippocrates blamed its development on excessive "black bile" in the body. Other early scientists blamed cancer development on trauma; still others on parasites.

At present, the understanding is that cancer is in fact caused by aberrations, either inherited or acquired, in the genes regulating the cell cycle. Genetic changes can disrupt the delicate balance of cell division and death. Stated simply, cancer is out-of-control cell reproduction.

Just as the understanding of cancer has evolved, so too have the methods for its treatment. In the early twentieth century, the only hope for curing cancer was in the context of small tumors that could be removed entirely by surgery. Later, chemotherapeutic drugs and radiation therapy were employed to target cancer cells beyond the reach of the surgeon.

At present, many more tools are available in our fight against cancer. Powerful new drugs have been developed (such as growth and angiogenesis inhibitors), which more specifically interfere with the biological functions of cancer cells. Surgical techniques have become more sophisticated and less invasive. Enormous technical advances in the field of radiotherapy have allowed for greater efficacy in destroying tumor cells, while sparing normal tissues. Recently, tumor ablation has become an important weapon in the armamentarium of cancer-fighting therapies. Ablation allows for the very localized destruction of cancers, which are inaccessible by surgery or radiation therapy, or for which these therapies are inappropriate. Often, combinations of therapies are employed to achieve cure or control of cancers previously deemed untreatable (Fig. 8.1).

For example, a patient with advanced colon cancer can be considered, which has also spread to the liver. Therapy at present might involve the surgeon removing the primary colon cancer, the interventional radiologist destroying the liver metastases with ablative techniques, and the oncologist administering chemotherapeutics to help destroy additional microscopic disease, elsewhere in the body. This multidisciplinary, multimodality approach has become the "state of the art" for many types of cancers.

Or, the example of a patient diagnosed with a small kidney cancer can be considered, who also suffers from heart disease and pulmonary disease and is thus considered at significant risk for undergoing major surgery to remove his tumor. Instead of surgical removal, tumor ablation has shown to be an effective and less-invasive way of destroying certain tumors.

Tumor ablation occupies an important role in the evolving field of "interventional oncology," a small, but dynamic and growing area in the complex field of oncology.

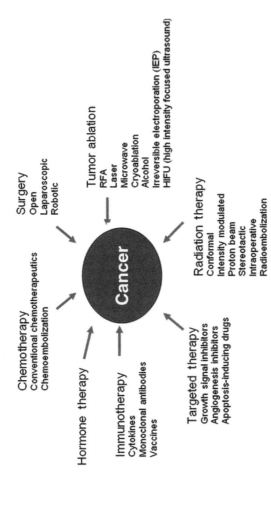

Figure 8.1 Evolving paradigms in cancer therapy.

8.3 BASIC ABLATION CATEGORIES AND NOMENCLATURE

"Ablation" refers to the localized destruction of target tissues by thermal or chemical means. Classically, these techniques have been placed into one of two categories: "thermal" techniques, which heat or cool tissues to achieve their destruction; or "chemical" techniques, in which caustic substances are injected into target tissues to achieve their destruction. A new technique, irreversible electroporation, utilizes electrical current to disrupt the permeability of cells to lethal levels and fits into neither category neatly (Table 8.1).

The device utilized to deliver the ablative technique is referred to as the *applicator*. The RFA and electroporation applicators are "electrodes," microwave applicators are "antennas," laser applicators are "fibers," and cryoablation applicators are "cryoprobes" (Goldberg et al., 2005).

Typically, the operator guides the applicator to the target tumor, with the aid of an imaging modality (i.e., ultrasound, CT (computed tomography), or MRI (magnetic resonance imaging)). The ablative technique is then applied, each technique achieving local tissue destruction in different ways (Fig. 8.2).

8.4 HYPERTHERMIC ABLATION

A number of ablative techniques use heat to destroy tissues, including RFA, microwave ablation (MWA), laser ablation, and high intensity focused ultrasound (HIFU). While the mechanism of tissue heating differs widely by the technology employed, at the cellular level, the outcome of cellular hyperthermia is the same—above threshold temperatures, cellular death is inevitable. As Goldberg et al. point out, "interventional radiologists (and tumor cells) do not care which energy source produces cytotoxic heat." (Goldberg et al., 2000).

In hyperthermic thermal ablation, energy deposition in target tissues results in local increases in tissue temperature. Injury to cells begins at 42 °C. At these relatively low temperatures, required exposure times are long, 3–50 h, to achieve cell death. As temperatures rise, the required exposure time drops off exponentially. Cellular lethality occurs by 8 min at 46 °C, by 2 min at 51 °C, and at temperatures

Table 8.1 Summary of Current Ablative Techniques

Ablative Techniques
Thermal: Radiofrequency ablation (RFA)
Cryoablation
Microwave coagulation therapy
Laser ablation
High Intensity Focused Ultrasound (HIFU)
Chemical: Alcohol or Acetic Acid Ablation
Electrical: Irreversible Electroporation (IRE)

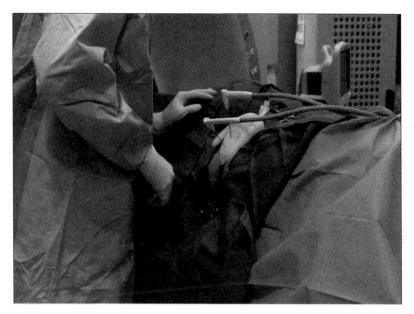

Figure 8.2 Intraprocedural photograph during performance of a thermal ablation. *Source*: Photograph courtesy of Fereidoun Abtin, MD.

above 60 °C, intracellular proteins become denatured, lipid bilayers melt, and cell death is inevitable (Curley, 2001). Optimal desired temperatures for ablation range from 50 to 100 °C (Ellis et al., 2004). Extremely high temperatures (higher than 105 °C) can result in tissue boiling, vaporization, and carbonization, which subsequently restricts total energy deposition by impeding the flow of current (Fig. 8.3) (Ellis et al., 2004; Goldberg et al., 2000).

8.5 FUNDAMENTALS OF *IN VIVO* ENERGY DEPOSITION

Unfortunately, predicting tissue heating turns out to be a very complex endeavor *in vivo*, with multiple physiological factors that must be accounted for.

The "bioheat equation" or the "Pennes equation" offers a starting point to discuss *in vivo* tissue heating. Described by Pennes in 1948, and derived from studying tissue heat flow in the human forearm, the complex bioheat equation attempts to account for primary factors that influence tissue heating *in vivo* (Pennes, 1948).

The Pennes "bioheat equation" is as follows:

$$\rho c \, \frac{\partial T}{\partial t} = \lambda T + \acute{\omega}_b c_b \rho_b \left(T_b - T\right) + q_m + q_r$$

In this equation ρ and ρ_b are the density of tissue and blood, respectively (kg/m^3), c and c_b are the specific heat of tissue and blood, respectively (J/kg/K), λ is the

(a)

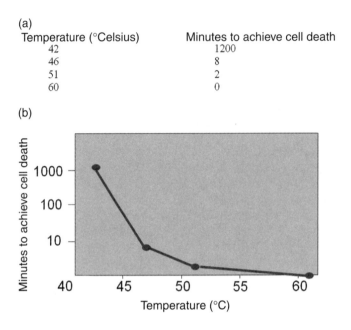

Temperature (°Celsius)	Minutes to achieve cell death
42	1200
46	8
51	2
60	0

(b)

Figure 8.3   (a) Table and (b) corresponding graph depicting the times required to achieve cell death as function of ablative temperatures.

thermal conductivity of tissue (W/m/K), $\dot{\omega}_b$ is the blood perfusion rate (kg/s^2/m^3), q_m is the excess heat generated by metabolism (W/m^3) (which is negligible and can be ignored), and q_r is the energy absorbed by tissue per unit volume per unit time (W/m^3/s) (Youjun et al., 2006).

Pennes made several assumptions in his work. He treated his study tissues as symmetric and perfectly cylindrical with uniform thermal properties and tissue constitution. Fundamentally, tissues are inhomogeneous, with varying tissue density and cellular composition, and thus thermal conductivity, electrical impedance, and other factors also vary. For example, in the case of hepatocellular carcinoma, a type of liver cancer, tumors have a tendency to form thick fibrous capsules (or "pseudocapsules") at their peripheries, which serve to trap heat, creating a so-called "oven effect" and resulting in higher temperatures than expected within the border of the capsule and lower than expected temperatures just outside the capsule (Fig. 8.4).

Or consider the case of a large blood vessel coursing near the edge of a target tumor. Circulating blood in the vessel serves to dissipate applied energy, effectively acting as a "heat sink" (or "cold sink" in the context of cryotherapies). As such, heating (or cooling) of tissues immediately adjacent to large blood vessels tends to be mitigated (Fig. 8.5).

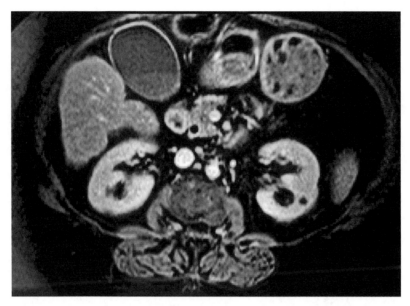

Figure 8.4   Axial T1W contrast enhanced MRI of the liver demonstrating a thin enhancing rim around a lesion in the liver, the so-called "pseudocapsule" sometimes seen with hepatocellular carcinoma (HCC).

Nonetheless, the Pennes equation offers a valuable theoretic starting point to discuss those factors that play a role in modulating tissue heating. Broadly, tissue heating is proportional to the energy absorbed by the tissue modulated by physiologic properties of the tissues (namely perfusion).

8.6   HYPERTHERMIC ABLATION: OPTIMIZING TISSUE ABLATION

Goldberg et al. greatly simplified the Pennes equation for the purpose of framing tissue ablation theory as follows:

"Coagulation necrosis (cell death) = energy deposited × local tissue interactions − heat loss" (Goldberg et al., 2000).

Using this equation as a framework, cell death can be increased by (i) increasing the amount of energy deposition, (ii) optimizing local tissue interactions, or (iii) decreasing heat loss. Most successful strategies for shifting the balance of this equation in favor of cell death have revolved around the first factor, increasing energy deposition (e.g., by increasing generator power, or designing ablation applicators to more effectively distribute energy), but strategies have also been devised to address the latter two factors as well. For example, several investigators have centered on reducing blood perfusion to ablated tissues, reducing perfusional "heat sink," and thus increasing the amount of coagulation necrosis. Mechanical or pharmacological techniques can be employed to reduce blood flow to target tissues

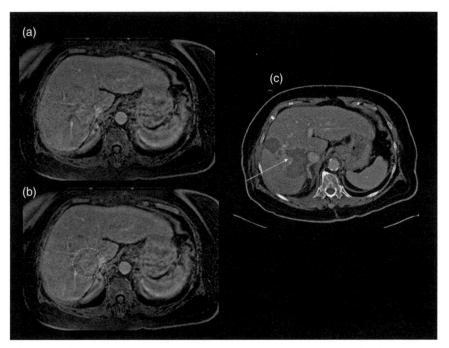

Figure 8.5 (a) MRI of the abdomen demonstrates a mass in the central liver, circled in (b). (c) Following thermal ablation of this mass, heat sink from a nearby blood vessel results in incomplete ablation at one of the tumors borders (arrow).

(Ellis et al., 2004). Additional specific strategies of optimizing tissue ablation are discussed in more depth in the sections dedicated to each ablative technology.

8.7 RADIOFREQUENCY ABLATION

French physiologist Jacques D' Arsonval first described the heating effects of alternating RF current on tissues in the late nineteenth century (Vogl et al., 2008). It was William Bovie and Harvey Cushing, however, who popularized the use of RF energy for medical applications in the 1920s by describing a monopolar, knifelike RF electrode, which could be utilized to cauterize or cut tissues within a few millimeters about the electrode tip (Cushing and Bovie, 1928). At present, it would be rare to find an operating room not equipped with this so-called "Bovie knife" (Fig. 8.6).

It was not until the 1990s, however, that RF systems were applied to the ablation of tumors. In 1990, two groups, McGahan et al. and Rossi et al. applied an alternating electrical current in the RF range (McGahan et al., 1990; Rossi et al., 1990) delivered via electrodes insulated except for their active tips, to achieve small ablations of normal liver tissue. In 1996, physicians Kenneth Tanabe and Nahum Goldberg performed the first RFA procedure on a patient with a liver tumor.

Figure 8.6 A typical electrosurgical device. *Source*: Courtesy of Megadyne Medical Products.

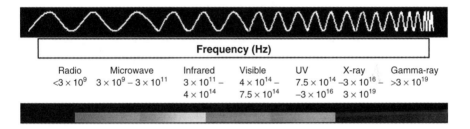

Figure 8.7 The electromagnetic spectrum.

At present, RFA is the most commonly performed percutaneous tumor ablation technique (Tatli et al., 2009).

8.8 RFA: BASIC PRINCIPLES

The RF energy occupies a well-defined portion of the electromagnetic spectrum (3 kHz–300 GHz) (Fig. 8.7).

The RFA applicators deliver alternating RF current, usually in the 450–500 kHz range, via needle electrodes to target tissues. Each ablation typically lasts between 8 and 20 min. Generator power typically ranges from 50 to 200 W. Alternating RF current passes from the uninsulated tip of the active RFA electrode into adjacent tissues. Typically, grounding pads, usually placed on the patient's thighs, complete the electrical circuit (Fig. 8.8).

The movement of ions, changing in concert with the applied alternating current (the so-called "ionic agitation"), results in the frictional heating of the tissue (Curley, 2001). Ultrasound imaging during the ablation allows the operator to monitor the size of the ablation zone, which will transiently appear echogenic (bright) as microbubbles are produced in treated tissues (Fig. 8.9).

8.9 RFA: *IN VIVO* ENERGY DEPOSITION

For RFA (or any thermal ablative technique), energy deposition, and thus tissue temperature, is highest closest to the applicator tip, falling off rapidly with increasing

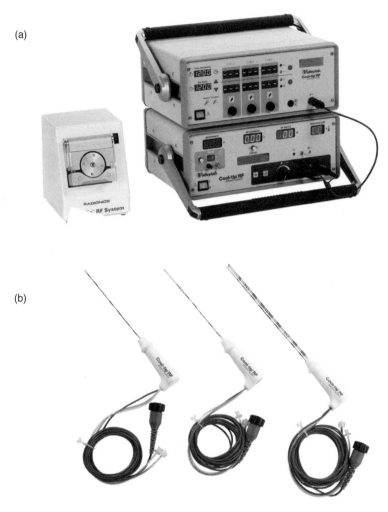

Figure 8.8   Radiofrequency ablation equipment. (a) An RFA generator. (b) RFA electrodes (*Source*: Copyright © 2012 Covidien).

distance from the applicator. For RFA, tissue temperature is roughly proportional to the inverse of distance to the tip to the fourth power ($T \approx 1/r^4$) (Vogl et al., 2008). However, excessive energy deposition results in extreme tissue temperatures near the applicator tip ($>105\,^{\circ}\text{C}$), which can result in tissue charring. This charring serves as an insulator that impedes thermal conduction, and thus also limits the volume of tissue ablation.

Thus, for any single, conventional RFA applicator, the diameter of the ablated tissue is limited. For example, a single, monopolar conventional applicator achieves

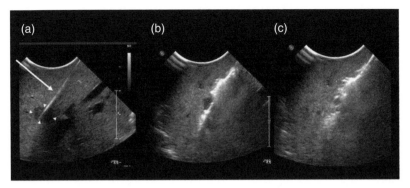

Figure 8.9 Ultrasound images during RFA of a liver lesion. (a) The RFA applicator (arrow) has been placed through the mass (arrowheads). Sequential ultrasound images during the ablation (b) and (c) demonstrate bright microbubbles, indicating the zone of ablation, engulfing the target mass.

ablations of up to approximately 1.6 cm in diameter (Ellis et al., 2004), though several strategies exist to increase ablation volumes (see the following text).

8.10 OPTIMIZING RFA

As we have discussed previously, there are several strategies to optimize hyperthermic ablation volumes. These involve increasing energy deposition, optimizing local tissue interactions, and decreasing heat loss.

Injecting saline into target tissues has been studied as one such strategy. Injected saline aids in heat conduction and also increases tissue ionicity, which enables greater current flow (Ellis et al., 2004). However, the distribution of the injected liquid is unpredictable, which can result in an irregular ablation zone (Livraghi et al., 1997), and diffusion of hot saline can cause burn injury to surrounding organs (Burdio et al., 2003). An expandable electrode using saline injection and multistep deployment of tines to a maximum theoretical diameter of 7 cm is commercially available. Other investigators have explored utilizing injected iron compounds to amplify current shifts (Ellis et al., 2004).

Most strategies for optimizing RF efficacy have revolved around strategies to increase energy deposition to target tissues. Most basically, increasing the power of the generator or the ablation time can result in increased energy deposition in target tissues (although again, tissue charring at the electrode tip will result in increased impedance about the active electrode and will impede further tissue ablation). Commercially available RF generators produce about 200 W of power and 2000 mA of current. Higher-power, higher-current generators (1000 W, 3000 mA) have been examined *in vivo* in animals and were found to enlarge the coagulation zone (Solazzo et al., 2007). Additional factors, including a slow increase in generator power and an increase in the exposed surface area of RF needle electrodes,

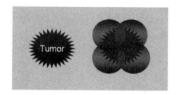

Figure 8.10   Overlapping ablations.

also have been shown to increase the volume of coagulated tissue (McGahan and Dodd, 2001).

Performing multiple *overlapping ablations* by repositioning a single applicator can also increase ablation volume, albeit a time-consuming and potentially technically challenging approach (Fig. 8.10).

Pulsing of energy, used with RFA (and also with laser ablation), has also resulted in increasing overall energy deposition. This technique employs periods of higher energy deposition alternating with periods of lower energy deposition. The periods of lower energy deposition allow for cooling at the applicator tip, without significantly affecting temperatures more distant from the tip (Ellis et al., 2004).

Finally, alterations in electrode design can effectively increase the amount of energy that can be deposited.

Expanding, or "umbrella" shaped applicators, consisting of small active tines, which can be deployed from a central cannula, also extends the ablation zone (Fig. 8.11).

To reduce charring at the applicator tip, *internally cooled applicators* have been designed. A chilled perfusate is circulated through the hollow core of the applicator to cool the applicator tip and the tissues immediately surrounding the tip. As such, greater energy can be applied overall without causing tissue charring (Fig. 8.12).

Multiprobe arrays, consisting of several, evenly spaced applicators deployed simultaneously, also achieve larger tissue ablation volumes. The RFA energy deposition is switched between the applicators, increasing the overall energy deposition, yet allowing for intermittent cooling at each electrode tip. This switching also prevents electrical interference that may occur if electrical current is simultaneously applied to neighboring electrodes (Fig. 8.13; Lee et al., 2003).

The use of *bipolar electrodes* instead of monopolar electrodes has been studied by some groups. These arrays use an "active" electrode and a nearby "grounding" electrode (the latter in lieu of grounding pads placed on the skin surface in conventional, monopolar systems). Current flows between the two electrodes, resulting in heating between them.

These techniques in various combinations have resulted in increases in the size of tumors, which can be effectively ablated. These are summarized in Figure 8.14. Novel designs employing new techniques, or combinations of the above techniques, are likely to emerge and hopefully will improve not only the ability to generate larger ablation zones effectively but also the reproducibility and reliability of ablation volumes and shapes (Fig. 8.14).

Figure 8.11   AngioDynamics StarBurst® XL RFA Device with expandable tines (*Source*: Photo courtesy of AngioDynamics, Inc.).

8.11 OTHER HYPERTHERMIC ABLATION TECHNIQUES

8.11.1 Microwave Ablation (MWA)

In 1979, Japanese surgeons described the use of a microwave device to achieve tissue coagulation and to control bleeding during hepatic surgery (Tabuse, 1979). This was the first clinical application of MWA. Later, intraoperative, laparoscopic, and subsequently percutaneous MWA techniques were developed in Asia to treat small liver tumors in the 1990s. Western countries were much later in adopting MWA techniques than their Asian counterparts. It was not until 2003 that the first commercially available MWA system was developed in the United States. Although still early in its evolution, MWA has already proven to be an important thermal ablative technique.

8.11.2 MWA: Basic Principles

Clinical ablation systems using this technology employ energy waves with frequencies ranging from 900 MHz to 2.45 GHz

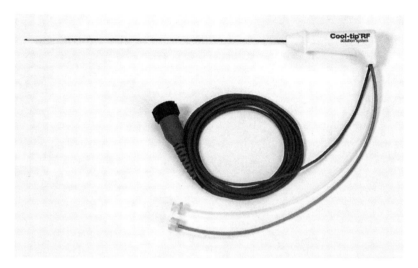

Figure 8.12 An internally cooled RFA applicator (*Source*: Copyright © 2012 Covidien).

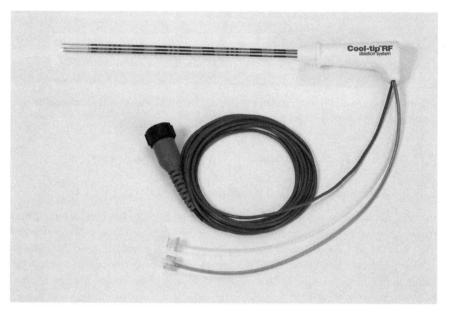

Figure 8.13 A (three applicator) multiprobe RFA array (*Source*: Copyright © 2012 Covidien).

All MWA systems are comprised of three basic elements—a microwave generator, a transmission line, and a microwave antenna (Liang and Wang, 2007. Microwave generators employ either a magnetron or solid-state amplifier to generate microwave radiation (Brace, 2009). The antenna (which is typically needlelike

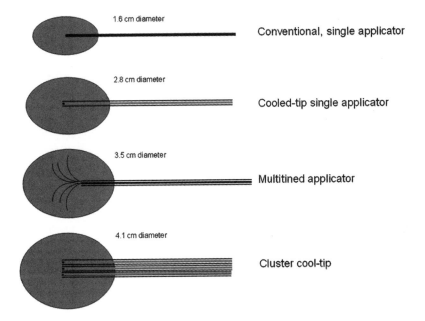

Figure 8.14 Schematic depicting RFA volumes achievable with different applicator types.

and inserted directly into the tumor under image guidance) is connected to the generator via the transmission line (typically low loss coaxial cable) and transmits microwaves to the target tissue.

Microwaves generate heat through a process known as *dielectric hysteresis*, whereby polar water molecules continuously realign with the applied electromagnetic field, producing heat by molecular friction (Fig. 8.15; Iannitti et al., 2007).

8.11.3   MWA: *In Vivo* Energy Deposition

MWA is a relatively new and promising ablation method, with several theoretical advantages over RFA. MWA has a much broader zone of active tissue heating than RFA. In RFA, active tissue heating is limited to a few millimeters about the active electrode tip, with the remainder of the ablation zone heated via thermal conduction (Simon et al., 2005), thus heating is predominately a passive process. This is not so with MWA, which can actively heat tissues up to 2 cm about the antennae.

Because of active rather than conductive heating, MWAs are less influenced by tissue charring and the resultant increase in electrical impedance about the applicator tip. In addition, microwave is less susceptible to "heat sink" created by nearby vessels than in RFA. This is because the cooling effect of blood flow is most pronounced in tissues heated by conductive rather than active means (Liang and Wang, 2007).

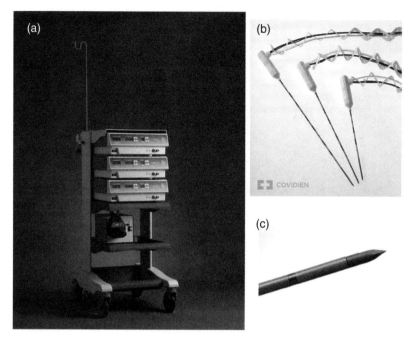

Figure 8.15   (a) Microwave ablation generator, (b) antennae applicators, and (c) applicator tip (*Source*: Copyright © 2012 Covidien).

The net result of these advantages is that intratumoral temperatures can be driven higher, faster, and larger, more consistent ablations, within a shorter time are achievable (Simon et al., 2005). However, by the same token, greater care must be taken in avoiding injury to normal peritumoral structures with MWA.

An additional advantage over RFA is that, unlike the electrodes used in RFA, which conduct electrical current, MWA antennae are electrically independent and are therefore amenable to synchronous ablations (Simon et al., 2006). This allows for the synergistic effect of multiple antennas acting in concert (in RFA, power in multiprobe arrays is switched between the deployed electrodes, rather than delivered to all electrodes simultaneously, to avoid electrical interference). Finally, MWA does not require grounding pads (as does RFA), decreasing patient preparation time and skin burns associated with these grounding pads.

8.11.4 Optimizing MWA

Most basically, increasing power output and ablation times will result in larger ablation volumes. However, several other techniques have been developed to optimize MWA ablations.

First, changes in microwave frequencies have achieved larger ablations. Historically, most systems have utilized 2450 MHz microwaves. However, recent evidence demonstrates that 915 MHz microwaves have a greater penetration depth, less energy attenuation (and thus greater conversion into heat energy), and are capable of producing larger ablation volumes. More recently developed MWA systems operate at 915 MHz.

The design of the MWA antenna plays a critical role in ablation efficacy. Most antennae are needle shaped, but antennas with novel designs, such as deployable loops, have also been developed. The loops can be deployed to encircle the target tumor, theoretically allowing for more effective and precise ablation of target tissues and less collateral damage to adjacent normal tissues.

To avoid overheating of the antenna shaft, cooled-shaft antennae have also been developed as in RFA. As shaft temperatures can be kept lower with this technique, higher and longer power output is possible, allowing for larger ablations without causing skin burns (Liang and Wang, 2007). The requirement for shaft cooling, however, usually necessitates increases in the diameter of the needle antenna. Designs that maximize energy delivery efficiency and minimize power loss via the antennae shaft may obviate the need for active shaft cooling.

Current systems, using generators capable of producing up to 60 W of power at a frequency of 915 MHz, and employing three simultaneous microwave antennas positioned in a triangular configuration (at 1.5 cm spacing), can achieve mean ablation diameters of 5.5 cm (with mean diameters of the ablated tumors being 4.4 cm) within 10 min (Simon et al., 2006).

8.12   LASER ABLATION

Laser ablation is the final hyperthermic ablative technique that is briefly discussed. Broadly, the term *laser ablation* (which goes by many names, including "laser coagulation therapy," "laser interstitial tumor therapy," and "laser interstitial photocoagulation") refers to the thermal destruction of tissue by conversion of absorbed light (usually at infrared or near-infrared wavelengths) into heat (Gough-Palmer and Gedroyc, 2008).

The optimal penetration of light in tissues occurs at wavelengths of 1060–1200 nm (Fig. 8.16) (vanSonnenberg et al., 2005).

The most widely used device in laser ablation is the neodymium:yttrium-aluminium-garnet laser (Nd-YAG) with a wavelength of 1064 nm. As light scatters within tissues, it turns to heat. Solid-state lasers, with power output of 30 W are available. Fibers, usually between 300 and 600 μm diameter (Gough-Palmer and Gedroyc, 2008), deliver light to target tissues (Fig. 8.17).

A single laser fiber is capable of producing a sphere of coagulative necrosis approximately 2 cm in diameter. Care must be taken to avoid overheating about the fiber, as tissue carbonization will reduce light penetration and will decrease the size of the ablation zone.

Laser equipment enjoys the advantage of being magnetic resonance (MR) compatible, because of the lack of ferromagnetic materials. Thus, MRI can be used to target the deployment of fibers and to monitor tissue heating (via MR thermal imaging). Unfortunately, the ablation zone created by lasers cannot be reliably visualized or monitored by ultrasound or CT scan.

Using a beam splitter, several evenly spaced fibers can be utilized simultaneously to achieve larger ablation volumes. The use of water cooled applicators allows for higher power output (up to 50 W compared with 5 W) while preventing carbonization (Vogl et al., 2004). Combining these techniques has allowed for ablative zones up to 8 cm in diameter (Gough-Palmer and Gedroyc, 2008). Laser ablation is the final hyperthermic ablative technique that is briefly discussed.

8.13   HYPOTHERMIC ABLATION

James Arnott, an English physician, is credited with being the first to use cold to destroy tissue in the 1800s (vanSonnenberg et al., 2005). He used ice and salt to treat superficial tumors. In the 1960s, Irving Cooper, an American neurosurgeon, described a closed cryoprobe, internally circulating liquid nitrogen, which was used to destroy tissues deep in the brain (vanSonnenberg et al., 2005). At present, "cryoablation," or ablation by means of tissue freezing, has become an important clinical tool in treating a number of types of tumors.

8.13.1   Cryoablation: Basic Concepts

Contemporary cryoablation employs surface, or penetrating, needle-shaped "cryoprobes," which are placed within target tissues. Thermally conductive material or "cryogens," circulated within the probe, results in cooling of the uninsulated probe tip (Fig. 8.18).

Cryoablation systems induce both intracellular and extracellular ice crystal formation and subsequent cell death. Intracellular ice crystals result in disruption of cell membranes and intracellular structures. Extracellular ice formation creates osmotic gradients that result in lethal cellular dehydration, as well as thromboses of small vessels and subsequent local tissue ischemia (Kahlenberg et al., 1998).

8.13.2   Cryoablation: *In Vivo* Considerations

Just as in hyperthermic ablation, there are critical temperatures at which cell death is achieved—for cryoablation, estimated to be in the range of -20 to $-50\,°C$, with certain tumor types and certain tissue types being more or less sensitive to cold than others. As with hyperthermic ablation, nearby large blood vessels can act as a "thermal sinks," mitigating cooling around these vessels.

8.13.3   Optimizing Cryoablation Systems

The first cryoablation systems utilized circulating liquid nitrogen. Argon gas-based systems, based on the Joule–Thomson principle (which states that rapid expansion of a pressurized inert gas will result in a temperature drop), were introduced in the early 1990s and offered several advantages over their liquid nitrogen counterparts. Argon is of relatively low viscosity and can be circulated very rapidly and at high pressure through smaller probes (vanSonnenberg et al., 2005). As such, procedures

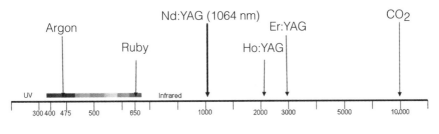

Wavelength (nm)

Figure 8.16   The light spectrum.

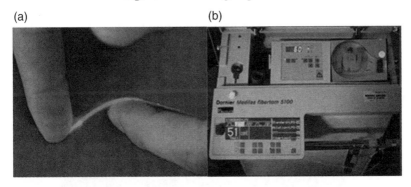

Figure 8.17   Laser equipment (*Source*: Courtesy of Dornier).

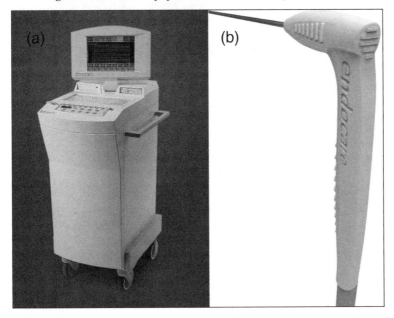

Figure 8.18   Cryoablation equipment. (a) Cryoablation generator and (b) cryoprobe (*Source*: Courtesy of EndoCare, Inc.).

can be performed via less-invasive means (laparoscopically or percutaneously) and tissue freezing is achieved more quickly.

Frozen tissue, or the "ice ball" around the probe tip is reliably and clearly visible by ultrasound, CT, or MRI, an important advantage over hyperthermic techniques, where the ablation zone is less clearly depicted by imaging. This advantage proves important when ablations are performed near critical structures (Fig. 8.19).

An additional advantage of cyroablation is the ability to "sculpt" the iceball. When using multiple cryoprobes, adjusting the flow of cryogen within each probe, allows the operator to shape the iceball to a desired configuration. The zone of complete tissue lethality, however, lies a variable distance inside the visible edge of the ice ball, at least 2–4 mm or more, meaning that the ice ball must extend well beyond the tumor to ensure a satisfactory treatment margin (Mala et al., 2001). Ablation zones of greater than 8 cm have been achieved with this technique (Callstrom and Kurup, 2009).

Experimental evidence suggests that multiple freeze-thaw cycles result in a greater likelihood of complete cell death (Neel, 1971). Typically, an initial freeze, lasting 5–15 min is followed by a spontaneous thaw and subsequently a second freeze, slightly larger than the first.

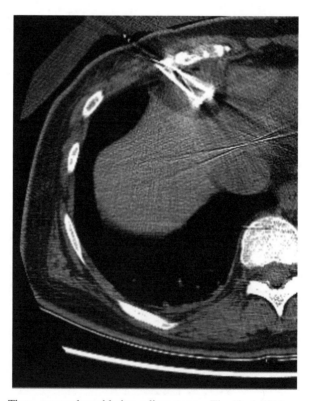

Figure 8.19   Three cryoprobes ablating a liver tumor. The "iceball" is represented by a lower density zone surrounding the cryoprobes (short arrows).

8.14 CHEMICAL ABLATION

Caustic substances such as alcohol have been used to ablate tumors for decades. Intratumoral injection of ethanol results in cytoplasmic dehydration and subsequent death of tumor cells. Chemical ablation also results in destruction of blood vessels feeding the tumor, thus potentiating the destruction of target tissues (Dodd et al., 2000). Although less widely utilized, acetic acid injection has been applied with similar effect. The typical setup for ethanol ablation is quite simple and inexpensive, including a 21- or 22-gauge needle, extension tubing, sterile 95% ethanol, and a syringe (Fig. 8.20).

Using image guidance, the needle (or needles if multiple injections are performed) is advanced into the lesion and intratumoral ethanol slowly administered. The total volume of ethanol to be injected can be estimated based on the size of the target lesion. For example, for a tumor of radius r, the necessary volume of ethanol can be approximated by estimating the volume of the lesion using the equation for a sphere: $4/3\pi\cdot(r\ \text{cm}+0.5\ \text{cm})^3$ (where 0.5 cm is added to achieve a small margin about the tumor). Numerous studies have demonstrated the safety and efficacy of alcohol tumor ablation, which has been used most extensively to treat hepatocellular carcinoma (HCC) in the liver. Chemical ablation is especially relevant in this context, because the capsule (or "pseudocapsule"), which often surrounds an HCC lesion, serves to trap and concentrate the injected caustic substance. Not uncommonly, adequate treatment requires several sessions, occurring on different days.

At present, alcohol ablation has largely been supplanted by more effective thermal ablative techniques. However, specific contexts still call for the use of chemical ablation. For example, when ablation is performed near critical structures susceptible to thermal damage (e.g., near bile ducts or bowel), chemical ablation may be the preferable ablation technique.

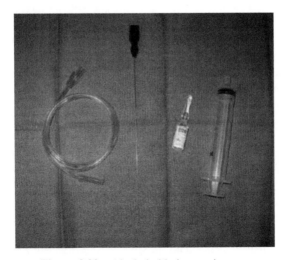

Figure 8.20   Alcohol ablation equipment.

8.15   NOVEL TECHNIQUES

8.15.1   High Intensity Focused Ultrasound (HIFU)

It may be more appropriate to refer to HIFU as a "rediscovered" rather than a "novel" technique. The ability of focused ultrasound energy to noninvasively heat tissues has been recognized for decades but is only recently emerging as a viable ablation technology. The renewed interest in focused ultrasound ablation is largely based on its recent coupling with advanced MR techniques for image guidance (the so-called *Magnetic Resonance-guided Focused Ultrasound or MRgFUS*). With HIFU, several hundred to thousands of ultrasound transducer elements are focused on a single target point under MR image guidance and tissues are "sonicated," resulting in tissue heating. MR thermography allows the operator to monitor this tissue heating. Sonic energy at intensities of approximately 10,000 times that used in diagnostic ultrasound is used in HIFU.

HIFU ablates tissues primarily via the thermal effects of molecular friction and the mechanical effects of acoustic cavitation (Kim et al., 2008). A single application of a typical 1.5 MHz HIFU transducer produces a thermal lesion measuring about 2 mm in width and 1.5–2 cm in length (terHaar, 2007). The target tumor is treated by creating multiple adjacent focused sonications to produce an overlapping ablation zone of the desired size. At present, HIFU has been used most commonly to treat uterine fibroids (the only current FDA approved application) but has also been used in the treatment of cancers of the brain, breast, liver, bone, and prostate.

The main advantage of HIFU compared to other ablative technologies is that it is completely noninvasive—the skin is not breached. HIFU, however, currently faces some limitations. Owing to the small size of each sonication, procedure times are long for large target lesions, sometimes lasting for several hours. Ultrasound beams are strongly attenuated in bone and reflected at gas interfaces, limiting their use in cases where bowel or bone is in the acoustic path of the target site. In addition, skin burns can occur if there is a poor acoustic coupling at the skin–transducer interface. Also, targeting accuracy must be continually updated for each sonication because of errors that could be introduced with tissue motion, thus increasing the challenge for targeting tumors in moving organs such as the liver.

The injection of intravenous microbubbles during HIFU application may be able to enlarge the resultant ablation zones. The microbubbles act as "seeds" for cavitation, decreasing the threshold and increasing the activity of acoustic cavitation (terHaar, 2007). This, in turn, leads to increased tissue temperatures with less sonication time, as demonstrated in animal models (Kaneko et al., 2005; Hanajiri et al., 2006), and may allow more rapid treatment compared to conventional techniques.

New geometries of HIFU transducers also promise to enlarge the coagulated volume and decrease the treatment time. A toric transducer, composed of eight ultrasound emitters with focal zones distributed over a 10 mm cone, has been described. Using sequential 5-s applications of each HIFU transducer in healthy pig liver, a coagulative lesion with an almost 2 cm diameter can be produced in only 40 s (Melodelima et al., 2009).

8.15.2 Irreversible Electroporation (IRE)

Irreversible electroporation refers to the process of permanent cell membrane permeabilization by the application of a high external electrical field. The transmembrane potential from the external electrical field is thought to cause the formation of innumerable nano-sized pores in the membrane and subsequent disruption of intracellular homeostasis. If the applied electrical field is below a certain threshold, this event is only temporary and the cell membrane returns to normal when the external electrical field is removed (reversible electroporation). This so-called "reversible electroporation" has become an important technique in molecular medicine, used to aid in the passage of drugs through a cell's membrane (Granot and Ribinsky, 2008). However, if the applied electric field exceeds a threshold, it will result in permanent disruption of cell membrane structures and intracellular homeostasis, leading to cell death (Lee et al., 2003). This latter effect has been recently seized upon as a novel technique for minimally invasive ablation.

There are two main types of IRE probes: monopolar and bipolar. The monopolar system requires placement of two probes into or bracketing the target. Applying 2500 V across two probes spaced 1.5 cm apart results in an ablation zone measuring approximately 2 cm × 3 cm × 3 cm (Lee et al., 2010). To achieve larger ablation areas, up to six monopolar probes can be placed simultaneously. The bipolar system consists of a single probe with two poles within its distal portion. Applying 2500 V across the two poles will create an ablation zone of approximately 2 cm × 2 cm × 3 cm (Fig. 8.21) (Lee et al., 2010).

Real-time ultrasound can be used for treatment monitoring, and accurately delineates the treatment zone, correlating well with the zone of cell death seen on immunohistopathology (Lee et al., 2007).

Irreversible electroporation offers some theoretical advantages over other ablation methods. The IRE has a very short ablation time, less than 1 min, to create an ablation area of approximately 3 cm in diameter (Lee et al., 2007). Because IRE is nonthermal, there is no "heat sink" effect, and complete cell death surrounding vessels can be produced (Lee et al., 2007; Al-Sakere et al., 2007; Edd et al., 2006; Rubinsky et al., 2007). While IRE ablates the living cells, it theoretically preserves the cellular matrix and pericellular structures, so large vessels and bile ducts remain structurally and functionally intact (Edd et al., 2006; Rubinsky et al., 2007). Finally, collateral damage to nearby structures when IRE is applied at the edge or dome of the liver is postulated to be less likely to occur compared to thermal modalities (Lee et al., 2010).

The IRE does have some disadvantages. Delivery of a high electrical field close to the heart (<1.7 cm) during IRE of lung lesions or high left lobe liver tumors has caused arrhythmias, including ventricular fibrillation in preclinical studies. The EKG gating of IRE application can be used to synchronize the application of electrical fields with the cardiac cycle and may overcome this disadvantage (Deodhar et al., 2009). Also, the high voltage used by IRE results in significant muscle spasm. General anesthesia and muscle relaxants are therefore used when performing IRE ablation to mitigate electric pulse-induced muscle contractions.

Figure 8.21 Irreversible electroporation generator (*Source*: Photo Courtesy of Angio-Dynamics, Inc.).

In vivo and clinical data using IRE is very limited at the time of this publication and so the efficacy and safety profile of IRE is not yet known.

8.16 TUMOR ABLATION AND BEYOND

The focus of this chapter has been on tumor ablation. Indeed, ablative techniques have become an indispensable tool in the management of many types of tumors, and new tumor applications are constantly being investigated. Some of the more common such clinical applications include (by no means a comprehensive list) the following:

- Both primary *hepatic* tumors (hepatocellular carcinoma) and secondary hepatic tumors (most commonly of limited metastases to the liver from colorectal, endocrine, and breast cancers, and also from other primary sites as well)
- Renal cell carcinoma of the *kidney*
- Both primary *lung* cancers (non-small-cell lung cancer, mesothelioma) and secondary lung tumors (most commonly of limited metastases from colorectal, breast, renal cell, head and neck cancers, and sarcomas, and also from other primary sites as well)
- Primary *bone* tumors (osteoid osteomas most commonly) and skeletal metastases for pain palliation

- A variety of *soft tissue* sites, with indications ranging from palliation of painful metastatic lesions to control of tumor recurrence in previously excised tumors
- *Prostate* ablation of select cancers
- *Uterine* fibroids
- Certain types of tumors of the *breast*, including fibroadenomas.

In addition to an ever-increasing list of tumors treatable by ablative techniques, adjuvant uses of thermal energy in oncology are being investigated. VanSonnenberg et al. (2005) summarize well some of these emerging paradigms. A few are given as follows:

- Thermal energy may one day be used in the context of heat-activated gene delivery.
- Hyperthermia is known to be a potent cellular radiosensitizer and so may be used synergistically with radiotherapy.
- Several studies have investigated the potential of using heat to release the contents of drug (chemotherapy)-laden liposomes in high concentrations into target tissues.

Beyond tumor ablation, clinical applications of tissue ablation are far-reaching. Aberrant electrical pathways in the heart, varicose veins, enlarged prostate glands, culprit nerves in chronic pain syndromes, the endometrial lining in patients with heavy menstrual bleeding (Lethaby et al., 2009), Barrett's esophagus (Zehetner and DeMeester, 2009), tendinopathy and sprained ligaments (Whipple, 2009), stress urinary incontinence, (Dillon and Dmochowski, 2009), and atria-esophageal fistula (Hazell et al., 2009) are just a few of the conditions that have been treated with ablative techniques. Whether in the context of tumor ablation or a myriad of other clinical scenarios, tissue ablative technologies represent a leap in minimally invasive methods for treating disease.

CHAPTER 9

Lasers in Medicine

ZACHARY TAYLOR

Center for Advanced Surgical and Interventional Technology (CASIT), University of California, Los Angeles, CA

ASAEL PAPOUR and OSCAR STAFSUDD

Department of Electrical Engineering, University of California, Los Angeles, CA

WARREN GRUNDFEST

Center for Advanced Surgical and Interventional Technology (CASIT), University of California, Los Angeles, CA

9.1 INTRODUCTION

Lasers and the manipulation of laser light have made inroads into countless medical specialties and subspecialties since Maiman unveiled the first working prototype in 1960. In many cases, lasers have initiated significant improvements to patient care, as is the case with optical coherence tomography (OCT), allowing diagnostics of the retina, which were previously unavailable. In other cases, lasers have completely revolutionized entire fields of medicine, such as dermatology and pathology, where lasers are able to target areas with minimal damage to surrounding normal tissue. As technology improves, costs are reduced, and acceptance by clinicians increases, lasers will continue to transform medicine.

This chapter discusses the history of the laser, the basic operation and types of lasers, laser–tissue interactions, and diagnostic and therapeutic applications of lasers. This chapter is intended to be a broad review of the use of lasers in medicine and in no way is intended to be a thorough treatise on the subject. The field is vast, and there are countless books and review articles on the subject. However, this chapter is broad enough in scope that it should provide the biomedical engineer

Medical Devices: Surgical and Image-Guided Technologies, First Edition.
Edited by Martin Culjat, Rahul Singh, and Hua Lee.
© 2013 John Wiley & Sons, Inc. Published 2013 by John Wiley & Sons, Inc.

or clinician a basic understanding of the role of lasers in medicine. References are provided for the interested reader and serve as a gateway to additional investigations.

Another goal of this chapter is to provide sufficient motivation for the use of lasers in medicine. Lasers have been explored for many applications, some which were not necessary or inappropriate. This chapter focuses on the more successful applications of lasers to medicine and provides details on the genesis of laser use and the advantages of the laser-based solution.

9.1.1 Historical Perspective

The acronym LASER (Light Amplification by Stimulated Emission of Radiation) dates to the late 1950s and is a modification of the term optical MASER, where the "M" in the acronym stands for microwave due to the original frequency of operation. The first theoretical foundations on lasers date back to a paper written by Albert Einstein in 1917, which predicted the absorption and stimulated emission aspects of laser operation (Heitler, 1984). These were followed by more theoretical works in the decades leading up to the late 1950s, including papers detailing operating characteristics and general properties of devices based on stimulated emission (Schawlow and Townes, 1958; Gordon et al., 1955; Shimoda et al., 1956). This work culminated in the first working MASER built simultaneously in 1953 by researchers in the United States and the USSR and operating at 24 GHz (Siegman, 2009). The first laser (optical MASER) was demonstrated by Theodore H. Maiman in 1960 at Hughes Research Laboratories in Malibu, CA. Maiman's laser was based on a ruby crystal pumped by an extremely powerful flashlamp wrapped around the cylindrical face of the ruby crystal rod and produced a near-infrared (NIR) beam at 694 nm. Maiman's laser was soon followed by more familiar laser types, including the gas laser (Javan et al., 1961), semiconductor diode laser (Hall et al., 1962), and excimer laser (Basov et al., 1971).

Lasers display a range of distinctive properties, including high temporal, spectral, and spatial coherency. Coherency is a unique feature of lasers and is the reason that beams can be propagated with low diverging beam (collimated) and focused to a very small area. Coherence means that every part of the beam's cross-section has a definite phase of the electromagnetic wave with respect to every other point. A pictorial demonstration of this concept is shown in Figure 9.1.

In light sources with low coherency or in lasers with poor beam quality, diffraction will limit the ability to form a tight focused spot and will cause the beam to diverge more rapidly, as interference of the beam within itself during propagation results in further degradation of the beam shape. For a laser beam to have high coherency, the laser must emit light/radiation in a very narrow band of wavelengths. This is why nearly all lasers, are close to being monochromatic. The narrow wavelength band they operate under is called the *linewidth* and represents the part of the spectrum in which the laser is emitting radiation.

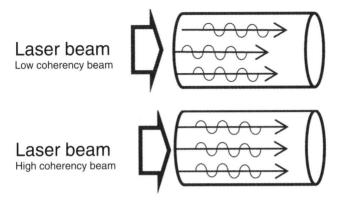

Figure 9.1   Electric component of the waves depicted as a sinusoidal shape.

Figure 9.2   Schematic representation of a laser cavity.

9.1.2   Basic Operational Concepts

To produce coherent, collimated laser radiation, a laser cavity must be built featuring a gain medium surrounded on both sides by specially designed mirrors. This cavity is typically called a *resonant cavity*, as it is designed to support a specific set of wavelengths by favoring a desired set of stimulated photons. This basic concept is displayed in Figure 9.2.

The gain medium can be solid (ruby laser), gas (CO_2 laser, water vapor laser, etc.), liquid (dye lasers), or semiconductor (diode lasers) and can be electrically excited through spark (electric discharge) or optically injected with a light source such as bulb, flashlamp, or another laser. The act of energizing the gain medium is also called *pumping*. The mirrors supply the necessary feedback to achieve a nearly monochromatic output. Once the gain medium is excited, radiation starts to build up in the cavity. Initially, only *spontaneous emission* occurs, which is characterized by radiation in random direction and a broad linewidth. This spontaneous emission starts a chain reaction of radiation in the desired direction, which resonates between the mirrors, and "forces" more radiation of the same propagation direction and

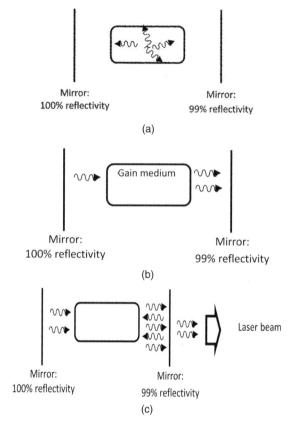

Figure 9.3 Laser startup dynamics: (a) Spontaneous emission in all directions, also called *fluorescence radiation*. (b) Stimulated emission starting to build up and get amplified, spontaneous emission not shown for simplification. (c) Stimulated emission amplified to greater intensities to produce the laser beam.

phase to be emitted. The resulting radiation is called *stimulated emission*. This process is displayed in Figure 9.3.

9.1.3 First Experimental MASER (Microwave Amplification by Stimulated Emission of Radiation)

In 1953, the first operational MASER was demonstrated and was able to amplify microwave radiation. Producing microwave radiation was a straightforward task, as radiation in the radio frequency (RF) domain was in technological use well before WWII. However, in order to achieve radiation amplification, the atoms/molecules in the gain medium must first be in an excited energy level. The gain medium can amplify radiation at a certain frequency, depending on the energy levels structure of that medium. In the first MASER, ammonia gas was introduced inside a resonant cavity to create feedback to produce coherent radiation at 1.2 cm in wavelength (Siegman, 2009). As radiation is reflected from the cavity walls, the excited medium is "forced" to emit radiation at the same frequency as in the incoming radiation, at the same phase, and the same direction. In case there is not enough feedback inside the cavity, the excited medium will relax into a lower energy level, in a random phase and direction, and will be considered as noise, which does not contribute to amplification.

9.2   LASER FUNDAMENTALS

One of the difficulties in creating light amplification is the need to achieve a state in which most of the gain medium is at an excited energy level without spontaneously relaxing into a lower state. This situation is called *population inversion*, meaning that the electrons in the medium reside in a higher energy level, and the lower level is considered empty. Population inversion is not trivial to sustain, however, and different types of lasers use varying schemes to achieve it.

9.2.1   Two-Level Systems and Population Inversion

In the simple case of a two-level system, the atoms are excited to a higher energy level. However, population inversion cannot be achieved because the rate at which the atoms are excited is equal to the rate at which they relax. The competing processes of excitation against spontaneous and stimulated emission make population inversion impossible, and no optical gain is possible.

In laser gain materials, when electrons decay from a higher energy state to a lower energy state, they emit a photon whose wavelength is determined by the difference between the two energy levels, as described in the following equation:

$$E_2 - E_1 = hv \tag{9.1}$$

where v (Greek, Nu) is the frequency of the photon (Hz) and h is a constant, Planck's constant 6.626×10^{-34} J S (joule × second). In nonlaser gain materials, energy is released as heat (Fig. 9.4).

The time constants in transitions between the energy states can reveal further the conditions toward a working laser. Spontaneous emission is occurring at a rate of about 10^{-8} s. However, exciting the electrons into a higher energy level, also called *pumping*, depends on the pump rate and is correlated to how much energy must be invested in the gain medium. In order to achieve a working laser, a high pumping rate is needed, coupled together with a slower mechanism of spontaneous relaxation. In multiple level laser systems, spontaneous relaxation can be on the order of few milliseconds (10^{-3} s), thus overcoming the limitations of a two-level system.

9.2.2   Multiple Energy Levels

A more realistic model of a working laser is the three-level laser. In this model, a nonequilibrium condition is achieved by exciting electrons to an energy level that

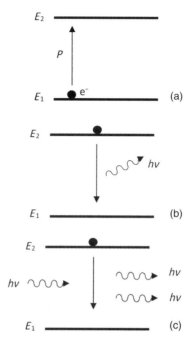

Figure 9.4 (a) Pumping in a two-level system. (b) Spontaneous emission: electron relaxes to lower energy level while emitting a photon in a random direction. (c) Stimulated emission: an incoming photon stimulates the electron to release energy by emitting a photon, with the same phase as the "stimulating" photon. All photons emitted/absorbed carry energy equivalent to the energy difference between E_2 and E_1.

indirectly contributes to a population inversion. A system is pumped to excite electrons from E_1 to E_3, and electrons from E_3 decay in a fast nonradiative transition (heat) to E_2, where they are held for a relatively long time. The slow transition rate from E_2 to E_1 increases the chance for a stimulated emission and also for a population inversion condition. Population inversion can subsequently be reached because electrons in E_1 are pumped immediately to E_3, thus creating enough inversion between E_1 and E_2. The disadvantage of a three-level laser is that it requires a strong pump, as every spontaneous and stimulated process will contribute to a decrease in the population inversion and is considered a highly inefficient system. An example of a three-level laser is the ruby laser, which was the first working laser, emitting red light at 694.3 nm (Fig. 9.5).

To reach a higher efficiency, more complex energy levels and transitions are needed and can be approximated well by a four-level laser system. In such system, pumping action excites atoms into E_4, then by a fast nonradiative transition to E_3, where a longer lifetime in this level accumulates the population in E_3. Spontaneous or stimulated emission causes decay to E_2 where another fast, nonradiative transition occurs at E_1. In this method, the short lifetime of E_4 and E_2 guarantees population inversion between the two lasing levels, E_2 and E_3 (Fig. 9.6).

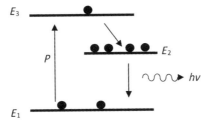

Figure 9.5 Schematic of a three-level system; majority of electrons now are in E2 thus population inversion condition is obtained.

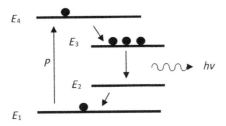

Figure 9.6 In a four-level laser system, photon emission is due to transition from E_3 to E_2.

The advantage of a four-level over the three-level system is the fact that the radiative decay is populating level E_2, and not the ground level E_1, ensuring that E_2 is always empty, thus achieving a steady-state population inversion. All lasers in use in medicine at present utilize four or more levels to achieve population inversion.

9.2.3 Mode of Operation

Lasers can be categorized into two distinct operational modes: continuous wave (CW) and pulsed wave. CW lasers emit beams of constant or quasi-constant amplitude. Because heating is a slow, time average process, many treatments that rely on photothermal effects (e.g., NIR treatment of choroidal melanoma) utilize CW lasers. Although most modern laser technologies can support CW, there are architectures and materials that cannot maintain this operation because the cavity cannot be pumped continuously without suffering from overheating effects and instabilities. For example, in excimer lasers such as the ones used in LASIK (laser-assisted *in situ* keratomileusis)/LASEK (laser-assisted subepithelial keratectomy) pulsed mode is the stable mode of operation and attempts to run these lasers in CW mode would result in catastrophic overheating (Krauss and Puliafito, 1995). Pulsed output allows instantaneous delivery of optical energy without causing large temperature rises in the tissue adjacent to the ablation area. This limits thermal damage in tissues when used for ablation.

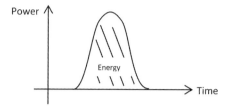

Figure 9.7 Pulse propagating in space.

A diagram showing the constituent parameters of a laser pulse is displayed in Figure 9.7. CW laser output is usually characterized with an average output power in units of joules per second or watts. Pulsed laser output can also be characterized with an average power but instead is often characterized with a peak power or energy per pulse (joule).

The total energy in a laser beam is computed as the power of the beam times the duration of the beam or, as is displayed in Figure 9.7, area under the curve. Similarly, the power is computed as the amount of energy in a given time interval divided by that interval as in Equation 9.2.

$$\text{Power}: P(\text{watt}) = \frac{\text{Energy (joule)}}{\text{Time (seconds)}} \tag{9.2}$$

In a pulsed, laser the output pulses are periodic and separated by an unchanging pulse period T_{rep} or pulse repetition frequency (PRF) f_{rep}. The average power and peak power of a pulse with known pulse energy pulse width is

$$\text{Average power}: P_{\text{avg}} = \text{Energy (joule)} \times \text{Repetition rate (Hz)} \tag{9.3}$$

$$\text{Peak power}: P_{\text{peak}} = \frac{P_{\text{avg}}}{f_{\text{rep}} t_{\text{pulse}}} \tag{9.4}$$

One major motivation for using laser technology in medical applications is the precise control over energy deposition in tissues in the volume, wavelength, and temporal regimes. These can be controlled with various laser system architectures and optics. For example, the ArF excimer laser used in LASIK procedures for corneal ablation operations is operating under relatively small energies of 3 mJ/pulse and repetition rates of 200–1000 pulses/s (Hz). Each photon is capable of electronic excitation of carbon–carbon bonds, which can lead to photodissociation in addition to photodisruption. Mid-IR lasers working with short pulse durations produce their effects through rapid photothermal decomposition or vaporization of the affected tissue. All short pulse lasers generate shockwaves during the process of tissue ablation. The solid and liquid portions of the tissue are vaporized over extremely short time scales, leading to ejection of material from the tissue surface at high velocity. These energized particles take much of the deposited energy with them, as they leave the tissue surface, thus limiting the transfer of

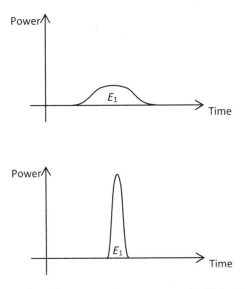

Figure 9.8 In lasers, where the purpose is to generate the highest peak power pulses, reducing pulse duration is preferable than increasing pulse energy.

energy (heat) to the adjacent tissue. Conversely, femtosecond lasers used for cutting corneal flaps in LASIK can have pulse energies down to nanojoule levels and repetition frequencies into the megahertz range.

An example of the energy power trade-off is displayed in Figure 9.8. The pulses in Figure 9.8a and b have the same energy (area under the curve); however, Figure 9.8b has a significantly shorter pulse width and hence higher peak power. Furthermore, if a pulse train is constructed of these waveforms, the PRF of the pulse train using Figure 9.8a can be increased sufficiently relative to Figure 9.8b to obtain equal average powers. Trade-offs between these parameters are utilized in for laser-based treatments depending on the desired laser tissue interaction. For example, in LASIK, the surgeon does not want to deposit heat in the cornea, thus he maximizes single pulse energy to achieve ablation yet reduces the PRF until the average power is extremely low. Conversely, in photothermal therapies that rely on hyperthermia, the surgeon's goal is to maintain a particular elevated temperature in the tissue while preventing and sorting of ablative or photomechanical effects that may damage the surrounding tissue. In this case, a very long pulse width, low peak power laser or CW laser is used.

9.2.4 Beams and Optics

9.2.4.1 Irradiance Profile The intensity profile of a laser beam projected on a flat surface orthogonal to the direction of propagation is called *irradiance distribution of irradiance profile*. Although a number of irradiance profiles exist (Siegman, 2009), the most common beam profile found in medical applications is the TEM_{00} Gaussian beam (see Siegman (2009) for an excellent overview of beam modes). In this distribution, the laser beam is radially symmetric about the optical axis and its intensity as a function of distance from the axis (r) is proportional to e^{-r^2}. A diagram of this irradiance and a 2D cut are displayed in Figure 9.9a and b, respectively.

Although lasers beams decrease in intensity proportional the Gaussian function, a good approximation to the laser peak intensity on target is:

$$I = \frac{2P}{\pi r^2} \left(\frac{W}{cm^2} \right) \qquad (9.5)$$

where P is the power (watts) and r is the beam radius defined as the radius in which the intensity drops to approximately 13.5% of the peak intensity (also known as $1/e^2$ value). Power can be replaced with energy depending on the desired calculation. The square radial dependence in the denominator of Equation 9.5 allows one to effectively control an arbitrarily energy/power density with an arbitrarily laser source. This is useful in certain settings, where the tissue treatment volumes are small and allow the clinician to achieve higher intensities by focusing a low powered, lower cost laser.

9.2.4.2 Divergence As a laser beam propagates in free space, its radius increases, the profile broadens while the irradiance (intensity) decreases. An exaggerated divergence of a laser beam is shown in Figure 9.10.

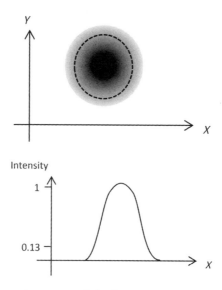

Figure 9.9 Irradiance—the power of the beam per unit area, measured in watts per square centimeter.

The radius r_1 is the minimum radius of the traveling laser beam assuming no lens or other optical components are in the beam path. The distance that it takes the beam central intensity to drop by half, z_0, is the Rayleigh length and the radius of the beam at z_0 is $r_2 = \sqrt{2}r_1$. Within the Rayleigh length, laser beam propagates as a collimated bundle of rays, while outside of this regime, the beam propagates similar to a modified point source with a constant divergence angle θ.

9.2.4.3 *Depth of Focus*

Laser beams are spatially and temporally coherent and can be focused to very small spots. Laser treatments and laser tissue interactions depend on wavelength and power/energy density. To fully characterize the laser beam energy density, one needs not only the transverse profile (irradiance) but also the axial variation. This parameter is often called the *depth of focus* (DOF), as it is modified by focusing lenses. Consider the focused beam displayed in Figure 9.11.

In this diagram, a collimated beam striking a convex lens is focused and then diverges. The beam waist is minimum at the beam focus, and the DOF describes an axial length corresponding to a given energy density. In other words, how long

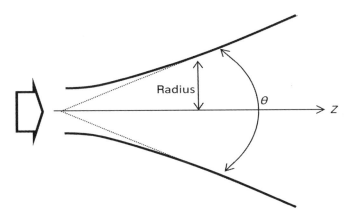

Figure 9.10 Laser beam divergence.

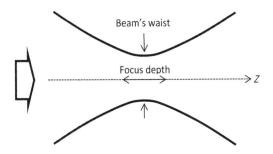

Figure 9.11 Laser beam depth of focus.

does the beam stay focused until it diverges? There are many definitions of DOF, but a commonly used metric is the Rayleigh length with the definition DOF $= 2z_0$.

Two examples of focused beams with varying DOFs are displayed in Figure 9.12 with equivalent collimated beam diameter sizes. In Figure 9.12a, a very short focal length, small radius of curvature lens is used for focusing, while a long focal length, large radius of curvature lens is employed in Figure 9.12b. This results in a tight focus with a small spot size and short DOF for Figure 9.12 and a larger spot size and longer DOF in Figure 9.12b. In general, the spot size and DOF are positively correlated.

9.3 LASER LIGHT COMPARED TO OTHER SOURCES OF LIGHT

Before studying laser operation and the biological and clinical applications, it is important to understand what makes laser light different from light emitted by other types of sources, such as light bulbs, the sun, or fire, and why these unique properties make lasers such an attractive and effective tool in a variety of diagnostic and therapeutic applications. Laser light from appropriately designed laser systems can have some of the following four properties, described in the following sections.

9.3.1 Temporal Coherence

Temporal coherence is the measure of the periodicity of the laser beam. To express this mathematically, we define the electric field of a laser beam as $E(t)$ in units

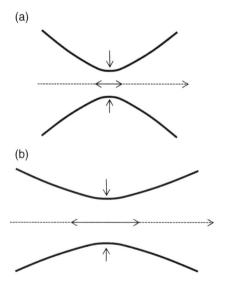

Figure 9.12 (a) Highly focused beam, narrow focus depth—sensitive to alignment. (b) Low focus, higher tolerance in positioning.

of energy per unit time. If the beam is periodic with period T then electric field obeys the following property:

$$E(t) = E(t + nT) \; \forall t, n \qquad (9.6)$$

where n is the integer. If time is fixed and the signal is compared with a copy of itself delayed or advanced by any number of periods, the same amplitude will be observed (note that the inclusion of n is unnecessary for a complete definition of periodicity but is done for the sake of clarity). In a laser beam with a high degree of temporal coherence, the beam can be compared to a copy of itself delayed/advanced by many periods and the same amplitude can be observed.

There are many methods to compute the level or degree of temporal coherence. The overall coherence of the beam is an important metric to establish the suitability of the beam for various medical applications. Several methods are available to provide precise analysis of the beam coherence. The overall metric of coherence time is the number of periods the beam can be advanced or delayed until the original and copy are no longer similar. Table 9.1 lists some typical coherence times for various light sources.

If appropriately designed, laser beams can be greater than 6 orders of magnitude more coherent than an equivalently powered source based on incandescent, halogen, or mercury arc lamps. This property is especially useful in frequency swept optical coherence tomography (OCT) systems whose axial resolution heavily depends on long coherence times.

9.3.2 Spectral Coherence (Line Width)

Spectral coherence, or linewidth, is measure of the wavelength/frequency or color content of a laser beam. In general, the time-varying response of the electric field of a laser beam can be written as

$$E(t) = p(t) \cos \left(\frac{2\pi c}{\lambda} \right) \qquad (9.7)$$

where c is the speed of light, λ is the center wavelength, and $p(t)$ is the envelope or modulus of the wave. Most laser beams encountered in biology and medicine (not including the superluminescent diodes found in time domain OCT imaging

Table 9.1 Lists of Typical Coherence Times for Various Light Sources

Source	Coherence Time, s	References
Light bulb	$\sim 3 \times 10^{-15}$	(Mandel, 1961)
LED	$\sim 33 \times 10^{-15}$	—
External cavity laser diode	$\sim 50 \times 10^{-9}$	(Boggs et al., 1998)
Frequency stabilized HeNe	$\sim 1 \times 10^{-9}$	—

systems) can be thought of having one color or wavelength. However, on closer inspection with a spectrometer or some other spectral sensing instrumentation, the irradiance of a laser beam is revealed to have a continuum of colors centered about some particular wavelength. For example, the spectrum of a green LED centered at 520 nm can be as wide as 20 nm, while the output of a HeNe laser running at 543 nm can be as narrow as 0.1 nm. The envelope term in Equation 9.6 thus describes the amount of spread (linewidth) existing in the output spectrum of a source.

To illustrate the concept of spectral coherence, the normalized output spectrum of a typical incandescent bulb is plotted in Figure 9.13a, with the spectra of the green LED and green HeNe lasers superimposed. Incandescent bulb and other thermal sources emit a very broad spectrum of light that extends deep into the near- and mid-IR. Lasers, however, are constructed with materials that offer gain (next section) at only a select range of wavelengths and are often mounted

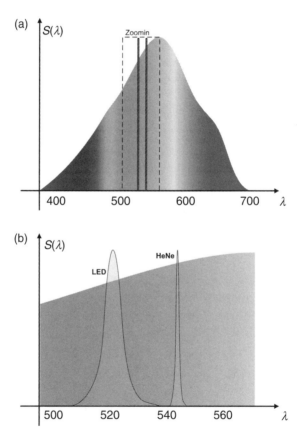

Figure 9.13 Spectra of different light sources. (a) Approximate spectra of incandescent source with superimposed LED and HeNe laser sources. (b) Zoom-in of area denoted by the dashed box in (a).

inside external cavities (next section) that reduce the operational wavelengths even further. Thus, it is possible to generate significant powers of light in a very narrow band of wavelengths (Fig. 9.13b). In contrast, bulbs cannot be designed to emit a narrow band of wavelengths and filters must be used if specific colors are desired.

9.3.3 Beam Collimation

Another significant aspect that differentiates laser light from other types of light is the high level of beam collimation. This means that the majority of radiation emitted by a laser travels in one direction with very little spread to other directions. As such, laser beam can be thought of and is often modeled as a parallel bundle of rays (plane wave). There are two major benefits of this narrow angular distribution of energy: (i) Laser light can be propagated between apertures with minimal energy loss making it possible to develop complex optical systems with high total throughput, and (ii) the plane wave nature of a laser beam approximates a perfect point source (located at infinity). This allows the beam to be focused down to a diffraction-limited spot with simple optics. Incandescent bulbs and other sources emit light evenly in all directions (isotropic). Optics and pinholes can be used to create a collimated beam filtering out unwanted directions; however, this method results in an arbitrarily high loss of power to yield power in a vanishingly small number of directions. Beam collimation is especially useful in microscopy and imaging, where high spatial resolution and throughput are requisite.

9.3.4 Short Pulse Duration

The final point that distinguishes laser light from all other sources of light is the pulse duration. Thermal sources such as incandescent and halogen bulb are limited in their ability to produce short pulse durations because of parasitic capacitance and inductance effects. The shortest pulses from traditional thermal sources are in the tens of milliseconds. Xenon lamps, which are driven by banks of capacitors and often used to pump solid state lasers such as ND:YAG, are limited by parasitic inductance down to a few milliseconds. The minimum pulse width of LEDs is generally constrained by parasitic capacitance of the junction and parasitic resistance of the resistive contacts. Under the right condition, LEDs can support pulse widths down to a few nanoseconds.

The above-mentioned sources are all driven electronically and are subject to pulse width limitations, which make it all but impossible to generate pulses in the picoseconds range. Lasers, however, are not limited by circuit effects and can achieve pulse widths of hundreds of picoseconds in the case of Q-switched lasers and hundreds of femtoseconds in the case of NIR mode locked lasers (Washburn et al., 2004). These short pulse widths allow the clinician to ablate tissues while minimizing thermal damage to adjacent tissues. A summary of typical pulse widths from a variety of sources is listed in Table 9.2.

Table 9.2   Typical Lower Limit Pulse Widths

Source	Pulse Width, s	References
Light bulb	$\sim$20 $\times$ 10^{-3}	(Land, 1975)
Xenon flashlamp	$\sim$10 $\times$ 10^{-6}	(Linford, 1994)
LED	$\sim$9 $\times$ 10^{-9}	(Authors' Experience)
Q-switched Laser	$\sim$100 $\times$ 10^{-12}	(Zayhowski and Dill, 1994)
Mode locked laser	$\sim$60 $\times$ 10^{-15}	(Spence et al., 1991)

9.3.5 Summary

Lasers offer a very efficient and straightforward manner of generating coherent radiation in a narrow band of wavelengths, directed in a small span of directions combined with the ability to modulate the amplitude on time scales many of orders of magnitudes shorter than conventional light sources. This combination of attributes makes lasers the perfect building block for a variety of diagnostic and therapeutic systems.

9.4 LASER–TISSUE INTERACTIONS

Lasers have allowed researchers and clinicians to explore many previously inaccessible light–tissue interaction physics because of the many orders of magnitude available in both power density and pulse width. While most conventional light sources are limited to coagulation or vaporization depending on the electrical driving systems and optical train design, lasers can provide incredibly high peak powers at vanishingly small pulse widths in a very limited number of spatial modes. With the appropriated optics, gigawatts to terawatts per squared centimeter of power density can be achieved for durations less than nanoseconds. This has enabled interaction modes such as ablation and photodisruption to be applied to treatments. There are essentially five types of interaction modes, and these are described in the following text and summarized in Figure 9.14.

9.4.1 Biostimulation

Cellular and biological effects have been observed at laser fluence (intensity) levels below that required for photodynamic therapy (PDT) (Morrone et al., 2000; Torricelli et al., 2001; Kertesz et al., 1982; Schwartz et al., 1987; Kubasova et al., 1995). Although the mechanisms behind the observed effects are not yet understood, this class of laser–tissue interaction is included in this chapter because it occupies the lowest level of fluence that has a measurable effect on tissue. Biostimulation has been proposed for the stimulation of the immune system (Kubasova et al., 1995), treatment of the optic nerve (Schwartz et al., 1987), and the treatment of cartilage tissue (Torricelli et al., 2001). Although many groups believe that power density and exposure duration are the important parameter at this

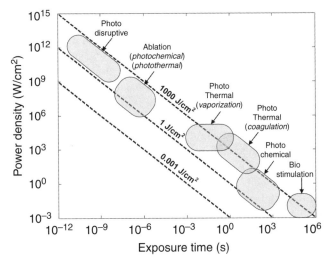

Figure 9.14 Different types of laser–tissue interaction modes as a function of exposure time and power density. The contours of constant energy are denoted by the dotted lines (Peng et al., 2008).

fluence level, some groups believe that the linear polarization of laser light is responsible for many of the observed surface effects through reorientation of the polar heads of the lipid bilayer in cell membranes (Kertesz et al., 1982). Whatever the case may be, biostimulation is still under investigation and its efficacy is unclear. The reader is cautioned that the efficacy of this technology has yet to be proven.

9.4.2 Photochemical Interactions

This class of interaction in the $10 \text{ mW/cm}^2 – 1 \text{ W/cm}^2$ range includes PDT, which relies, in most cases, on the generation of reactive oxygen species (Peng et al., 2008; Toyokuni, 1999; Carbonare and Pathak, 1992; Zorov et al., 2000; Halliwell, 1992; Bayr, 2005; Krötz et al., 2004; Henderson and Dougherty, 1992). Although proposed for a wide variety of treatments, the PDT and other photochemical interactions are most commonly used for cancer treatments, where the lasers light can be used to precisely to treat tumors with specific drugs (Pass, 1993; Brown et al., 2004; Overholt et al., 1999; Dolmans et al., 2003). The PDT has been used for treatment for skin cancer, Barrett's esophagus, lung cancer, and kidney cancer. In addition, fluence levels in this range have been used to treat age-related macular degeneration (AMD) (Bressler, 2001). It is important to note here that the fluence levels of appropriately administered PDT induce chemical reactions in either the tissue or injected photosensitizer and are usually not high enough to cause thermal damage in the irradiated tissue.

9.4.3 Photothermal Interactions

Precise and selective heating of tissue volumes is a widespread use of medical laser technology. In this regime, the fluence is sufficiently high either to cause thermal necrosis, protein denaturation, and coagulation or to excite cavitations (air pockets, or bubbles) in the tissue (Steger et al., 1989; Utley et al., 1999; Fitzpatrick et al., 1996; Kahn et al., 1996; Wyman et al., 1992). Fluences corresponding to the higher power density levels in the photothermal range in Figure 9.14 are often referred to as *photocoagulative*, due to common applications of this regime. Although photothermal effects can be induced with flashlamps and other traditional light sources, lasers are ideal for this application because the beams can be focused to arbitrarily small spot sizes, and specific wavelengths can be selected to maximize absorption in a particular volume while leaving the surrounding tissue untouched. While this effect forms the basis of treatment for various ocular (Framme et al., 2009; Bird and Grey, 1979) and oncologic (Steger et al., 1989) applications, it is most commonly used in dermatology to remove port-wine stains (PWSs) and other vascular malformations (Anderson and Parrish, 1983; Jacques et al., 1993; Anvari et al., 1995; Chang and Nelson, 1999; Nelson et al., 1996; Dierickx et al., 1995; Nguyen et al., 1998; Jasim and Handley, 2007).

9.4.4 Ablation

At sufficiently high peak power densities, the instantaneous electric field of the laser pulse is strong enough to vaporize material (photothermal ablation) and/or disrupt chemical bonds (photochemical ablation) of the material's molecules (van Leeuwen et al., 1991; Dougherty et al., 1994; Munnerlyn et al., 1988; Brinkmann et al., 1996; Walsh and Deutsch, 1988; Walsh et al., 1988). When performed at a wavelength with sufficiently high tissue absorption coefficients, laser pulses can precisely remove volumes of tissue with little to no thermal effects in the surrounding tissue volume. UV lasers of 193 nm are often used for tissue ablation, given the large absorption coefficient of both proteins and water at these wavelengths and the large volume fraction of water in most tissues. UV-based ablation methods form the basis of most refractive surgeries, including LASIK (Marcos et al., 2001; Moreno-Barriuso et al., 2001; Scerrati, 2001; Guell and Muller, 1996), LASEK (Taneri et al., 2004a, 2004a; Dastjerdi and Soong, 2002; Tychsen and Hoekel, 2006; Anderson et al., 2002; O'Doherty et al., 2007; Kornilovsky, 2001), and photorefractive keratectomy (PRK) (Munnerlyn et al., 1988; Fantes et al., 1990; Seiler et al., 1994; Seiler et al., 2000; Hersh et al., 1998). UV lasers offer an interesting comparison of ablation dynamics. At 193 nm, both proteins and water display extremely high absorption coefficients and ablation of tissues. Furthermore, the photon energy at 193 nm (6.4 eV) is significantly higher than the carbon–carbon bond energy (3.6 eV) (Garrison and Srinivasan, 1985). Thus, ablation in tissues, such as cornea, occurs through processes of material/water vaporization and photochemical bond disruption. Conversely, laser pulses centered at 1064 nm, such as those from a Q-switched Nd:YAG laser, have photon energies in the 1.17 eV range.

This energy is too low to disrupt chemical bonds, and thus, ablation at these wavelengths is a consequence of photothermal effects. Ablation in the NIR range thus produces much more intense mechanical shockwaves than ablation using sources in the far UV range.

9.4.5 Photodisruption

At extremely intense fluence level electric fields are so large that the irradiated molecules are torn apart and optical breakdown creates micro plasmas in the adjacent tissue areas. As these micro plasmas progress and expand, they generate mechanical shockwaves that propagate throughout the tissue. The current set of practical applications for this amount of laser fluence is narrow, but the most common use is in ophthalmology, where the mechanical shockwaves are utilized in conjunction with cataract surgery (Thompson et al., 1992).

9.5 LASERS IN DIAGNOSTICS

The unique properties of laser light enumerated in previous sections, including resolution, penetration, and hyperspectral sensitivity, have enabled a wide range of diagnostic instrumentation that was either not possible before the advent of the laser or was practically/financially infeasible. In some cases, such as blood oxygen monitoring, lasers have increased both measurement sensitivity and specificity. In other cases, such as OCT in ophthalmology, lasers have completely revolutionized the clinician's diagnostic tool box. In this section, some of the more common laser-based diagnostics are motivated and discussed.

9.5.1 Optical Coherence Tomography

OCT is a technique where the layers of the tissue are imaged by interfering the probing beam with a reference beam (usually an advanced/delayed copy of itself) in a typically fiber-based interferometer. An interferometer is an optical system consisting of a beam splitter, two mirrors, and a detector. This system allows one to interfere an incoming light beam with a delayed copy of itself, thus enabling a measurement of the envelope of the light beam of interest.

Transverse scanning is performed using galvo mirrors or translation stages, and axial scanning is performed with an adjustable mirror on the reference arm to control the relative delay in the resulting interferogram. Although some of the first *in vitro* and *in vivo* OCT images were published using superluminescent diodes (Ko et al., 2004; Goldberg and Mehuys, 1994), which are technically not lasers in the context of the definitions given in Section 9.3, many systems now employ either femtosecond lasers or swept source NIR lasers for illumination. These systems may offer advantages in signal-to-noise ratio (SNR) and sensitivity.

A block diagram of a basic OCT setup is given in Figure 9.15a and an image of the Zeiss Cirrus OCT system is displayed in Figure 9.15b. In this diagram,

(a)

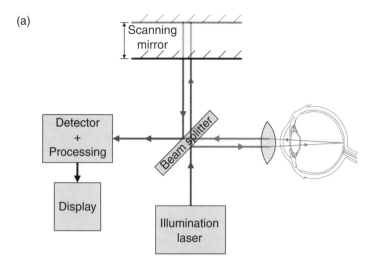

(b)

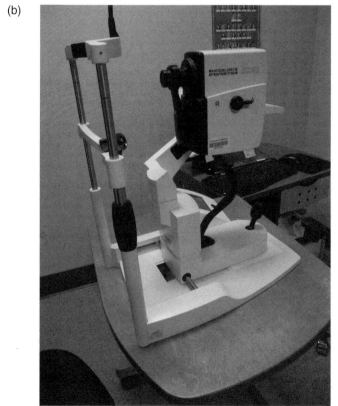

Figure 9.15 (a) Operational principle of optical coherence tomography. (b) OCT system.

the broadband light source is coupled into a fiber and then sent through a 50/50 coupler, which sends half the energy to the scanning arm and the other half to the reference arm. Images can be acquired with either of the two methodologies described in the following two sections.

9.5.1.1   *Low Coherence or Pulsed Illumination*   To acquire an image with this modality at a particular depth, the reference arm is held fixed and the probe scanning module sweeps the probe beams across the 2D plane of interest. Then the moving mirror in the reference arm is translated, and the 2D scan is initiated again. Because of the low coherence or pulsed nature of the source in this setup, the instrument is sensitive to constructive interference that occurs when effective optical path length of the reference beam and the probe beam are equal. Adjustment of the reference beam mirror allows scanning in the longitudinal/axial dimension.

9.5.1.2   *Swept Source Illumination*   This technology replaces the pulsed/low coherence source with a wavelength tunable laser source. Axial scans are performed by sweeping the wavelength of the laser and recovering the amplitude and phase of the reflected wave by modulating the optical path length of the reference arm. A Fourier transform (mathematical computation) is then applied to the magnitude and phase data to extract the depth profile of the tissue. Transverse scanning is performed in the same manner as the low coherence setup. This method offers some advantages in terms of axial resolution and SNR when compared to low coherence OCT imagers and is gaining increase use at ophthalmology clinics.

9.5.1.3   *Retinal Imaging*   Laser-based OCT has revolutionized retinal diagnostics, as it has provided clinicians with the ability to create detailed 3D reconstructions of the macula, fovea, and many other structures in the retina. Before the advent of OCT, optical imaging of the retina was much more difficult and time consuming and less predictable. When using an ophthalmoscope, the pupil must be dilated and the retina illuminated by an external source to maximize the imaging illumination that makes it to the back of the eye. This method allows for superficial visualization of the macula and optic nerve but does not permit 3D imagery of retinal structures.

The OCT is performed with NIR radiation in the 800–1100 nm wavelength band. This wavelength is ideal, as it suffers extremely low loss through the aqueous and vitreous humor and displays significant tissue penetration. Furthermore, these wavelengths cannot be detected by the patient, and therefore patient discomfort is reduced. Newer systems based on swept source technology have more refined image processing algorithms and increasingly high SNR, and often do not require pupil dilation for high resolution results. With these advantages, OCT has become a commonly used technique in the ophthalmology clinic. OCT has revolutionized the detection and diagnosis of macular degeneration, macular pucker, and macular holes. It has enabled a noninvasive virtual biopsy of retinal tissue that was completely inaccessible before the advent of this technology.

9.5.2 Fluorescence Angiography

Angiography is the study and imaging of blood vessels and vessel network through the introduction and detection of a contrast agent. In fluorescence angiography, a visible or IR fluorescent dye is injected into the blood stream. The area of interest is illuminated with narrow band light of the selected wavelength, and the emission is detected with blocking, long-pass filters, and a CCD. A block diagram of this process is given in Figure 9.16a.

The most common dyes used in medical fluorescence angiography are indocyanine green (ICG) and fluorescein. These dyes are known to cause slight discomfort to a minority of patients but display no long-term side effects and are completely removed by the body's renal system after 2–3 days. Furthermore, the fluorescence yield of these molecules is quite high and thus low, eye-safe fluences are more than sufficient for high contrast imaging.

9.5.2.1 Retinal (Fluorescein) Angiography

One of the most common uses of fluorescence angiography is retinal imaging, where the technique is used to visualize the blood vessel network in the retina and assess perfusion throughout the macula and beyond. This is especially important for detection of ischemia, retinal detachment, and diabetic retinopathy. Before the advent of retinal angiography, ophthalmologists were limited to various types of ophthalmoscopes to produce rather subjective assessments of retinal perfusion.

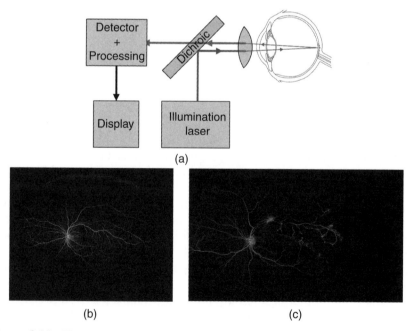

Figure 9.16 Fluorescence angiography. (a) Basic block diagram of a fluorescence angiography system, (b) fluorescence angiogram of a normal retina, and (c) fluorescence angiogram of a retina with a vein occlusion.

Retinal angiography typically uses fluorescein dye administered intravenously roughly 10 min before imaging. The patient is then placed in front of a specially designed ophthalmoscope and the retina is illuminated with 494 nm (turquoise) light. Fluorescein emits at 520–530 nm, which is long enough in wavelength to use standard dielectric stack filters for the dichroic mirror (figure) yet short enough for silicon CCD of CMOS detection. Thus, images can be acquired in real time.

The absorption spectrum of fluorescein is quite broad, and an LED or halogen lamp with appropriate filters is sufficient to excite emission from this contrast agent. However, laser diodes are more efficient than the aforementioned sources and are often used as the illumination mechanism. Furthermore, the narrow output spectrum of a laser diode minimizes the probability of illumination light leaking through the blocking filter, thus maximizing contrast in the detected image.

9.5.2.2 *ICG Angiography* The ICG is another dye used in medical diagnostics. It has a significantly different molecular structure than fluorescein and thus can be used to image a different structure in conjunction with the same administration and imaging techniques used in fluorescein angiography. ICG's absorption spectrum peaks in the 790–800 nm band, and it emits in the 830–840 nm band. These NIR wavelengths are almost ideal for deep penetration, and therefore systems based on ICG angiography allow imaging of the choroidal circulation below the retinal pigment epithelium (RPE). This contrasts with fluorescein angiography, which captures blood flow above the RPE, highlighting retinal vessels and capillary networks against the dark background of the RPE.

The NIR operation has enabled the assessment of vascular networks and the measurement of blood perfusion in a variety of applications, including plastic surgery, abdominal surgery, general surgery, and internal medicine. Before the advent of this technique, it was nearly impossible to image small vasculature *in vivo*.

9.5.3 Near Infrared Spectroscopy

As discussed above, NIR lasers exhibit deep penetration depths due to the relatively low absorption by water and the low level of scattering as compared with visible light. This property combined with the absorption spectrum of oxygenated and deoxygenated hemoglobin makes near infrared spectroscopy (NIRS) an ideal technology for noninvasive blood oxygenation monitoring, especially in infants. Infants are at a much higher risk of developing complications from abnormal blood oxygen levels than adults and their reduced body weight and blood volume necessitates the use of noninvasive techniques. If exposed to very high concentrations of oxygen, an infant can develop acute toxicity (Bert effect), which results in respiratory stress, and can lead to retinopathy of prematurity (ROP). This is caused by vasoconstriction, especially in the temporal portion of the retina and is the single largest cause of blindness in childhood. High oxygen concentrations may also damage the pulmonary epithelium and can lead to atelectasis, the collapse or closure of alveoli in the lungs.

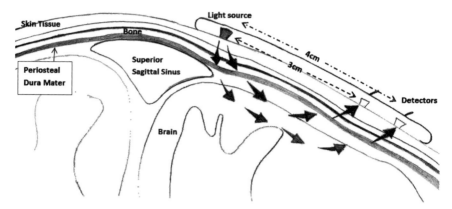

Figure 9.17 NIRS device for neonatal oxygen monitoring.

A typical NIRS system for neonatal blood oxygen monitoring is displayed in Figure 9.17 and consists of two to three laser diodes emitting radiation in the 700–900 nm range with an array of detectors located at a range of distances away from the emitter diodes. The NIR radiation is launched into the cranium of the neonate at an angle. The radiation undergoes multiple scattering and absorption events until it finally makes its way to the detectors. As can be seen from the figure, detectors closer to the emitter probe the more superficial layers, while detectors on the periphery probe the deeper layers. Using different emitter wavelengths and signals from different detector positions, an accurate noninvasive assessment of blood oxygen can be acquired.

9.6 LASER TREATMENTS AND THERAPY

Laser beam properties allow for the precise deposition of energy in tissue with high spatial and temporal precision. This has facilitated the development of numerous techniques that were not feasible before the clinical application of laser technology. This is readily apparent in the ophthalmologic and dermatologic applications, where even slight damage to tissue adjacent to the treatment area is not tolerable due to aesthetic or functional constraints.

An exhaustive treatment of the current therapeutic uses of lasers is beyond the scope of the chapter. Instead, a brief overview is provided, followed by four specific topics, one for each of the final four laser–tissue interactions overviewed in Figure 9.14. These are retinal PDT (photochemical), transpupillary thermal therapy (TTT) (photothermal), vascular birth mark removal (photocoagulation), and laser-assisted corneal refractive surgery (ablative).

9.6.1 Overview of Current Medical Applications of Laser Technology

9.6.1.1 Urology One of the most successful uses of lasers in urology (Mulvaney and Beck, 1968; Johnson et al., 1992; Michael, 1996; Stein and Kendall,

1984; Turek et al., 1992) is laser lithotripsy (Bagley, 2002; Watterson et al., 2002; Razvi et al., 1996), where high energy pulsed lasers are used to fragment kidney stones through photochemical or photothermal means. This procedure originated with photomechanical disruption using fiber coupled pulses from a 504 nm dye laser. In today's procedures, the dye laser has been replaced by a Ho:YAG laser operating at ~2120 nm. Pulses from these lasers are coupled directly to the stone using a fiber delivery in specially designed endoscopes and fragment the target area through photothermal effects.

9.6.1.2 Ophthalmology Ophthalmology has arguably been more positively impacted by the advent of laser technology than any other specialty in medicine. The optical properties of ocular tissue allow treatment of targeted areas with less damage to adjacent structures than most other applications. For example, numerous laser-based treatments can be applied to the retina without any damage to the cornea, iris, or lens. This level of noninvasiveness has revolutionized many ophthalmologic applications that were either highly invasive or unfeasible.

Refractive error surgeries, including LASIK, LASEK, and PRK, are performed with a 193 nm excimer laser. Precise volumes of tissue can be ablated (photomechanical effect) using these high energy pulses to reshape the cornea and correct refractive errors such as myopia, hypermetropia, and astigmatism. UV light at 193 nm suffers significant absorption in both water and corneal proteins, resulting in high axial resolution and short ablation depths, therefore making the procedure safe for the retina.

Numerous retinal surgeries are performed with visible and IR lasers. The TTT utilizes NIR laser energy in the 800–900 nm band for the treatment of retinal tumors through heat-assisted protein denaturation (Korver et al., 1992; Shields et al., 1996; Oosterhuis et al., 1995; Reichel et al., 1999; Shields et al., 2002). Retinal PDT uses an NIR laser to activate blood vessel coagulating drugs for the treatment of wet AMD (Bressler, 2001; Dhalla et al., 2006; Schlötzer-Schrehardt et al., 2002; Miller et al., 1999; Schmidt-Erfurth et al., 1999). In addition, lasers ranging from 500 to 700 nm depending on the depth of treatment are used to treat retinal detachment (Hunter and Repka, 1993; Yannuzzi et al., 1992) and retinal ischemia (Stefánsson, 2001; McAllister et al., 1998; Adamis et al., 1996).

9.6.1.3 Lasers in Otolaryngology The primary laser technologies used in head and neck surgery are CO_2 laser-based surgical systems, which operate at 10,600 nm. This mid-IR wavelength displays extremely high absorption in water, making it ideal for intraoperative cutting, ablating, and coagulation. In addition, the large wavelength allows for efficient fiber coupling, enabling the use of convenient handheld devices, and the diameter of the delivery fibers can be adjusted to fit the desired outcomes. Although adoption of CO_2 lasers is a relatively recent occurrence, this technology is now routinely used in a number of otolaryngological subspecialties, including head and neck oncology (Vilaseca-González et al., 2003; Gallo et al., 2002; Holsinger et al., 2006), otology/neurotology (Vernick, 1996; Haberkamp et al., 1996; Antonelli et al., 1996), laryngology/airway (Cotton and

Tewfik, 1985; Sosis, 1989; Fried, 1983), pediatric otolaryngology (Senders and Navarrete, 2001; Schraff et al., 2004; Pasquale et al., 2003), and rhinology (Şapçi et al., 2003).

9.6.2 Retinal Photodynamic Therapy (Photochemical)

Retinal PDT (Bressler, 2001; Dhalla et al., 2006; Schlötzer-Schrehardt et al., 2002; Miller et al., 1999; Schmidt-Erfurth et al., 1999) is commonly used to deliver targeted energy in the choroidal neo vascular membrane (CNVM) to treat diseases and pathologies such as AMD, high myopia, and choroidal tumors. AMD is a deterioration or breakdown of the macula, the small area in the retina that is responsible for central vision and the resolving of fine details. AMD can be broken down into two types: dry AMD and wet AMD. The dry AMD cellular deposits, known as *drusen*, accumulate beneath the retina, leading to retinal detachment in more severe cases. In wet AMD, abnormal blood vessels begin to grow in the choroidal layer and can eventually break through the retinal pigment layers and leak blood into the photoreceptor layer, resulting in significant photoreceptor damage and loss of vision. Cross-sectional views of a normal retina and AMD are given in Figure 9.18a and b, respectively.

As in choroidal melanoma, the tissue requiring treatment lies adjacent to many sensitive structures, and damage to these can lead to significant loss of functionality. However, unlike TTT (the choroidal melanoma), which uses thermal effects to deliver treatment, the PDT uses photochemical effects and enhances and directs energy absorption through the stimulation/activation of a photosensitive dye injected into the patient before treatment. A special dye (e.g., verteporfin), which has a high affinity for the abnormal blood vessels associated with wet AMD, is injected in the patient's vein before treatment. After injection, a diode laser in the 800–900 nm range is focused on the affected area in the retina, with the aid of an external lens. The illumination activates the dye bonded to the unwanted blood vessels and leads to vessel closure and blood coagulation. The dye is designed to selectively bond to the abnormal blood vessels associated with AMD. This combined with the minimum absorption in the retina to NIR and the reduced laser power as compared to TTT results in minimal or no damage to the adjacent normal tissue. While promising, this technique has been largely supplanted by antiangiogenic therapy (Hlatky et al., 2002; Spaide et al., 2006).

9.6.3 Transpupillary Thermal Therapy (TTT) (Photothermal)

Choroidal melanoma is the most common primary malignant intraocular tumor (Char, 1978; Diener-West et al., 1992). The choroid lies beneath the retina and its pigment epithelium, and its primary function is to provide oxygen and other nourishment to the retinal photoreceptors. Choroidal melanomas affect the RPE as they push against it and deprive it of circulation, resulting in areas of atrophy, drusen, and localized pigment epithelial detachments. Choroidal neovascularization can occur as the tumor grows, leading to ischemia, subretinal exudation, hemorrhage,

(a)

Outer
nuclear →
layer

Photo
receptors →

Retinal
pigment →

Choroid →

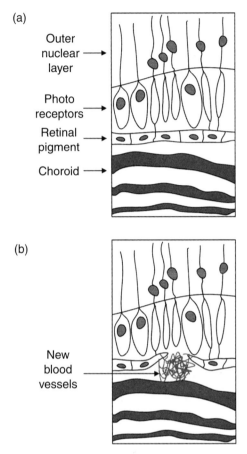

(b)

New
blood
vessels

Figure 9.18   Wet age-related macular degeneration. (a) Normal cross-section of the macula showing photoreceptors retinal pigment layer, and choroid. (b) Cross-section of ocular wall geometry.

and fibrous plaque formation. Owing to the location of the choroid, tumors in this tissue are inaccessible without highly invasive procedures and are poor candidates for mechanical removal, as they lie directly behind the fragile retinal layers. These factors make laser treatment a desirable option for choroidal melanoma.

The TTT is a therapeutic method used to deliver heat to specific volumes of the wall of the eye using NIR laser irradiation. This technique is employed primarily in the treatment of retinal tumors. The TTT typically uses narrow band diode lasers in the 700–1000 nm focused onto intraocular tumors to induce tumor cell necrosis by hyperthermia (Shields et al., 1996; Oosterhuis et al., 1995; Reichel et al., 1999; Shields et al., 2002). Pigment in the tumor is responsible for the NIR absorption and resulting heat generation that destroys the tumor cells. Tumor cell necrosis is caused by cell membrane damage, protein denaturation, chromosomal damage, and the breakdown of mitochondria because of heat deposition (Korver et al., 1992).

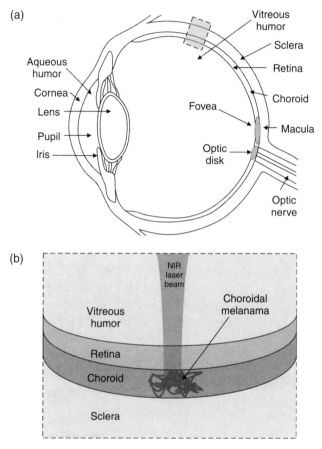

Figure 9.19 Location of choroidal melanoma. (a) Structure and geometry of the eye. (b) Cross-section of ocular wall geometry.

The location of the choroidal tumors with respect to the adjacent structures in the eye is displayed in Figure 9.19a and b. This stratified media poses much the same problem as mentioned in the vascular lesion application, where the overlying tissue must remain unaffected by treatment of the deeper layers. Wavelengths from the NIR band are poorly absorbed by the retina (the human eye cannot detect NIR wavelengths) and suffer reduced scattering from retinal structures as compared to the visible band. An NIR laser beam can be focused at the lesion in the choroid while suffering minimal absorption, and thus heating damage, in the retina.

9.6.4 Vascular Birth Marks (Photocoagulation)

The PWS is a congenital vascular malformation of the dermis. This anomaly is typically visible at birth as a light red or pink discoloration on the skin. These discolorations arise from a dense branching of blood vessels and capillaries in the upper dermis. In infancy, PWSs are typically limited to discoloration; however, if left untreated, they can elicit morphological changes in the surrounding tissue. Starting in adolescence and continuing into older age, the blood vessels grow progressively thicker and change color from a light pink to red to dark purple. These changes often coincide with a thickening of the afflicted tissue and benign masses can form from the underlying dermis.

Approximately 60% of PWSs occur on the face, and untreated malformations can lead to vision impairment depending on the location. Another consequence of the high facial incidence is the prevalence of psychological problems. Personality development in children is adversely affected in nearly all cases, and many patients report on the negative reaction of others to a "marked person" (Fig. 9.20).

The most successful treatment of PWS has been the selective photothermolysis of the abnormal blood vessels with high powered pulsed dye lasers. In this process, laser radiation is focused on the vascular abnormalities, and the pulse width and energy density are adjusted to elicit irreversible heat damage to the blood vessels and to essentially fragment them. The body then clears these fragments away with its normal injury response. The precision in both energy and spatial distribution needed to disrupt just the abnormal vascular area in addition to the need to reduce scaring that makes this the perfect application for laser surgery.

The treatment of PWS with laser irradiation presents a unique challenge to the dermatologic surgeon as the treatment mechanism, heat, can induce intense scaring in the epidermis and dermis. To effectively couple laser energy to the offending blood vessels, the surgeon must overcome absorption and scattering by tissue above the PWS and prevent excessive absorption by tissue below the PWS. This represents a difficult dosimetry problem (Fig. 9.21).

To overcome these problems, PWS is currently irradiated with yellow (575–585 nm) high pulse energy from a dye laser. This wavelength is selected because absorption by hemoglobin is quite high while absorption by other tissue constituents is comparably low. In addition, the microsecond pulse duration from dye lasers is optimized for maximal thermal damage to the abnormal vessels with minimal thermal damage to adjacent tissues and overlying skin. Recent advances in laser treatments have added active spray cooling capability to this procedure (Nelson et al., 2000; Majaron et al., 2001; Ahčan et al., 2004) to reduce skin heating, and thus significantly reduce tissue scarring in the affected area.

9.6.5 Laser Assisted Corneal Refractive Surgery (Ablation)

The most common type of laser surgery by patient volume is LASEK (Taneri et al., 2004a, 2004b; Dastjerdi and Soong, 2002; Tychsen and Hoekel, 2006; Anderson et al., 2002; O'Doherty et al., 2007; Kornilovsky, 2001). In this procedure, a pulsed UV excimer laser is used to precisely ablate extremely small volumes of the corneal tissue (typically the stroma) to change the geometry/shape of the cornea, thus correcting refractive errors. LASEK and its variants are used to treat hypermetropia (farsightedness), myopia (nearsightedness), and astigmatism. The LASEK procedure evolved from two earlier procedures called *photorefractive deratectomy* (PRK)

(a)

(b)

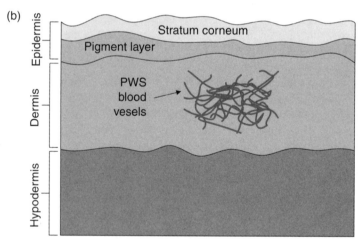

Figure 9.20   (a) Port-wine stain on the head of former Soviet leader Mikhail Gorbachev. Retrieved from http://en.wikipedia.org/wiki/File:Gorbatschow_DR-Forum_129_b2.jpg. (b) Location of PWS blood vessel plexus within the skin layers.

(Munnerlyn et al., 1988; Fantes et al., 1990; Seiler et al., 2000; Seiler et al., 1994; Hersh et al., 1998; Oshika et al., 1999; Pallikaris and Siganos, 1994) and laser-LASIK (Marcos et al., 2001; Moreno-Barriuso et al., 2001; Scerrati, 2001; Guell and Muller, 1996). LASEK surgery has combined the benefits of both PRK and LASIK and is now the procedure of choice for refractive vision correction because of advantages in healing and its superior corrected visual acuity.

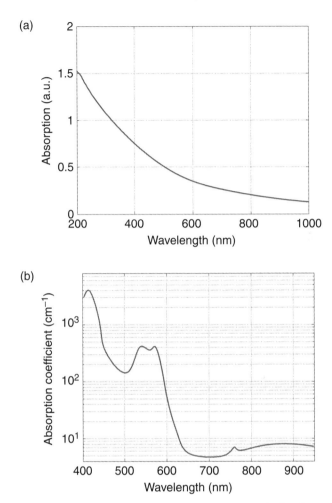

Figure 9.21 Skin constituent absorption parameters. (a) Melanin absorption and (b) hemoglobin absorption. The goal in this surgery is to maximize hemoglobin absorption while minimizing melanin absorption.

Hypermetropia (Fig. 9.22b), or farsightedness, is the increased difficulty focusing on objects near the observer as compared to focusing on distant objects. This is the result of the virtual image created by the eye's optics being focused behind the retina rather than directly on it and may be caused by the eyeball being too short in the axial dimension or insufficient refractive power from the cornea or lens. Conversely, myopia (Fig. 9.22c), or nearsightedness, is the increased difficulty focusing on objects farther away from the observer. In mypoia, the virtual image is formed before the retina, resulting in blurring of the observed scene. This can be caused by excess refractive power from the cornea and/or lens or if the eyeball is too long in the axial dimension. Astigmatism is a condition where the refractive power

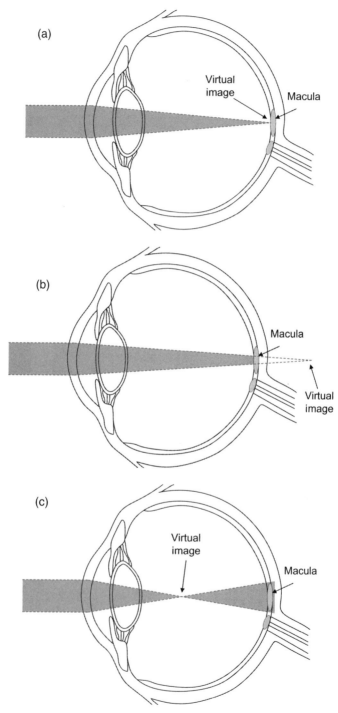

Figure 9.22 Ocular refractive errors. (a) Normal cornea, (b) hypermetropia, and (c) myopia.

of the eye varies in the two corneal axes (meridians). In other words, the focal point in one plane aligned with the eye's optical axis occurs at a different distance from the lens as it does in the orthogonal plane, resulting in differing amounts of blur in the horizontal and vertical axes of the imaging plane. Because of the large index mismatch between air and the tear film, the cornea provides roughly 66% of the total refractive power of the eye, and thus is the tissue of choice for refractive error treatments (Taneri et al., 2004a, 2004b). Although refractive errors can be treated with additive methods (Taneri et al., 2004a, 2004b; Lovisolo and Fleming, 2002), incisional methods (Koch et al., 1989; Waring et al., 1994; Applegate et al., 1998), and thermal methods (McDonald et al., 2002a; McDonald et al., 2002b; McDonald et al., 2004), excimer laser-based ablative techniques are employed in the vast majority of refractive error correction surgeries.

In PRK, the epithelium is removed with mechanical means to expose Bowman's layer, a smooth collagen layer that helps the cornea to maintain its shape (Fig. 9.23). Excimer laser pulses at 193 nm are then applied to the top volume of the stroma,

(a)

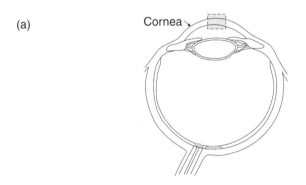

(b)

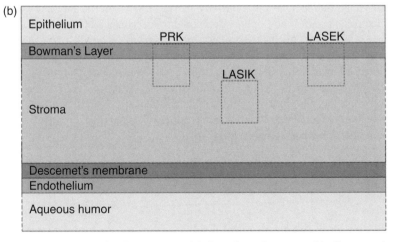

Figure 9.23 Corneal refractive surgery. (a) Location of cornea. (b) Cross-section of corneal layers showing ablation areas for PRK, LASIK, and LASEK.

and tissue is ablated until the geometry corresponding to the desired refractive power is achieved. While this technique is effective for patients with thin corneas, the eye must regrow the epithelium, and the patient can experience pain and dry eyes post surgery. Furthermore, the mechanical removal methods employed in this procedure can damage Bowman's layer, leading to increased discomfort. In LASIK, a femtosecond laser or microkeratome is used to cut a flap off the top surface of the cornea consisting of the epithelium, Bowman's layer, and $\sim$80 μm of stroma. A 193 nm laser is then used to ablate the deeper stroma for the desired refractive properties. The surgery finishes with a returning of the flap to the normal position. LASIK has excellent outcomes and significantly reduced discomfort (as compared with PRK) but is not feasible for patients with thin corneas because of the ablation of deeper stromal layers. Finally, in LASEK, alcohol (typically ethanol) is used to remove the epithelium from the top layer and excimer laser pulses are used to ablate the stroma. This procedure usually results in less discomfort than PRK, because the removal process does not damage Bowman's layer and the epithelium is reattached following stromal ablation.

While each of these techniques is used to treat various corneal refractive imperfections and differ in execution, they all rely on the short pulse widths, large pulse energies, and the short wavelength of UV excimer lasers. Although corneal tissue is removed in discrete steps, the high volumetric resolution achievable with UV excimer lasers and the appropriate optics result in continuous reshaping that significantly improves the vision of the patient.

9.7 CONCLUSIONS

Although the current use of lasers in medicine is broad and varied, the technology is still in the beginning stages of its development as a diagnostic instrument, treatment modality, and therapeutic tool. New advances in laser materials, optical packaging, optoelectronic control, and optical component performance are enabling scientists, engineers, and clinicians to take research systems from the bench to the bedside with ever decreasing development times. In addition, advances in optical modeling and the increased availability of low cost, high performance computational capabilities are allowing researchers to integrate sophisticated postprocessing into instruments, thus opening the possibility of real-time laser fluence calculations of a number of treatment systems and real-time image reconstruction and display in OCT and other tomographic optical imaging systems.

PART IV

IMPLANTABLE DEVICES AND SYSTEMS

CHAPTER 10

Vascular and Cardiovascular Devices

DAN LEVI

Department of Pediatric Cardiology, University of California, Los Angeles, CA

ALLAN TULLOCH

Center for Advanced Surgical and Interventional Technology (CASIT), University of California, Los Angeles, CA

JOHN HO

Department of Pediatric Cardiology, University of California, Los Angeles, CA

COLIN KEALEY

Center for Advanced Surgical and Interventional Technology (CASIT), University of California, Los Angeles, CA

DAVID RIGBERG

Department of Surgery-Vascular, University of California, Los Angeles, CA

10.1 INTRODUCTION

A wide range of devices are used to repair the heart and vascular system. The mainstay of vascular repair is the stent—a metal tube used to support the patency of vascular structures. Vascular stents can be self-expandable or balloon-expandable. They can also be covered or coated with biocompatible materials and drugs. These stents can be placed in the coronary arteries, carotid arteries, peripheral arteries, the aorta, and even venous structures. While entire books have been written on stents, this chapter attempts to describe the full range of devices used for both structural heart disease and vascular abnormalities. This also includes devices designed to close holes in the heart or other vascular structures and transcatheter heart valves. Because all intravascular devices must interact with the blood and vascular system,

Medical Devices: Surgical and Image-Guided Technologies, First Edition.
Edited by Martin Culjat, Rahul Singh, and Hua Lee.
© 2013 John Wiley & Sons, Inc. Published 2013 by John Wiley & Sons, Inc.

we also consider the types of materials used in these devices and the requirements for biocompatibility and thrombus resistance.

The first cardiovascular intervention was a pulmonary valvuloplasty performed in 1956. This procedure involved the use of catheters to intentionally damage blocked tricuspid valves. The balloon atrial septostomy described by Rashkind and Miller was performed in 1966 for children with transposition of the great vessels. This procedure was designed to make a hole in the heart of children in order to allow the flow of deoxygenated blood to the lungs and oxygenated blood to the body. Portsmann and Wanke performed transcatheter closure of a patent ductus arteriosus (PDA) with a device designed to plug this structure when it persists beyond the newborn period. The first published transcatheter closure of a hole in the heart was performed in 1976 by King and Mills [this defect was an ASD (atrial septal defect)]. Although shape memory alloys such as nickel-titanium (Ni-Ti) were discovered in the 1960s, their use in medical devices was delayed until the 1990s. In the 1990s, Dr. Kurt Amplatz exploited the unique shape memory and pseudoelastic properties of nitinol to design ASD and PDA closure devices. The success of these devices ignited an explosion of transcatheter interventions in congenital heart disease.

In general, this chapter discusses several types of devices: (i) devices designed to open blocked vessels, including stents and angioplasty balloons; (ii) devices deigned to block or occlude unwanted vascular structures, such as a hole in the heart; (iii) devices designed as internal band aids to tamponade a bleeding vessel or to wall off an aneurysm, the best example of which is the covered stent; and (iv) the emerging field of catheter-based valve replacement technologies, as well as endografts and vascular filters.

10.2 BIOCOMPATIBILITY CONSIDERATIONS

Biocompatibility is a term used to describe the interaction of a foreign material with host tissue. Broadly speaking, there are six components of biocompatibility: (i) hemocompatibility, (ii) direct cytotoxicity, (iii) systemic toxicity, (iv) carcinogenicity, (v) immunologic sensitization, and (vi) material fatigue. Placement of a permanent, indwelling medical device requires an understanding of the interactions of the biomaterial with the biological properties of the host. For the sake of this discussion, we focus on the biocompatibility issues as they relate to implantable cardiovascular devices.

For an endovascular medical device, hemocompatibility, which is the interaction between blood and a biomaterial, is the most important parameter. Indeed, platelet-mediated thrombosis, that is, an acute blood clot forming on the device, is the most common source of early graft failure (Kannan et al., 2005). To minimize this complication, many researchers have sought to identify ways to modify biomaterial surfaces to prevent platelet adhesion. One strategy is to mimic the microenvironment of vascular endothelial cells that form the inner lining of blood vessels. As early as the 1950s, it was shown that native blood vessels carry a net negative charge. When vessel wall injury occurs, the native blood vessel charge at the area of injury turns positive, preferentially attracting negatively charged platelets to the site of injury (Sawyer et al., 1953). Later work demonstrated the importance of a material's surface energy on platelet adhesion (Van Kampen and Gibbons, 1979). Surface energy is used to quantify how hydrophilic or hydrophobic a material is. In essence, the surface energy refers to the amount of energy needed to expel a water molecule from the surface (i.e., more hydrophilic materials demonstrate higher surface energy). Platelets are naturally attracted to hydrophobic surfaces. This point cannot be overemphasized as studies have found a correlation between the hydrophilicity of a material and its thrombogenicity (Xu and Siedlecki, 2007; Michiardi et al., 2007).

This interplay between electrostatic charge and surface energy represents an important factor in determining the thrombogenic profile of indwelling medical devices. For example, expanded polytetrafluorethylene (ePTFE) is commonly used to cover endovascular stents and for artificial surgical bypass grafts. The surface of ePTFE is electronegative but hydrophobic, which has been hypothesized to cause thrombosis in low flow states. To decrease the rate of acute thrombosis associated with ePTFE, researchers have bonded heparin, a commonly used anticoagulant, to this material's surface. The Viabahn Endoprosthesis (Gore Medical, Flagstaff, AZ) made use of this strategy and has demonstrated improved patency in clinical use. This effect is at least partly because of heparin's hydrophilicity (Walluscheck et al., 2005; Laredo et al., 2004). Other experimental surface treatments for vascular grafts to improve hydrophilicity, such as polyethylene glycol and polyethylene oxide, have also been shown to prevent platelet adhesion and increase these materials' hemocompatibility (Karrer et al., 2005). Unfortunately, these polymers bond poorly to grafts and have so far been relegated to laboratory science.

The degree of electronegativity and hydrophilicity is not the only determinant of surface thrombogenicity. Surface roughness and the binding of other thrombogenic blood products such as fibrin and leukocytes must also be taken into account (Kallmes et al., 1997; Van Kampen and Gibbons, 1979). For metal alloys such as nitinol, the quality of the surface oxide layer is also important since it acts as a semiconductor preventing the release of electrons at the surface that has been shown to decrease platelet adhesion *in vitro* (Huang et al., 2003). Mechanical factors related to the deployment of the stent and design elements, such as the shape and size of the struts themselves, are also critical in preventing thrombosis. It is well established that turbulent blood flow and shear stress promote platelet activation and graft thrombosis (Karino and Goldsmith, 1984). In order to minimize flow disturbances, proper sizing of the stent as well as the anatomic location must be taken into account. Stents that cross joint spaces undergo a higher degree of mechanical stress, which can also contribute to thrombotic complications.

The other components of biocompatibility (i.e., direct cytotoxicity, systemic toxicity, carcinogenicity, immunologic sensitization, and material fatigue) are largely due to the physical and mechanical properties of the materials used to construct the device. In the case of implantable cardiovascular devices, these tend to be metal alloys. Currently, the most common alloys used to construct these devices are stainless steel 316 (SS316L), chromium–cobalt (Cr–Co), and nickel–titanium (nitinol) (Mani et al., 2007). Corrosion is defined as the degradation of a metal over time because of oxidation or chemical reaction. When these metals come into contact with blood and other tissues, the potential for corrosion exists. As the device corrodes, undesirable by-products such as metal ions may be released into the blood and surrounding tissues, possibly leading to local and systemic toxicities. Corrosion will also reduce the life of the implant and is of particular concern in areas where implants undergo a high degree of mechanical stress, such as a device placed across a joint (Bates et al., 2008).Thus a significant amount of research has focused on defining these implants' corrosion profiles over time.

Longer term biocompatibility issues with cardiovascular implants revolve around the process of neointimal hyperplasia (NIH). This issue is most relevant to stents but applies to any foreign object placed intravascularly. The intima is the innermost lining of the vessel wall. Placement of an intravascular device inevitably damages this intimal lining and a new or "neo" intima must regrow to cover the device and restore vascular homeostasis. While the process of neointimal regrowth is necessary, it can also be maladaptive. When the neointima grows too much, it can occlude the lumen of the vessel, a process termed *neointimal hyperplasia*. A large number of strategies have been developed to stop or reduce the process of NIH. Among these are the use of drug-eluting stents (DESs), covered stent grafts, surface modification of the stent to increase biocompatibility, and seeding of stents with endothelial cells before placement (Dichek et al., 1989; Shirota et al., 2003; Palmaz et al., 2002; Clowes et al., 1987; Golden et al., 1990). Of all the strategies developed to date, DESs are far and away the most successful.

A DES is a typical stent that has been coated with a drug-containing polymer. The most commonly used drugs are immune modulators such as sirolimus or antiproliferative drugs such as paclitaxel. These stents have been extremely successful in the prevention of in-stent restenosis of the coronary vessels, which has led to their off-label use in the peripheral vasculature. However, current DES are not specific for smooth muscle cells that are the primary cell type that causes NIH. As a result, these stents also inhibit growth of the endothelial cells that form the vessels' inner lining. Inhibition of these cells may put patients at risk of late thrombotic complications. Therefore, there is considerable interest in new strategies to reduce the rate of neointimal and thrombotic complications following placement of endovascular medical devices.

10.3 MATERIALS

As mentioned above, the materials used to make endovascular devices play an important role in both the mechanical and the biocompatibility characteristics of the device. While a detailed discussion of all the materials used to construct intravascular medical devices is beyond the scope of this chapter, the following section provides an overview of the three most commonly used materials for these devices.

10.3.1   316L Stainless Steel

316L stainless steel is the most commonly used material for the manufacture of stents (Mani et al., 2007). 316LSS is well suited for stents because of its high tensile strength and excellent corrosion resistance. Disadvantages of stainless steel include its ferromagnetic nature and low density; thus, these devices are not MRI compatible and are poorly visible under fluoroscopic guidance. Despite these limitations, stainless steel is both strong and biocompatible and is therefore used for a large number of endovascular applications. It should be noted that all stents constructed from stainless steel are balloon expandable. This means that they are packaged over a balloon that is then advanced to the desired location and expanded to appose the stent against the arterial wall. While this approach gives clinicians excellent control over stent placement and final diameter, it means that these devices are relatively inflexible once placed. This makes them susceptible to crushing or permanent deformation owing to an outside force. Therefore, the use of these devices is generally limited to internal vascular beds such as the coronary arteries where they cannot be easily crushed by external forces. Another concern with 316LSS devices is the leaching of other metal ions. In the case of SS316L coronary stents, Cr, Ni, Mn, and Mo ions have been demonstrated to be released when the metal is placed in the coronary vessels (Gutensohn et al., 2000).While this is not thought to be clinically significant at this time, manufacturers must strive to achieve the highest possible alloy purities.

10.3.2   Nitinol

Nitinol is an acronym for nickel-titanium, developed by the Naval Ordinance Laboratory in the early 1960s. Initially, nitinol's pseudoelastic and shape memory properties were used solely for industrial applications; however, it was later found to be biologically inert. This later discovery, coupled with its pseudoelastic response, has led to its use in a variety of medical applications. Specifically, both the elastic and shape memory properties of nitinol make it ideal for transcatheter devices where the device is compressed into a small tube (i.e., the catheter) and then expanded on delivery. In direct contrast to stainless steel, nitinol-based devices will undergo this expansion without the application of an outside force. This obviates the need for balloon-based delivery systems and makes these devices better suited for certain applications such as the peripheral and carotid arteries where their innate recoil provides protection against device crushing.

Another advantage of nitinol is its fairly unique thermal/mechanical-induced phase transformation. This response gives rise to nitinol's shape memory and pseudoelastic capabilities. In the low temperature martensite phase (monoclinic state), nitinol is very malleable because of reversible twin boundary motion (unlike dislocation motion in most metals) and can be compressed into catheters. On heating (in many cases simply to body temperature), nitinol transforms to the austenite phase, which has a cubic crystal structure. The practical result of this behavior is that nitinol can be compressed and deformed in the martensite phase and then recovers its original shape during its austenitic transformation. This original shape

Table 10.1 Mechanical Properties of the Metals Commonly Used to Make Intravascular Medical Devices

Metal	Elastic Modulus (GPa)	Yield Strength (MPa)	Tensile Strength (MPa)	Density (g/cm^3)
316L stainless steel	190	331	586	7.9
Nitinol	83 (austenite phase)	195–690 (austenite phase)	895	6.7
Cobalt-chromium	210	448–648	951–1220	9.2
Mg alloy	44	162	250	1.84
Pure iron	211.4	120–150	180–210	7.87

can be set into the material by simply heat treating it in a constrained shape near its crystallization temperature (500 $^\circ$C) for a few minutes. Therefore, nitinol can be "taught" to recover virtually any shape as long as the deformations are below 10%. For comparisons, other competitive elastic materials such as stainless steel and Ti are limited to 1% elastic deformations (Henderson et al., 2011).

10.3.3 Cobalt–Chromium Alloys

Cobalt–chromium (Co–Cr) alloys have been used in dental and orthopedic applications for decades. More recently, these alloys have been used to make stents. One of the biggest advantages of Co–Cr is its excellent radial strength. As seen in Table 10.1, the radial strength of Co–CR is approximately double that of stainless steel, making it the strongest of the commonly used stent materials. This is advantageous, as one of the primary goals in endovascular device design is to reduce the device's profile as much as possible. Co–Cr is favorable in this regard, as it allows manufacturers to significantly reduce the size of the device while still maintaining an adequate radial force. This allows the devices to be packaged in smaller delivery systems, allowing physicians to access more distal vascular beds. In addition, Co–Cr-based devices are innately radiopaque and are thus readily visible under standard fluoroscopic guidance. Because Co–Cr-based stents lack the pseudoelastic response of nitinol-based devices, these stents require a balloon delivery system like stainless steel devices (Mani et al., 2007).

10.4 STENTS

Stents are small metal tubes that are designed to prop-open or "stent" blood vessels. Stents are most commonly used for treatment of atherosclerotic arterial disease. This occurs when arteries become blocked by accumulation of fatty plaques in the vessel wall. This limits blood flow to tissue served by the artery beyond the level of the blockage and can have severe clinical consequences such as a myocardial infarction, known more commonly as a *heart attack*. Stents were first introduced into clinical practice in the 1980s with the introduction of the Palmaz–Schatz stent and have since dramatically changed the practice of medicine throughout the world. Before the introduction of stents, the only treatment for severe atherosclerotic disease was an open surgical bypass operation. With the introduction of stents, a minimally invasive alternative is now available, and the number of open surgical procedures being performed has declined dramatically. While there is still considerable debate over when stents should and should not be used, they undoubtedly represent one of the most important medical advancements of the past 20 years (Palmaz, 2007). In this section, we explore the basics of stent design and how they are used throughout the body.

As discussed above, there are two basic types of stents: the balloon-expandable and the self-expanding stents. The former has traditionally been made of 316LSS (e.g., the original Palmaz stent) and uses an angioplasty balloon to expand and place the stent within the arterial segment. By contrast, self-expanding stents utilize the unique shape memory properties of nitinol or the weave of the stent to assume a preconfigured shape within the vessel lumen. Because of the radial force required to hold diseased arteries open, balloon-expandable stents have been the device of choice for atherosclerotic arterial disease. In contrast, self-expanding stents are used in situations where less radial force is required or where crushing is a concern. This includes neurological stents designed to be placed over the necks of intracranial aneurysms as well as stents placed over joints and close to the surface of the skin.

One of the more important advances in stent technology has been the introduction of DESs. As mentioned above, the primary limitation of stent technology is the tendency of the vessel to grow over the stent and leading to restenosis over time. This process is known as *neointimal hyperplasia* (NIH). NIH is currently an area of intense research but is generally regarded as an inflammatory process whereby the smooth muscle cells in the medial layer of the vascular wall undergo excessive proliferation, causing them to overgrow the stent and occlude the vascular lumen. DESs were a response to this observation and involve coating the stent with a drug that inhibits smooth muscle cell proliferation. DESs are most commonly used to treat coronary artery disease and are coated with immune-modulating drugs such as sirolimus or a cell-cycle inhibitor such as paclitaxel. Use of DES has dramatically decreased the rates of restenosis and in the coronary arteries, now represent the majority of devices placed (Kabir et al., 2011).

Just as in the coronary arteries, recommendations on the treatment of peripheral arterial disease (PAD) include an expanding role for endovascular procedures instead of open bypass surgery to revascularize ischemic limbs (Norgren et al., 2007). However, it has proven more difficult to demonstrate the benefits of endovascular technology in the peripheral system. For instance, several randomized control trials comparing self-expanding nitinol stents to more traditional percutaneous transluminal angioplasty (PTA) for the treatment of femoropopliteal disease showed no significant differences in clinical outcomes between the two groups. In fact, the data suggested that patients receiving balloon-expandable stents in the lower extremity fared worse because of excessive NIH (Schillinger et al., 2002; Becquemin et al., 2003). In addition, the success at preventing restenosis with DES achieved in the coronary circulation has been difficult to repeat in the periphery as well (Oliva and Soulez, 2005). Therefore, use of these devices for PAD is less-established than in the coronary circulation and is currently an area of ongoing investigation.

One vascular region outside the coronary arteries where stent technology has dramatically changed clinical practice is in the treatment of abdominal aortic aneurysms (AAAs). An AAA is an outpouching of the aorta as it traverses the abdominal cavity. This outpouching is an area of weakness and can burst, leading to catastrophic complications and oftentimes death. In the past, the only treatment option was an invasive open surgery where the affected segment of the aorta was removed and an artificial conduit was sewn in as a replacement. While this procedure is both effective and durable, it is a large operation and involves significant risk to the patient. This state of affairs changed dramatically in 1990 when Parodi et al. performed the first endovascular AAA repair (Veith et al., 2005). In this procedure, a self-expanding nitinol stent backbone is covered circumferentially with a biocompatible polymer, most commonly expandable polytetrafluoroethylene (ePTFE), to create a covered stent graft. The graft is then placed over the neck of the aneurysm to redirect blood flow away from the aneurysm sac. This isolates the aneurysm from the normal circulation and causes the stagnant blood inside to clot. These stent grafts are now used routinely for the treatment of both AAAs and thoracic aneurysms and have dramatically changed clinical treatment patterns worldwide (Parodi et al., 1991; Makaroun et al., 2008).

Endografts have now supplanted open surgical repair as the first-line treatment for thoracic and abdominal aneurysms. Because of the high flow and large diameter of the aorta, thrombosis and NIH are less of a concern than in other vascular territories. However, because these endografts are large in diameter, they require large delivery catheters anywhere from 18 to 21 Fr (3 Fr is 1 mm in diameter) for AAA repair and up to 26 Fr for thoracic aneurysms. This large delivery catheter size can limit its utilization and can make arterial access through tortuous and calcified vessels extremely difficult. The stent covering makes up approximately 50% of the predeployment profile. Decreasing the size of these grafts and their delivery systems is an area of active investigation. Other unique design elements that are already being seen in clinical trials include both fenestrated and branched endografts to treat aneurysms involving the renal and visceral vessels.

10.5 CLOSURE DEVICES

Although the first open heart surgeries were performed to close holes in the heart, modern devices (made primarily with a nitinol weave) have largely replaced open heart surgery as the treatment of choice for most ASDs (holes between the upper two chambers of the hearts) and even for many ventricular septal defects (VSDs—holes between the bottom two cardiac chambers). These transcatheter devices have disks or umbrellas that sit on either side of the hole to allow for stable closure of the hole even in a beating heart. Pediatric devices are also available to close the PDA, a vessel communication between the pulmonary artery and the aorta. As with stents, shape memory alloys are ideally suited for use in this sort of device because they can be "trained" to form certain shapes at body temperature, while also capable of being cooled to the martensite shape and compressed into very small catheters. These devices have allowed tens of thousands of children and adults with congenital heart disease to avoid open chest and heart surgical interventions.

The ASD, PFO (patent foramen ovale), VSD, and PDA nitinol closure devices manufactured by AGA Medical (Golden Valley, MN) were originally designed by a radiologist, Kurt Amplatz, MD (Fig. 10.1). These devices have significantly improved the treatment of these basic congenital heart defects. All of the devices are built from mesh tubes of woven nitinol wires as small as 0.003 in in diameter and are shape set to form whatever general shapes are needed to close a given congenital heart defect. For example, the Amplatzer Septal Occluder for ASD closures forms two disks with a waist in between and can be easily delivered so that it is stable even in a beating heart. The ASO devices can be delivered in sheaths as small as 6 Fr and up to 12 Fr with sizes ranging up to devices appropriate for closure of 35–40 mm ASDs (Ryhanen et al., 1997; Cheatham, 2001) (Fig. 10.1). Most of the Amplatzer devices are designed such that the heart can adapt to the shape of the device.

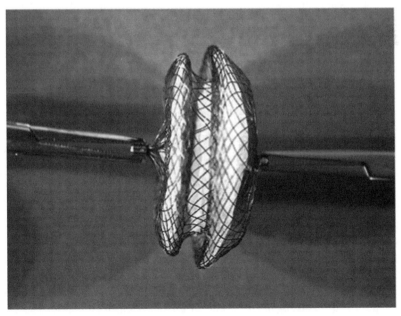

Figure 10.1 AMAPLATZER™ septal occluder. St. Jude Medical is a trademark of St. Jude Medical, Inc. (*Source*: Reprinted with permission of St. Jude Medical™, © 2012 All rights reserved).

ASD, VSD, PFO, and PDA closure devices have allowed over 50,000 patients with congenital heart disease to undergo nonsurgical closure of these significant congenital heart defects. In addition to AGA Medical, many other companies have now employed a range of different designs for ASD, PFO, and PDA closure devices. Nearly all of these devices utilize nitinol's unique shape memory properties (Thanopoulos et al., 2007; Pass et al., 2004; Du et al., 2002; Chessa et al., 2002; Hong et al., 2003; Chan et al., 1999). A nitinol coil manufactured by *pfm* Medical (Cologne, Germany) is now being used for closure of both PDAs and smaller VSDs (Lorber et al., 2003).

10.6   TRANSCATHETER HEART VALVES

An emerging field in interventional cardiology is the development of transcatheter heart valve replacement. There is a wide variety of prosthetic valves designed to replace failing heart valves. These valves typically involve prosthetic animal leaflets combined with a collapsible stent. The device is delivered through a catheter just like a conventional stent and is placed over the defective valve, pushing it out of the way, and offering a replacement without a major surgery. Thus far, this approach has been particularly useful in those patients older than 75 years and suffering from severe aortic stenosis who are at high risk for open surgery (Nkomo et al., 2006). In the near future, however, this technology is likely to be extended to children, adolescents, and adults who need pulmonary valve replacements. Although the concept for a transcatheter heart valve has been around for decades, this field was made a reality in 2000 by Dr. Philip Bonhoeffer with studies performed with pulmonary valve replacement in lambs (Lurz et al., 2008; Bonhoeffer et al., 2000).

The Melody® valve (Fig. 10.2) is a prosthetic pulmonary valve developed by Medtronic and based off of the work done by Dr. Bonhoeffer. The valve is an 18-mm modified bovine jugular vein valve mounted in a platinum-iridium stent. The stent is 28 mm in length and is crimped down to 6 mm. This balloon-expandable stent can also be reexpanded to 18 or 22 mm (Brinkman and Mack, 2009). The Melody heart valve was recently approved by the FDA for use in the United States, making it the first transcatheter heart valve approved for use. Early outcomes with the Melody valve in patients with right ventricle to pulmonary artery conduits with insufficiency have been encouraging (Zahn et al., 2009). It is ideal to prevent the need for high risk surgery and reduce the total number of open heart surgeries a person may need in their lifetime (Lurz et al., 2008; Khambadkone and Bonhoeffer, 2004).

The SAPIEN transcatheter heart valve (Fig. 10.3) is an aortic valve replacement designed by Edwards, Inc. The valve is made from bovine pericardium mounted in a stainless steel or cobalt-chromium stent. Since the SAPIEN valve is engineered for aortic valve replacement, it is able to tolerate greater pressures. The current generation of Edwards SAPIEN valve consists of bovine pericardial leaflets, a cobalt-chromium frame, and a sealing cuff on the inflow aspect of the stent to prevent leakage around the valves (Brinkman and Mack, 2009). The SAPIEN valves

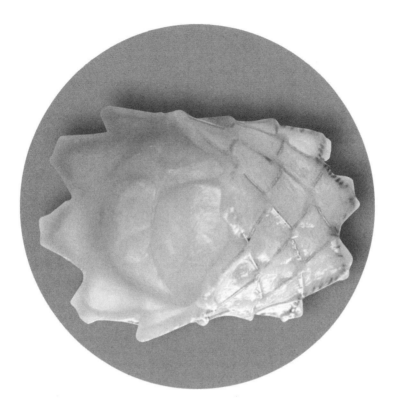

Figure 10.2 Melody® transcathether pulmonary valve. Melody® TPV is approved in the United States as a humanitarian device. It is authorized by Federal law (USA) for use in pediatric and adult patients with a regurgitant or stenotic right ventricular outflow tract (RVOT) conduit. The effectiveness of this device for this use has not been demonstrated (*Source*: Courtesy of Medtronic, Inc.).

are pretreated with an anticalcification technique known as *Thermafix*. This process involves the removal of glutaraldehyde and phospholipids from the device (Schoen and Levy, 2005). Studies by Edwards have shown a reduction in calcification for valves treated with this proprietary method. This valve has undergone a few structural and material changes to allow for tighter crimping of the valve without loss of stability or radial force. The SAPIEN XT aortic valve will be available in 20, 23, 26, and 29 mm external diameters (Brinkman and Mack, 2009). The SAPIEN valve is somewhat stiffer than the Melody valve and requires a larger delivery sheath. This device is currently under investigation in the PARTNER (placement of aortic transcatheter valve) Trial for aortic valve replacements in high risk adult patients. A trial for pulmonary valve replacements has also begun. This trial focuses on patients with congenital heart disease with failed right ventricular to pulmonary artery conduits.

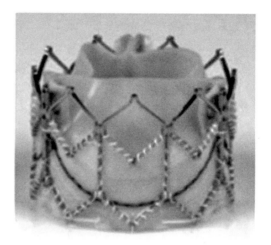

Figure 10.3 Edwards SAPIEN XT transcatheter heart valve (*Source*: Courtesy of Edwards Lifesciences LLC, Irvine, CA).

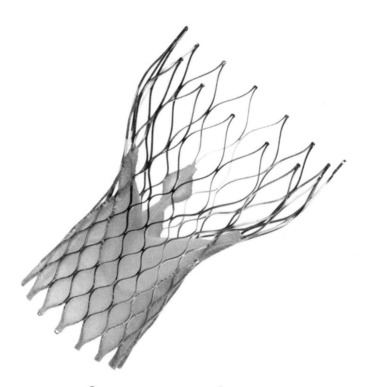

Figure 10.4 CoreValve® System. The CoreValve® System is an investigational device and is limited by US law to investigational use. CoreValve is a registered trademark of Medtronic CV Luxembourg S.a.r.l. (*Source*: Courtesy of Medtronic, Inc.).

The Core Valve® (Fig. 10.4) is another device designed for aortic valve replacement. The major advantage of the Core Valve® is that its self-expanding nitinol frame does not require ballooning (both the Melody TPV® and SAPIEN® valves are balloon inflatable). Within its frame are three porcine pericardial leaflets. The Core Valve® length is 55 mm for the 26-mm valve size, 53 mm for the 29-mm valve size, and 52 mm for the 31-mm valve size. The device is designed with a waist in the middle and the frame experiences variable radial forces. The inflow portion of the frame has a high radial force. The mid portion, which contains the valve, remains fixed to allow blood to fill the coronary arteries. The upper portion of the stent with modest radial forces then enables proper orientation (Brinkman and Mack, 2009; Nietlispach et al., 1998). Early data from implantation in Europe has been encouraging in a large number of patients.

The MitraClip device (Abbot Vascular) valve (Fig. 10.5) is a device designed to reduce mitral valve insufficiency by providing a nonsurgical alternative to the Alfieri stitch, a surgical technique used to sew the mitral valve leaflets together in select cases of mitral valve insufficiency. The device is covered with polyester and has two arms that are designed to capture the two leaflets of the mitral valve. The delivery of this device is complex and requires traveling across the atrial septum through a naturally occurring or catheter-created hole. It is important to deploy this clip in the correct alignment with the center of the mitral valve opening in order to maximize its effectiveness (Herrmann and Feldman, 2006; Gillinov and Liddicoat, 2006). Once deployed, the device effectively creates a double orifice to reduce valve leakage. The EVEREST II (Endovascular Valve Edge-to-Edge Repair Study) randomized clinical trial is ongoing to assess the safety and effectiveness of this device.

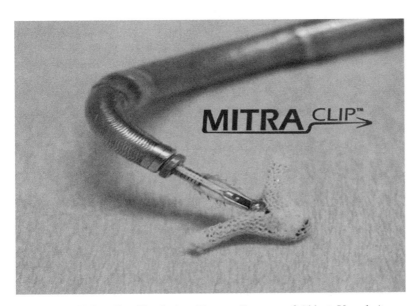

Figure 10.5   MitraClip device (*Source*: Courtesy of Abbott Vascular).

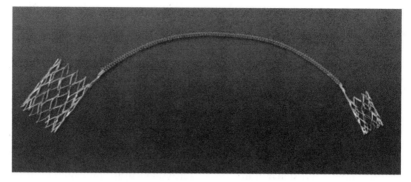

Figure 10.6 Edwards MONARC Mitravl Valve system (*Source*: Courtesy of Edwards Lifesciences LLC, Irvine, CA).

An alternative method for managing mitral valve insufficiency is through coronary sinus annuloplasty. This technique is based on the anatomic relationship of the mitral valve and the coronary sinus, a vein that circles the mitral valve. By placing a device in the coronary sinus and constricting the orifice via a device that sits in the coronary sinus, mitral valve leakage should improve. One such device is the MONARC Edwards system (Fig. 10.6). First, the coronary sinus is accessed through the right internal jugular vein. Two stents joined by a connecting bridge are then deployed into the coronary sinus. This connecting bridge is a spring, which is held open by bioabsorbable material. The spring tightens as its bioabsorbable suture coat is degraded. As a result, two stents are pulled closer together and the mitral annulus tightens. This process occurs over a 3- to 6-week period (Brinkman and Mack, 2009; Gillinov and Liddicoat, 2006; Piazza et al., 2009). The MONARC system is currently in phase II clinical trials in the United States. Although efficacy data are encouraging, coronary compression and anchor separation are important safety concerns (Piazza et al., 2009). A nonrandomized, multicenter, prospective safety and efficacy study (EVOLUTION phase II) is planned for the near future (Piazza et al., 2009). This device has been placed in more than 80 patients in studies outside the United States (Brinkman and Mack, 2009).

10.7 INFERIOR VENA CAVA FILTERS

Inferior vena cava (IVC) filters are a minimally invasive approach to the problem of pulmonary embolism. Certain health conditions predispose patients to forming blood clots in the veins of the lower extremities. When these clots dislodge, they can travel to the lungs, via the IVC, and cause severe consequences such as pulmonary hypertension and death. For at-risk patients, a filter has been designed to catch these clots. These devices are placed in the IVC and act as a filter to prevent the clot from traveling any further. The Greenfield filter is the prototypical example of such a device (Fig. 10.7), although many different devices and design innovations are available. These devices can be permanent or retrievable and have been

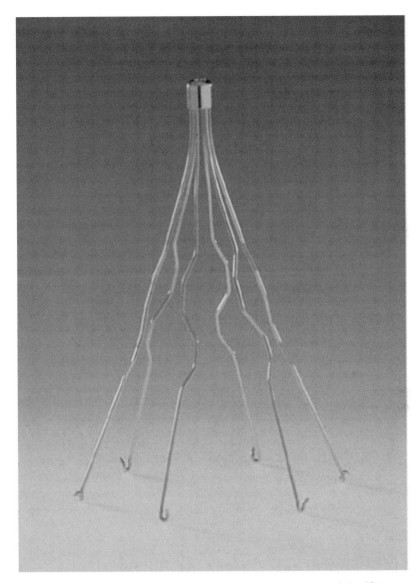

Figure 10.7 Greenfield filter (*Source*: Courtesy of Boston Scientific).

constructed using SS316 (e.g., the original Greenfield filter), nitinol (e.g., Simon Nitinol, Bard Inc., TempeAZ), and titanium (e.g., new generation Greenfield filter). Most retrievable filters are fitted with a device that allows them to be pulled back into the catheter and retrieved usually from the jugular vein. For example, the Gunther Tulip Filter (Cook Medical, Indianapolis, IN) has a hook eye at the cephalad aspect that permits repositioning or removal by the transjugular route (Fig. 10.8). Removal of filters is important in cases of filter malposition or clotting. Many

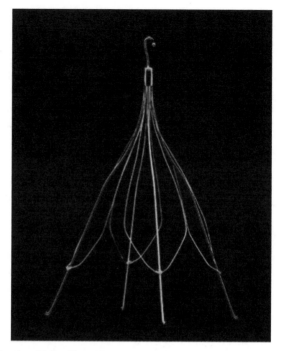

Figure 10.8 Gunther Tulip filter (*Source*: Permission for use granted by Cook Medical Incorporated, Bloomington, Indiana).

patients with trauma are young and despite the long term patency of IVC filters (~98%), up to 44% develop long term sequelae of venous stasis (Greenfield and Michna, 1988). Explanation of the filters can be difficult because of endothelialization of the filter, which can occur as early as 12 days after placement (Burbridge et al., 1996). Thus, IVC filters are commonly used endovascular devices that have protected innumerable patients from potentially devastating pulmonary emboli.

10.8 FUTURE DIRECTIONS — THIN FILM NITINOL

Currently, the most widely used application of nitinol has been in the design of self-expanding stents for stenotic or blocked blood vessels. These stents are constructed from bulk nitinol and are generally cut from nitinol hypotubes using advanced laser cutting technology. In its smallest form, bulk nitinol is approximately 50 μm in thickness. Recently, there has been increasing interest in thin film nitinol (TFN)-based endovascular devices. Thin films are defined as being about 1–15 μm thick—thinner than a human hair. During the past two decades, researchers have attempted to fabricate TFN with marginal success. The manufacturing of TFN has been attempted by vacuum evaporation, flash evaporation, ion beam sputtering, and

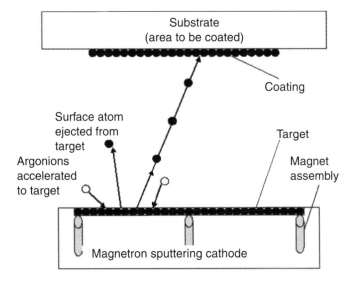

Figure 10.9   Vacuum sputter deposition.

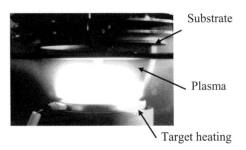

Figure 10.10   Vacuum sputter deposition with target heating.

laser ablation (Busch et al., 1990; Sekiguchi et al., 1983; Makino et al., 1998; Lehnert et al., 1999; Ikuta et al., 1994; Wu, 2001). Most of these fabrication approaches have been unsuccessful at producing medical grade thin film. Vacuum sputter deposition, which involves knocking atoms from a target material and directing them to form a thin film on a substrate in an extremely low pressure vacuum, has been used in attempts to produce high quality TFN (Figs 10.9 and 10.10). This approach is currently the preferred method for thin film fabrication because of process controllability and consistency. Because sputter deposition "nanosynthesis" requires an ultrapure vacuum chamber (less than 10^{-12} Torr), there are very few contaminants introduced during the fabrication process.

Although the general process of sputter deposition of various materials has been successfully used by many institutions, only a handful of teams have been able to produce high quality TFN using a sputter deposition approach. Recently, it was discovered that target heating during the sputtering process creates uniform films not present in conventional sputtering processes (Ho and Carman, 2000). Expertise with this process produces precise control of the film composition and uniformity. The target composition and annealing temperatures required to produce nitinol thin films with transition temperatures in the desired temperature range for medical applications has also been determined (Stepan et al., 2005). These advancements should facilitate the introduction and subsequent use of TFN in biomedical devices.

Although traditional stents are nonocclusive metal scaffoldings, the treatment of many disease processes relies on the ability to use a "covered stent," able to both open vessels and provide a circumferentially occlusive boundary between the stent and the vessel. The prototypical example is for the treatment of AAA discussed above. The next generation of covered stents could be significantly improved by utilizing TFN technology. This is because TFN could allow for production of very low profile covered stents because of its extreme thinness. Current materials used to create covered stent grafts such as ePTFE are approximately 150 μm thick and add substantial bulk to the catheter delivery system. Use of TFN coverings could dramatically decrease device bulk and allow for treatment of not only AAA but also intracranial aneurysms where current covered stents are far too bulky to access. Although thin film stent applications will certainly be seen first in adult patients, many niche applications will become available in pediatric patients. These applications include stenting of the PDA in symptomatic newborns with congenital heart disease (Gewillig et al., 2004), treatment of arteriovenous malformations, and pulmonary vein stenting. These applications are more suited to TFN because of the fact that thin film can be collapsed into much smaller diameter catheters than "bulk" materials, and smaller catheters are a requirement in the pediatric community.

10.9 CONCLUSION

This chapter focused on basic implantable devices. There are many other devices including angioplasty balloons, steerable catheters, atherectomy catheters, radiofrequency ablation catheters, perforation catheters, steerable wires, and even magnetic wires able to be controlled with magnets external to the patient. The range of therapeutic transcatheter devices is large and constantly expanding. Because of the rapid growth of this field, cardiologists, radiologists, and vascular surgeons are able to accomplish more without resorting to risky open surgical procedures.

As many surgical procedures have been eliminated, these devices have been a huge benefit to patients and to the health-care systems. Especially, in the case of stents and closure devices, hundreds of thousands of patients have been able to avoid surgery and its associated medical and cosmetic morbidities. In most or all cases, the results with catheter-based interventions have been demonstrated to be at least as good if not better than surgical interventions. As evidenced by the development of percutaneous heart valves, this trend is likely to continue and even to accelerate. Many investigators are already working on biocompatible polymers to improve drug delivery, special surface coatings, nanotubes, and other advanced materials such as TFN that will comprise the next generation of transcatheter devices, stents, filters, and valves.

It is important to note that as the complexity of these devices increases, there will be a need for increased collaboration between engineers, biological scientists, and clinicians. Current devices have an origin that is largely mechanical in nature. Future devices are likely to be hybrid in design and combine advances in materials science with those in molecular biology to rationally interact with the vessel wall in a predetermined manner. DESs are the first widely used example of such a device. This trend is likely to continue and calls for the training of scientists well versed in these disparate fields.

■ CHAPTER 11

Mechanical Circulatory Support Devices

COLIN KEALEY

Center for Advanced Surgical and Interventional Technology (CASIT), University of California, Los Angeles, CA

PAYMON RAHGOZAR and MURRAY KWON

Department of Surgery-Cardiothoracic, University of California, Los Angeles, CA

11.1 INTRODUCTION

Heart failure is the result of chronically reduced blood flow to the heart. It may develop acutely after a heart attack, or more slowly in response to a variety of diseases that compromise cardiac blood flow. Heart failure is both common and deadly. In 2006, an estimated 5.7 million Americans suffered from heart failure, and 500,000 new cases are being diagnosed each year. The aggregate 5-year survival from this disease is less than 50%, and the economic costs in the United States were estimated to be $37.2 billion in 2009 (Lloyd-Jones et al., 2009). Since the inception of the NIH artificial heart program in 1964, large sums of both public and private money have been devoted to developing mechanical circulatory support (MCS) devices that augment or replace the function of a failing heart. This effort was stimulated by the increasing prevalence of end-stage heart failure and its grave prognosis. While considerable progress has been made, heart failure continues to be an extremely deadly disease and the idea of a fully mechanical replacement for the human heart is still many years out from realization. Treatment options for heart failure remain limited. Patients with mild to moderate heart failure have been shown to benefit from pharmacologic management. There are few options, however, for those in the advanced stages of the disease. Historically, cardiac transplantation has been the only modality to provide substantive benefit. Transplantation is not a

Medical Devices: Surgical and Image-Guided Technologies, First Edition.
Edited by Martin Culjat, Rahul Singh, and Hua Lee.
© 2013 John Wiley & Sons, Inc. Published 2013 by John Wiley & Sons, Inc.

panacea, as the number of organs and the complications are limited and the average cost is approximately $75,000 (Cope et al., 2001). Thus, there has been a long-term concerted effort to develop artificial circulatory support devices to benefit patients with heart failure.

Currently, MCS is used primarily as a bridge to cardiac transplantation. These devices are placed in patients in the advanced stages of heart failure as a temporizing measure until an organ becomes available. In 2001, the Randomized Evaluation of Mechanical Assistance for the Treatment of Congestive Heart Failure (REMATCH) study showed that patients receiving MCS, who were not candidates for cardiac transplantation had significantly better 2-year survival than those receiving optimal pharmacologic management only (Rose et al., 2001). These results were very encouraging for advocates of MCS as a definitive or destination therapy and sparked new interest in devices capable of reducing or eliminating the need for cardiac transplantation.

11.2 HISTORY

In 1953, Dr. John Gibbon performed the first successful open heart surgery using a machine to oxygenate and circulate the blood, thus bypassing the heart and lungs (Romaine-Davis, 1991). What followed was a rapid advancement of cardiopulmonary bypass (CPB) technology, and by the early 1960s, CPB was considered a safe modality to facilitate cardiac surgery. The inability to wean some patients from CPB after operation, however, fueled interest in more long-term approaches to MCS.

Early efforts to create an MCS device centered on the creation of an artificial heart. In 1969, Dr. Denton Cooley implanted a highly experimental artificial heart into a 47-year-old patient who could not be weaned from CPB following repair of a left ventricular aneurysm (DeBakey, 2000). The artificial heart sustained the patient for 64 h until a transplant could be performed. This experience generated increased interest in artificial hearts, and in 1982, a team at the University of Utah implanted the first permanent artificial heart, the Jarvik 7. The patient survived the operation but died 112 days later after battling numerous complications (Devries, 1988). The subsequent experience with permanent artificial hearts has been disappointing. The longest survival of a recipient was 620 days, and to date, there are no artificial hearts approved by the FDA for permanent implantation.

Concurrent with setbacks to the development of permanent MCS devices in the early 1980s was an increase in the number of cardiac transplantations being performed. Transplantation was generally a successful therapy, but up to 30% of candidates were dying while waiting for an organ. In 1984, a group at Stanford became the first to perform a successful cardiac transplantation after temporarily supporting the patient's cardiac function with a ventricular assist device (VAD) before the operation (Portner et al., 1985). VADs differ from an artificial heart in that they support only the ventricles. The most commonly used VADs support the left ventricle only, though biventricular or solitary right ventricular VADs are used as well. These devices proved vastly more successful than total artificial hearts, and in the early 1990s, the FDA sponsored clinical trials on the use of VADs as a bridge to transplantation. In 1994, the HeartMate left ventricular assist device (LVAD) became the first implantable device to be approved by the FDA as a bridge to transplantation (Frazier et al., 1995). Since that time, there has been a marked increase in the use of VADs around the world. They are now well established as a bridge therapy before transplantation and as permanent "destination" therapy in patients not suitable for transplant (Rose et al., 2001).

11.3 BASIC PRINCIPLES

11.3.1 Biocompatibility and Mechanical Circulatory Support Devices

11.3.1.1 Overview Biocompatibility refers to the interaction between host tissue and any foreign material it is exposed to. Table 11.1 gives an overview of the components of biocompatibility for any given material. In the context of MCS, the primary issue is hemocompatibility. Hemocompatibility is a subset of biocompatibility concerning the interaction between the host's blood and foreign materials. Hemocompatibility is an extremely complex and in completely understood phenomenon. For MCS devices, considerations run from the microscopic interactions at the interface between blood and material, to the macroscopic considerations of pump design, flow rate, and types of valves used. Devices with poor hemocompatibility profiles will quickly form blood clots that can interfere with the pumping mechanism and cause significant complications for the patient, such as strokes and even death. The vast majority of devices in the market at present require that the patient receive pharmacologic anticoagulation to minimize their blood's reaction to the device. While use of anticoagulation is generally well tolerated, these therapies predispose patients to bleeding complications and cause a significant number of deaths each year (Odén and Fahlén, 2002). Other biocompatibility concerns include device placement, size, predisposition to infection, and compression or damage to adjacent organs. There is no MCS device that achieves perfect biocompatibility. Biocompatibility is therefore an area of active research in MCS devices.

Table 11.1 Considerations When Determining the Biocompatibility of an Implanted Material

Components of Biocompatibility
Hemocompatibility
Direct cytotoxicity
Systemic toxicity
Carcinogenicity
Immunologic sensitization
Material fatigue

11.3.2 Hemocompatibility: Microscopic Considerations

Hemocompatibility on the microscopic scale involves a complex series of interactions between the biomaterial surface and blood. When the material first contacts blood, protein adsorption and desorption begin. This is a dynamic process where a layer of protein coats the biomaterial's surface. The composition of this layer changes over time and the factors that influence composition are multiple. A large body of research has identified surface charge, hydrophobicity, and surface topography as being important determinants of this interaction (Dee et al., 2002). It has also been shown that native vascular endothelium, the cells lining blood vessels, is hydrophilic with a negative surface charge (Sarkar et al., 2007).

Attempts to apply the principles of biocompatibility discussed above have yielded mixed results. Biomaterials that are negatively charged, hydrophobic, and smooth (i.e., less surface topography) tend to lessen activation of the body's blood clotting cascade, which leads to thrombosis and accumulation of clot (Fig. 11.1) (Jones et al., 2000). While most MCS devices make use of these principles, some devices contain an intentionally roughened blood-contacting surface. Experience has led some investigators to believe that this tends to activate

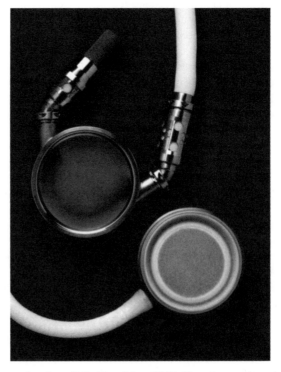

Figure 11.1 Inner chamber of the HeartMate XVE. Note the roughened surface, thought to promote formation of a stable coagulum. *Source*: Reprinted with the permission of Thoratec Corporation.

less overall blood clot formation. The rationale for this is not well understood, but it has been postulated that the roughened surface encourages the formation of a stable coagulum that inhibits further clot formation (Zhuo et al., 2005). Clearly, the principles that dictate the hemocompatibility of biomaterials are not fully understood. There appear to be considerable differences between static and dynamic systems, and different aspects of the clotting cascade are activated by different types of materials and conditions. Basic research on this topic will continue for some time.

11.3.3 Hemocompatibility: Macroscopic Considerations

Macroscopic hemocompatibility in MCS is primarily an issue of flow and wall shear stress. These are defined through Poiseuille's law, which relates the wall shear stress (τ_s) of a fluid flowing through a pipe to flow rate (Q), fluid viscosity (μ), and the pipe's radius (r), as

$$\tau_s = \frac{4\mu Q}{\pi r^3} \text{ (Poiseuille's law)}$$

Blood is a fragile substance, and it has been shown that blood is most stable at wall shear stresses between 10 and 70 dyn/cm^2. Below this range, blood products tend to deposit, and these regions are prone to a buildup of fatty deposits that can clog the vessel in a process known as *atherosclerosis*. Above this range, the blood begins to exhibit shear-induced thrombosis and activation of the clotting cascade (Malek et al., 1999). Therefore, the design of MCS devices must carefully consider the path of blood, size of the conduit, and expected velocity. Engineers must also consider the type of pump and valves, if any, that are to be used with the system. A variety of techniques have been used to study and predict the macroscopic interactions of blood, as it travels through a device. Computational fluid dynamics (CFD) is often a first step. The CFD is useful for pinpointing areas of high wall shear stress and turbulence. Once a prototype has been constructed, *in vitro* flow loops of varying design can be used to identify areas of thrombus formation. Finally, large animal studies are necessary to test the device *in vivo*, with careful attention being paid to areas of consistent thrombus accumulation.

11.4 ENGINEERING CONSIDERATIONS IN MECHANICAL CIRCULATORY SUPPORT

11.4.1 Overview

The design and manufacture of a modern MCS device is a complex, multidisciplinary task (Table 11.2). The typical human heart maintains a flow of approximately 6 l/min at a pressure of 120 mmHg. Some MCS devices are designed to function for 5–10 years. If one assumes a median lifespan of 7.5 years, then approximately 24 million l of blood will be pumped. Table 11.3 gives a list of common

Table 11.2   Overview of Engineering Considerations

Basic Engineering Considerations
Materials used
Type of pump
Path of blood flow
Use of valves
Fatigue/reliability
Anatomic position
Control system
Energy storage
Alarms

Table 11.3   Common Mechanical Circulatory Support Devices

Common Mechanical Circulatory Support Devices
Aortic balloon pump
Left ventricular assist device (LVAD)
Right ventricular assist device (RVAD)
Biventricular assist device (BiVAD)
Extracorporeal membrane oxygenation (ECMO)
Total artificial heart (TAH)

MCS devices. The most common devices used at present for long-term circulatory support are LVADs. Figure 11.2 gives an overview of a basic LVAD configuration. Generally speaking, LVADs involve the placement of an outflow cannula in the apex of the left ventricle. Blood is then shunted out the left ventricle, through a pump and into the patient's ascending aorta, reentering systemic circulation.

Consider also that these systems must function almost maintenance-free and that failure of the device is most likely catastrophic. Further complicating the problem is the inherent variability in the rate of flow. Starling's law of cardiac function dictates that the normal heart continuously adjusts its output to equal its input in order to avoid large swings in blood pressure, so too must the MCS device. This necessitates complex control systems to modulate the pump's function in response to normal physiologic changes. It is also necessary to consider the device's power source and method of energy storage. They must be robust and portable and generate only a small amount of excess heat. Finally, the anatomic position of the device is of vital importance. Today's devices are both intra- and extracorporeal and can provide univentricular or biventricular support. If intracorporeal, these devices must fit easily within the body and must not compress vital internal organs. Extracorporeal devices should still afford the patient a degree of mobility, and large lines traversing the skin must be carefully monitored for infection. MCS devices are therefore highly complex machines, whose construction requires careful consideration of both biologic and engineering issues by multidisciplinary groups.

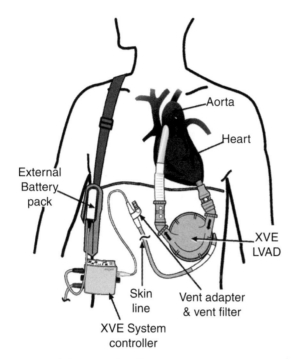

Figure 11.2   A typical left ventricular device (LVAD) setup. *Source*: Reprinted with the permission of Thoratec Corporation.

11.4.2   Pump Design

Current MCS devices generally use one of the two pump designs. These are positive displacement, or pulsatile, pumps, and rotary, or turbodynamic, pumps.

11.4.3   Positive Displacement Pumps

Positive displacement pumps are similar to the native heart in that they propel blood forward by cyclically changing the internal volume of a pumping chamber. Output is dependent on the pump's stroke volume and rate of activation. The volume of blood pumped is equal to the volume displaced inside the pumping chamber. This pumping mechanism results in a pulsatile pressure wave, akin to that created during the normal cardiac cycle. This configuration requires one-way valves to ensure that the pumped blood does not reverse its forward flow. In addition, the pump must have a method for quickly and reliably regulating the chamber's internal volume. These pumps have traditionally used a compressible sac coupled to a gas reservoir (Loisance, 2008).

There are several drawbacks to the positive displacement pump design. Perhaps, foremost among these is the use of valves to ensure one-way blood flow. Experience with these devices has shown that valves are commonly the source of mechanical failure and thromboembolic complications. Positive displacement devices tend to be large and heavy. If implanted intracorporeally, this necessitates large incisions and abdominal wall trauma. This may lead to chronic infections and recurrent bleeding episodes. Likewise, the inflow and outflow cannula are large, thus their implantation involves traumatic surgery to an already damaged heart muscle. Bleeding complications after the placement of positive displacement LVADs occur in approximately 60% of recipients, and chronic infection as result of using these devices is still a major hurdle to their long-term use. The flexible diaphragm, or compression sac, has also posed significant engineering problems in the past. If external, they require large percutaneous lines prone to infection that are tethered to gas sources. If internal, they become inaccessible, or very difficult, to maintain and suffer from a loss of gas over time. Despite these drawbacks, positive displacement pumps are a mainstay of MCS technology and their use will continue for some time (Mesana, 2004).

11.4.4 Rotary Pumps

Rotary or turbodynamic pumps are composed of a rotating shaft aligned parallel to the direction of blood flow. The rotation of the shaft produces torque, and impellers protruding from the shaft convert this to radial blood flow. The radial flow is then converted to axial flow by stationary blades (stators) that catch the swirling fluid, thereby generating pressure and forward blood flow (Fig. 11.3) (Loisance, 2008). This relationship can be described by the following equation (Frazier and Kirklin, 2006):

Torque $\times$ rotational speed (rate of energy supplied)

$$= \text{Pressure} \times \text{flow (rate of energy provided)} + \text{Energy wasted}$$

The output of a rotary pump is therefore a function of the speed of the rotor and the pressure difference between the pump's inlet and outlet. Therefore, if the speed, or revolutions per minute (RPM) of the rotor is held constant, the pump's output will be a function of the pressure difference between the outlet and the inlet. As this difference increases, the pump's output will decrease and vice versa. Conversely, if the pressure difference across the outlet and inlet is held constant while the speed of the rotor is increased, output will increase as well (Butler et al., 1999). This equation also demonstrates the primary disadvantage of rotary pumps, where the rate of flow is sensitive to the pressure gradient across the inlet and outlet. This tendency to decrease flow at higher outlet pressures is contradictory to the body's intrinsic demands and must be overcome with complex control systems (Konishi et al., 1996).

When using a rotary pump device in a patient, two variables must be optimized to insure adequate cardiac support. The first is pump speed quantified as the RPM to which the rotor is set. The second is pump flow. This is equivalent to the pump's "cardiac output." If this parameter is too low, the patient's heart will not

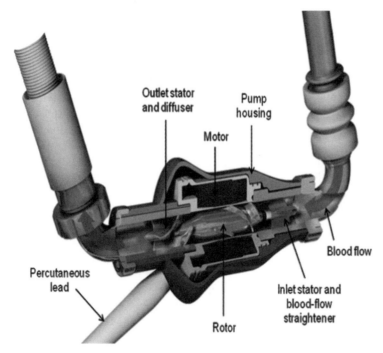

Figure 11.3 Schematic for a rotary MCS pump. *Source*: Reprinted with the permission of Thoratec Corporation.

be sufficiently unloaded, and they may begin to experience systemic symptoms such as syncope and fatigue. If it is set too high, it may cause severe ventricular suction and potentially even ventricular collapse (Henein et al., 2002). The optimum settings vary from patient to patient and as well as within a single patient, as their physiologic state changes. For instance, a larger patient may require a higher RPM to ensure sufficient flow. If, however, their blood pressure is significantly decreased, either pharmacologically or otherwise, then their flow will be higher at that same level of RPM and the rotor's speed may be decreased.

The primary advantage attributed to rotary pumps is their inclusion of only one moving part, namely, the rotor shaft. The use of a single moving part suggests an increased lifespan and maintenance-free operation. Even so, there is still the problem of fatigue on the bearings facilitating the shaft's rotation. Devices making use of a magnetic field to levitate the rotor, thus obviating the need for mechanical bearings, are currently undergoing clinical trials and will likely be available in the near future (Esmore et al., 2008). Another interesting feature of rotary pumps is that they do not contain valves. As stated earlier, valves are a significant source of both mechanical failure and thromboembolic complications in positive displacement pumps. The absence of valves is a double-edged sword in the case of pump failure, however, as there is no mechanism to prevent backward flow into the ventricle.

Other advantages of rotary pumps include the compact size of these devices, reduced hemotrauma, and their relatively noiseless function (Mesana, 2004).

11.4.5 Pulsatile Versus Nonpulsatile Flow

In addition to the mechanical considerations mentioned above, there is controversy regarding the type of flow provided by the two pump designs. Advocates of positive displacement pumps point to the more physiologic nature of the pressure waveform that is generated. Advocates of rotary pumps say that the effects of nonpulsatile flow are negligible. Historically, this controversy has its roots in the use of CPB. A review of the literature from 1952 to 2006 concluded that pulsatile flow was superior to nonpulsatile flow during CPB. This was evidenced by decreased pulmonary vascular resistance, decreased levels of inflammatory mediators and decreased end-organ damage with pulsatile perfusion systems. The authors also state that investigators advocating a nonpulsatile approach to CPB were able to show equivalency to pulsatility in some circumstances but not superiority. (Ji and Ündar, 2006)

Experience with LVADs has generally shown that nonpulsatile rotary systems produce either equivalent or superior results to pulsatile systems. A study of 70 patients receiving either pulsatile or continuous LVAD support for a mean of 370 days showed equivalent outcomes in terms of both survival and end-organ function (Radovancevic et al., 2007). Other studies have suggested that both operative CPB time and duration of ICU stay are shorter in patients receiving continuous-flow LVADs. (Feller et al., 2007) Perhaps, most convincingly, the results of a large randomized trial of 200 patients receiving either a pulsatile or nonpulsatile device were published in 2009. All patients enrolled were deemed ineligible for transplant. The study's primary endpoint was 2-year survival free from disabling stroke or reoperation to repair the device. Among those patients receiving the pulsatile device, 11% of patients achieved this primary endpoint, while 46% of patients receiving the nonpulsatile device achieved this endpoint. The data was statistically significant with $p < 0.001$. (Slaughter et al., 2009) This supports the idea that nonpulsatile flow is compatible with life and that the current generation of nonpulsatile devices are as good as, and perhaps better than, their nonpulsatile counterparts.

11.5 DEVICES

11.5.1 The HeartMate XVE Left Ventricular Assist System

11.5.1.1 Overview The HeartMate XVE is a left ventricular assist system (LVAS) manufactured by Thoratec Corp. (Pleasanton, CA). It is the result of a series of modifications to the HeartMate LVAD system and currently represents the state-of-the-art offering in positive displacement VAD technology from the Thoratec Corporation (Fig. 11.4). The XVE is an intracorporeal positive displacement pump and is placed intracorporeally in either a preperitoneal or intraabdominal position

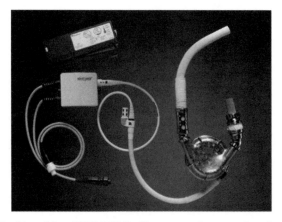

Figure 11.4 The HeartMate XVE LVAD. Shown are the pump housing, cannulae, percutaneous driveline, controller, and batteries. *Source*: Reprinted with the permission of Thoratec Corporation.

(Fig. 11.5). This sets it apart from some of its peers, where the pump housing is placed extracorporeally.

The XVE can produce a maximum flow of 10 l/min at a mean arterial pressure of 120 mmHg, a fill pressure of 20 mmHg, and a rate of 120 beats/min. The pumping chamber is composed of a pusher-plate actuator coupled to a flexible polyurethane diaphragm. The actuator causes the diaphragm to expand and contract, generating both positive and negative pressure within the pumping chamber. The inlet cannula is placed in the apex of the left ventricle while the outlet cannula empties into the ascending aorta. The cannulae are made of Dacron, a woven polymer commonly used for blood-contacting surfaces, and each one contains a porcine valve to prevent the backward flow of blood (Fig. 11.4). The pump is powered via a percutaneous driveline and may use either pneumatic or electrical power sources. The driveline also serves to vent the air that is displaced. To reduce the risk of infection, the driveline is covered with woven polyester. This encourages tissue ingrowth and formation of a less permeable barrier. Power is supplied by external batteries connected to a system controller. The system controller is a microprocessor-based unit that regulates the pump's function and serves as the primary interface for user interaction. (HeartMate XVE LVAS Operating Manual, 2008)

One unique feature of the HeartMate XVE LVAS (and all HeartMate VADs) is that blood-contacting surfaces are textured to encourage the formation of a stable clot on the device's blood-contacting surfaces. It is postulated that this "stable coagulum" is less thrombogenic than other biomaterials and surfaces typically found in VADs. This theory supported by clinical studies of HeartMate VADs have shown significantly less risk of thromboembolic complications on minimal anticoagulation therapy. Investigations of the stable coagulum have indicated that it contains pluripotent hematopoietic cells, monocytes, and macrophages. (Spanier et al., 1999) Given the inflammatory nature of this cellular population, some investigators have

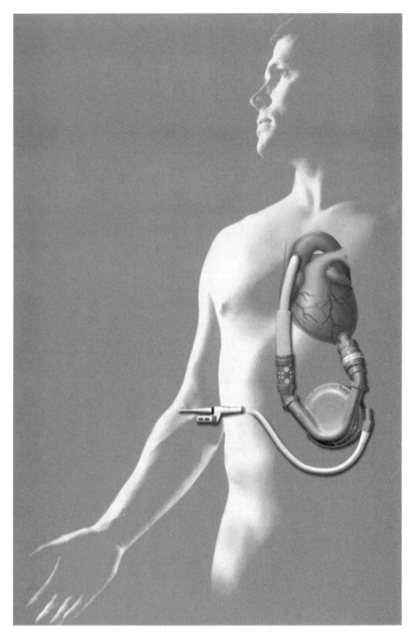

Figure 11.5 Diagram of an implanted HeartMate XVE LVAS. The inlet cannula exits the apex of the left ventricle while the outlet cannula empties in the ascending aorta. Also note the percutaneous driveline. *Source*: Reprinted with the permission of Thoratec Corporation.

hypothesized that recipients of these VADs may suffer from increased immunoreactivity and are thus less suitable for transplant. However, subsequent investigations have not shown that other VAD devices without the roughened surfaces show comparable levels of immunoreactivity (Kumpati et al., 2004).

11.5.1.2 *Device Limitations* Bearing wear, motor failure, and driveline fracture are the most significant sources of problems with the XVE device. This device is actually a refinement of a previous VE model. Most of these changes were in response to durability issues with the VE model. The most important changes were a repositioning of the diaphragm, additional support of the valves in the cannula, and modification of the percutaneous driveline to prevent kinking. An analysis of these changes showed a significant improvement over the VE model. Durability issues persist, however, as experienced clinicians report signs of bearing wear and pump fatigue after 18 months of use. (Martin et al., 2006)

11.5.1.3 *Clinical Outcomes* A study published in 2005 looked at outcomes from use of the XVE. The findings support the view that it is a significant improvement over the previous VE model. The authors examined 42 patients deemed ineligible for transplant who received the XVE. They found that a 40% lower rate of death than those reported in the 2001 Randomized Evaluation of Mechanical Assistance for the Treatment of Congestive Heart Failure (REMATCH) trial when the VE model was in use. Patients were also 2.1 times less likely to experience an adverse event, such as device failure, sepsis, or thromboembolic complications. (Long et al., 2005) Given that the patients in this study were comparable to those chosen for the REMATCH trial, it seems that there have been significant improvements to the design of this device.

11.5.2 The HeartMate II Left Ventricular Assist System

11.5.2.1 *Overview* The HearMate II LVAS is a second generation rotary pump system manufactured by Thoratec Corp. The design and manufacturing of rotary pump devices was in response to the large size and weight of positive displacement VADs, as well as lingering questions regarding their long-term viability. The HeartMate II is an intracorporeal device designed for long-term, outpatient circulatory support. It is placed just below the margin of the left rib cage in a preperitoneal position that is below the skin but not inside the abdominal cavity. A percutaneous driveline then exits at the skin 3–4 cm below the placement site.

The HeartMate II is capable of producing a maximum flow rate of 10 l/min at an average pressure of 100 mmHg. The rotor's operating range is 6000–15,000 RPM. The pump is approximately 7 cm in length and 4 cm at its largest diameter, making it significantly smaller than the HeartMate XVE. The pump itself is contained

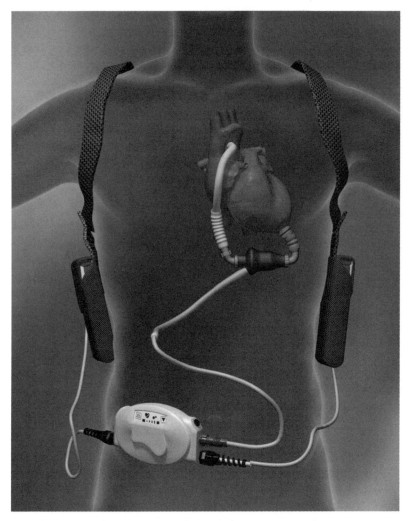

Figure 11.6 Diagram of an implanted HeartMate II LVAS. Seen here is the complete system with wearable battery packs and external controller. *Source*: Reprinted with the permission of Thoratec Corporation.

within a titanium housing. Attached to the housing is a percutaneous driveline that provides both power and control over the pump's function. The external controller is connected to the system via the percutaneous driveline and may be powered by two wearable batteries or a stationary power supply (Fig. 11.6) (HeartMate II LVAS Operating Manual, 2008).

An electromagnetic motor drives the pump. The motor's coils are contained within the pump housing and wrap circumferentially around the rotor. Similar to the HeartMate XVE, the inlet and outlet cannulae are made of woven Dacron. The outlet cannula is placed at the apex of the left ventricle and the inlet cannula is placed in the ascending aorta. Unlike the Heartmate XVE, this system does not contain any valves and has only one moving part (the rotor). This is thought to reduce mechanical wear and increase the pump's operational lifespan. Also, similar to the HeartMate XVE is the HeartMate II's use of roughened blood-contacting surfaces. As discussed earlier, this is thought to reduce the risk of thromboembolic complications.

11.5.2.2 *Device Limitations*

Control of blood flow through a rotary pump is a challenging problem. Unlike positive displacement pumps, there is no set stroke volume and few easily observable parameters. Current designs, including the HeartMate II, have no direct method to measure flow through the pump. Instead, they utilize the motor's use of electrical power over time in conjunction with the rotor's RPMs to estimate flow (Konishi et al., 1996). Under normal operating conditions, these algorithms work well. At the extremes of flow, however, these relationships break down, and more invasive medical testing is necessary to quantify the pump's function.

One of the primary indicators of pump malfunction is the sudden change in power consumption. In the case of thrombus accumulating on the rotor, drag will increase and so will power consumption. Likewise, one of the most feared complications of rotary pumps is the potential for a suction event. In this case, the pump may pull too much blood from the ventricle, causing suction and potentially leading to a ventricular collapse. The control system will register this as a marked decrease in power consumption and automatically decrease the RPM. This allows the patient to seek medical consultation.

As discussed earlier, one of the primary advantages of the rotary pump design is that it contains only a single moving part. The rotor for the HeartMate II is suspended by two bearings, and future modifications may use a magnetic field to hold the rotor in place, thus eliminating this final source of mechanical friction. Another improvement likely to be implemented in the near future is the elimination of driveline with the use of transcutaneous power delivery instead. The driveline is a common source of infection and even device failure in the case of fracture or kinking. Transcutaneous technology may eliminate the need for this line and allow for a fully intracorporeal system.

11.5.2.3 *Clinical Outcomes*

The HeartMate II was first implanted in a human in Israel in 2000. In 2005, it received approval for clinical use in Europe. Following this, a number of European centers began to implant the device. A retrospective analysis of the first 101 patients to receive the device in Europe found a one-year survival of approximately 65%. The main causes of death were multiorgan system failure and cerebrovascular accidents. Of the five cerebrovascular deaths that occurred, four were in the first 9 days after surgery (Strüber et al., 2008). These results are encouraging and significantly better than those found in the landmark REMATCH trial.

In 2007, a study of 133 patients in the United States receiving the HeartMate II was performed. The primary outcome was the proportion of patients who had undergone transplantation, were going through cardiac recovery, or had ongoing mechanical support but remained eligible for transplantation. Seventy-five percent of the patients met this primary outcome, and there was a 68% 12-month survival among those still receiving mechanical support (Miller et al., 2007). The results of this study paved the way for FDA approval, and in April of 2008, the HeartMate II was approved for commercial use in the United States as a bridge to transplant therapy. Finally, a recent study of 134 patients compared outcomes of the HeartMate XVE to those of the HeartMate II. Patients receiving the HeartMate II had significantly increased 2-year actuarial survival rates (58% vs. 24%), and lower numbers of adverse events and device replacements (Slaughter et al., 2009). This data suggests that rotary-based devices have reached a point in their development, where they are as good as, and perhaps better than, positive displacement systems. This will undoubtedly be an area of intense future investigation.

11.5.3   Short-Term Mechanical Circulatory Support: The Intraaortic Balloon Pump

11.5.3.1   Overview   Another device frequently utilized for mechanical support of cardiogenic shock is the intraaortic balloon pump (IABP). This device is commonly employed to stabilize patients with heart failure to restore blood flow on a temporary basis. It can be very beneficial in centers that do not have a cardiac surgeon, where a patient arrives in acute myocardial failure in order to stabilize with an IABP and transfer the patient to a tertiary care center where surgery can be performed. The IABP provides physiologic assistance to the failing heart by decreasing myocardial oxygen demand and improving coronary perfusion. A flexible catheter with a polyethylene balloon mounted on its end is inserted percutaneously, typically through the femoral artery with the balloon situated just distal to the left subclavian artery (Fig. 11.7).

Once inserted into the aorta, the balloon provides "counterpulsation" by inflating during left ventricular diastole and deflating during left ventricular systole. Inflation during left ventricular systole pumps blood throughout the arterial system while the left ventricle is filling, and deflation during left ventricular systole increases aortic compliance and lowers aortic pressure. The IABP timing is performed with an electrocardiogram (ECG) or arterial waveform. From the ECG, inflation is set for the peak of the T-wave at the end of systole, and deflation is set just before or on the R-wave. If the timing is set from arterial waveform, inflation should occur at the dicrotic notch with deflation just before the onset of the aortic upstroke. The balloon can be set to inflate with varying ratios to the heartbeat. A typical waveform is shown in Figure 11.8.

By unloading the heart, the IABP improves the myocardial oxygen supply and demand requirements, thus allowing injured myocardium to recover. It decreases myocardial oxygen demand by unloading the failing ventricle and increases the

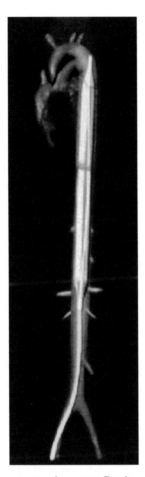

Figure 11.7   Intraaortic balloon pump placement. Retrieved from http://en.wikipedia.org/wiki/File:Dscp_Iab.JPG (Papaioannou (2005)).

oxygen supply by increasing the diastolic coronary perfusion pressure, improving coronary flow in a patient with otherwise diseased vessels. In addition, use of an IABP has been shown to salvage ischemic and reperfused myocardium. In an experimental model of ischemia-reperfusion, the use of IABP resulted in a smaller infarct area 1 h after reperfusion (Smalling's RW Circulation).

11.5.3.2   Device Limitations   Unfortunately, the IABP has its limitations. Specifically, it cannot be utilized in patients with aortic insufficiency, as diastolic augmentation would be impossible with an incompetent valve. Other contraindications include aortic dissection and severe aortic and peripheral vascular atherosclerosis. Furthermore, the inflating and deflating of the balloon is entirely dependent on the input tracing of the ECG or arterial waveform. Thus, patients who are being paced by unipolar atrial pacing may produce a large atrial spike that

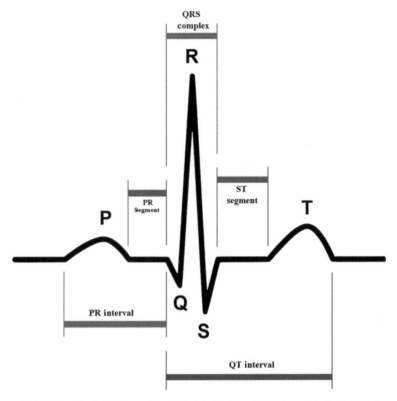

Figure 11.8 A typical ECG tracing with labels. The IABP is set to inflate at the peak of the T wave (left ventricular diastole) and deflate during the R-wave (left ventricular systole). Retrieved from http://en.wikipedia.org/wiki/File:SinusRhythmLabels.svg.

can be interpreted by the IABP console as a QRS complex leading to inappropriate inflation. Also, the development of atrial or ventricular ectopy (arrhythmias) can disrupt the normal inflation and deflation patterns.

Rapid heart rates can also be problematic. Some balloon consoles cannot inflate and deflate fast enough to accommodate heart rates over 150. Moreover, there is always a risk of balloon rupture and escape of gas (typically helium) into the bloodstream. If balloon rupture occurs, the device must be removed immediately to prevent thrombus formation. Difficulty in removal may occur if thrombus has already been formed within the balloon before removal. Thrombus formation is always a risk with the presence of an IABP. Embolization to visceral vessels (mesenteric and renal arteries) can occur in the presence of significant aortic atherosclerosis.

Distal ischemia is the most common complication of indwelling balloons occurring in approximately 5–10% of patients. Ischemia is most likely to occur in a patient with a small body surface area and small femoral vessels or severe iliofemoral occlusive disease. The presence of distal pulses or Doppler signals must be assessed regularly in all patients with an IABP and anticoagulation should be considered with intravenous heparin to keep the partial thromboplastin time at 1.5–2 times the control.

11.5.3.3 *Clinical Outcomes* Unfortunately, despite the important and well-established therapeutic advantages of the IABP, there are no good randomized clinical trials evaluating its use. Most of the current data are from retrospective analyses. In the National Registry of Myocardial Infarction (NRMI)-2, the largest registry of acute myocardial infarction included 23,180 patients. IABP was used in 31% of patients with cardiogenic shock. Patients who received thrombolytic therapy and an IABP had an in-hospital mortality of 49% versus 67% for those patients who did not receive an IABP ($p < .001$). The patients who had percutaneous revascularization showed no survival benefit with the use of IABP (45% vs 47%) (Barron et al., 2001). Two other large retrospective trials include the GUSTO I, and the SHOCK trial registry support the conclusion that IABP groups who experience survival benefit were those that had eventual revascularization (Sanborn et al., 2000). Therefore, it seems that the most important role of the IABP in the era of mechanical revascularization is to serve as a bridge to revascularization.

11.5.4 Pediatric Mechanical Circulatory Support: The Berlin Heart

Pediatric cardiac transplantation has a generally excellent long-term survival approaching 70% at 10 years (Kirklin, 2008). However, the donor pool is quite limited necessitating the need for miniaturized MCS for the pediatric population. As of 2003, the extracorporeal membrane oxygenator (ECMO) was essentially the only support technique for myocardial failure in the pediatric population. ECMO is a method for oxygenating and circulating blood similar to the machine used during cardiac surgery. ECMO is not, however, a permanent solution and is typically effective for only 1–2 weeks. Therefore, the National Heart Lung and Blood Institute (NHLBI) established a program for the development of pediatric circulatory devices in 2004, which would have specific applications for infants and small children.

The Berlin Heart EXCOR is a device used clinically that has drawn attention in the past few years. It is a paracorporeal pump with an electropneumatic drive system (Fig. 11.9) and includes pediatric blood pumps available in sizes from 10 to 80 ml depending on the size of the patient. There is a trileaflet inflow and outflow polyurethane valve in the pumps. Heparin coats the blood-contact surfaces of the valves to prevent thrombus formation. As the pumps are intended to be used in infants, the drive unit must be able to provide high positive systolic pressures (up to 350 mmHg) and negative diastolic suction pressures (up to 100 mmHg) at rates as high as 140 beats/min. The blood chambers and polyurethane ports are transparent to allow direct visualization for thrombus and chamber filling and emptying. As the Berlin Heart EXCOR is a paracorporeal pump, bedside pump exchanges can

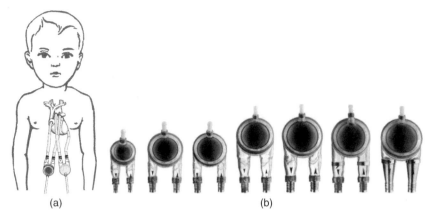

(a)   (b)

Figure 11.9 (a) The Berlin Heart EXCOR BVAD pediatric ventricular assist device for both right and left ventricular assistance. (b) EXCOR blood pumps. *Source*: Reprinted with permission from Berlin Heart, Inc.

be performed in the event that a thrombus is detected on a valve or within the chamber.

Selection of pump size is critical because a pump that is too small will result in limited filling and inadequate cardiac output, and may induce hemolysis (destruction of red blood cells) if run at an excessive pump rate. Blood pumps that are too large for the patient will have adequate pump filling, at the expense of a slower pump speed, increasing the likelihood of thrombus formation. Oversized pumps have also been associated with systemic hypertension. Pump size can be estimated from a customized chart used to match pump size to patient weight. The targeted flow for infants and children is 120–150 ml/kg/min and in adolescents, 100 ml/kg/min. Thus, a 10 ml pump is appropriate for infants up to 9 kg, a 25–30 ml pump for those between 10 and 25 kg, and a 50 ml pump for children and adolescents greater than 30 kg.

According to the German Heart Institute's experience, since 1999, 73% of patients have survived to hospital discharge, via either transplantation or recovery with device explantation (Kirklin, 2008). The major causes of mortality were multiorgan failure, sepsis stroke, and hemorrhage. The North American experience from 2000 to 2007 indicates that approximately 80 patients have been supported up to 234 days with the smallest patient being 3 kg and 15 days old. Fifty-five percent of the patients underwent transplantation, 13% were weaned from the device, and mortality was approximately 25%.

The major limitation in the use of the Berlin Heart is the incidence of neurologic events. Intracerebral and thromboembolic episodes are a major cause of morbidity and mortality with an estimated incidence being as high as 7–8%.

11.6   THE FUTURE OF MCS DEVICES

The following is a brief description of some newer devices in development and undergoing clinical trials. Improvements sought include development of a compact, light, efficient, and nonthrombogenic device.

11.6.1   CorAide

The Arrow CorAide (Cleveland Clinic, Lerner Research Institute, Cleveland, Ohio) LVAS is an electrically powered third-generation centrifugal pump. The pump is made of titanium with a biocompatible coating over the blood-contacting surfaces. The pump and electrical cable weigh 303 g, which is significantly lighter than most devices. It operates at 2000–3000 RPM providing flow rates of 1.5–8.0 l/min. Its rotating assembly uses a combination of magnetic and hydrodynamic forces for its suspension and function. What makes this device unique is the design of the inverted, blood-lubricated fluid film bearing. It supports the rotating assembly radially with a stable hydrodynamic fluid film resulting in no surface contact in the bearing. A noncircular profile stabilizes the radial hydrodynamic forces acting on the rotating assembly, allowing an axial passage through the center of the pump for blood circulation. The CorAide pump is therefore suspended, and there is no contact of the rotating element with either of the other two stationary components. This contrasts with the second-generation axial pumps, and thus, there is no wear and theoretically less potential for thrombus deposition.

11.6.2   HeartMate III

The HeartMate III LVAS (Thoratec, Pleasanton, CA) is a centrifugal pump with a "bearingless" magnetically levitated rotor design intended for long-term use. The device also has textured surfaces to reduce anticoagulation requirements, an induced pulse mode to achieve a level of pulsatility with continuous-flow assistance. It also has a sensorless flow estimator and a modular connection to allow percutaneous to transcutaneous upgrades. The pump's impeller draws blood from the ventricle and places an angular acceleration to the flow. The energy from this acceleration is converted to pressure, as it is delivered to the aorta. There are large gaps above and below the rotor to wash surfaces outside the main flow path to reduce the risk of thrombogenesis and hemolysis. Motor function and magnetic levitation are achieved in a single integrated unit that is incorporated with the pump's lower housing with all the control electronics. Another unique feature of this device is that it has sintered titanium to create texture on all the surfaces of the pump that come into contact with blood, which reduces thrombogenecity, and in turn anticoagulation requirements.

11.6.3 HeartWare

The HeartWare LVAD (HeartWare Limited, Sydney, Australia) is a centrifugal flow pump that resides in the pericardial space. The apical inflow cannula is an integral part of the pump itself. It weighs 145 g and displaces a volume of 45 ml and can generate flows up to 10 l/min. The pump's only moving part is the wide-bladed impeller which contains magnets within impeller blades. Electricity is not required to induce the device's magnetic effect eliminating the need for wires and connections. A hybrid magnetic and hydrodynamic bearing system holds the impeller blades in position.

11.6.4 VentrAssist

The VentrAssist LVAD (VentraCor Ltd, Sydney, Australia) is a centrifugal pump manufactured from titanium and a titanium alloy consisting of hydrodynamic bearings and an electromagnetically driven impeller. It weighs 298 g and is 67 mm in diameter. The VentrAssist has a conical rotor suspended in the blood by hydrodynamic forces generated by the tapered edges of the blades. The wide spaces between the blades create an "open" nature to the rotor to ensure that all areas of the pump are washed completely with high flows eliminating the occurrence of thrombosis. Its small size allows the pump to be positioned within the thoracic cavity or below the diaphragm within the extraperitoneal cavity behind the rectus sheath. The pump can sustain a flow rate of 2–10 l/min with a pressure range of 50–160 mmHg.

11.7 SUMMARY

MCS devices are an invaluable option for heart failure given the limited number of organs available for transplant. Perhaps the most important principle for the design of MCS devices is that of biocompatibility. These devices must be biocompatible on both a microscopic and macroscopic level. Clearly, all MCS devices are foreign bodies and their implantation results in a significant inflammatory response. Efforts to blunt that inflammatory response and prevent the dreaded, and often fatal, complication of thrombosis are constantly being investigated in the development of new devices. A multitude of mechanical considerations such as type of pump, energy storage method, anatomic position, path of blood flow, use of valves, and durability must also be considered. Current devices are approaching outcomes comparable to those achieved with cardiac transplantation and the near future may bring a device that obviates the need for transplant altogether in some patients.

CHAPTER 12

Orthopedic Implants

Sophia N. Sangiorgio

Biomechanical Engineering and Surgical Research Facility, Santa Monica UCLA Medical Center and Orthopaedic Hospital, Santa Monica, CA

Todd S. Johnson

Global Extremities Development, Reconstructive Division, Zimmer, Inc., Warsaw, IN

Jon Moseley

Implant Technology, Wright Medical Technology, Inc., Arlington, TN

G. Bryan Cornwall

Research and Clinical Resources, NuVasive, Inc., San Diego, CA

Edward Ebramzadeh

Biomechanical Engineering and Surgical Research Facility, Santa Monica UCLA Medical Center and Orthopaedic Hospital, Santa Monica, CA

12.1 INTRODUCTION

12.1.1 Overview

Collectively, the technology involved in the conception, design, development, evaluation, manufacturing, and marketing of orthopedic implants constitutes a multibillion dollar industry. Every year, more than $17 billion is spent on reconstructive, fracture and soft tissue repair devices, spine implants, orthobiologics, and other related products in the United States alone (Englehardt et al., 2009). The use of joint replacement implants, particularly for the hip and knee, has grown to account for the largest share of the orthopedic device market, representing approximately one-third of total orthopedic sales (Fig. 12.1). Similarly, the market

Medical Devices: Surgical and Image-Guided Technologies, First Edition.
Edited by Martin Culjat, Rahul Singh, and Hua Lee.
© 2013 John Wiley & Sons, Inc. Published 2013 by John Wiley & Sons, Inc.

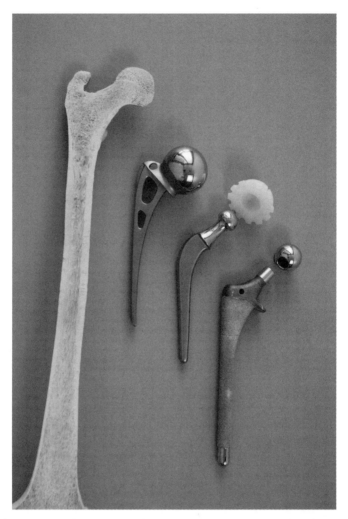

Figure 12.1 From left to right, a frontal plane cross section of a femur is shown, revealing the hollow medullary canal. For a hip replacement, the femur is cut at the neck, removing and replacing the femoral head. The implant on the left is a press-fit, Austin Moore stem design with a ball diameter similar in size to the natural femoral head. In the middle, the original Charnley femoral stem and polyethylene cup, both of which are cemented into place to achieve immediate initial fixation and to fill the gaps between the implants and the natural bony anatomy. The stem on the right is a more contemporary noncemented stem with a porous ingrowth surface. The larger size is intended to fill the medullary canal and the femoral ball is modular to allow customization of sizes, offsets, and materials.

for spine implant technology has increased dramatically over the past two decades. Specifically, between 1990 and 2003, the number of primary cervical and lumbar fusions with instrumentation increased by 170%. Globally, the market for spine implants, particularly motion-preservation technology as an alternative to vertebral fusion, has been estimated to be growing at a rate of 20% per year since 2000 (Kurtz and Edidin, 2006).

In general, orthopedic device technology can be subdivided into the same categories as the major clinical orthopedic subspecialties:

- adult hip and knee reconstruction;
- sports medicine and soft tissue joint repair;
- hand and upper extremity reconstruction;
- foot and ankle reconstruction;
- trauma (fracture repair);
- pediatric orthopedics;
- spine.

While there are many advantages to subspecializing for clinical practice, from an engineering point of view, it is important to be familiar with the basic technology and innovations in all areas of orthopedics. For example, modern hip replacements predate knee replacements by approximately 10 years; innovators of total knee replacements leveraged the engineering knowledge gained through the failures associated with using various materials for the hip. Conversely, one must be wary of translating innovations from one area to another. To illustrate this, most modern highly polished metal-on-metal (MOM) hip replacement bearings have been relatively successful at reducing wear debris; consequently, designers adapted this technology for articular lumbar spine disk implants. While both technologies are based on ball-and-socket constructs, the loading and lubrication conditions are not the same, resulting in different performance characteristics for the lumbar spine (Mathews et al., 2004; Lee et al., 2008). This emphasizes the importance of understanding the specific biomechanical conditions a given implant will encounter, as well as the major historical successes and failures related to similar devices used in other anatomical locations.

12.1.2 History

As early as the sixteenth century B.C., external splints were used for fracture repair by ancient Egyptians. Protective casts were composed of materials such as the midrib of the date palm leaf, spongy strips of wood or bark from the acacia tree, or bundles of straw to treat fractures (Salib, 1960). Later, in the fifth century B.C. in ancient Greece, Hippocrates detailed the importance of achieving proper anatomical alignment of fractured bones, how to apply compression correctly when bracing a fracture, and how to avoid leg length discrepancies when treating a fractured hip. He classified fractures by severity, explaining why injuries of long bones that ruptured the skin rarely healed, while those that did not rupture skin were more successfully treated. (Today, these fractures are known as open and closed, respectively.) Most importantly, Hippocrates applied logic and reason to the art of healing, separating religion and superstition from medicine.

There have been many advances in fracture repair since Hippocrates; however, for the vast majority of closed fractures that are mechanically stable, a short period of casting followed by bracing is widely used with excellent outcome (Charnley, 1950; Sarmiento, 1987; Sarmiento, 2006). Most open fractures are treated surgically today and are able to resume activity within a short period of time after surgery, whereas, as recently as a century ago, these patients would have been considered untreatable. The introduction of aseptic techniques in the mid-nineteenth century enabled the development of new and successful orthopedic surgical procedures. The introduction of biocompatible materials has enabled orthopedic surgeons to treat fractures internally as opposed to simply casting or splinting. In fact, today, a long bone that is abnormally short because of a congenital disorder or traumatic injury can even be lengthened using technology adopted for fracture treatment, through a process of regrowth known as distraction osteogenesis (Ilizarov, 1990). For this process, a long bone is carefully cut and held by an external brace, with a 1-mm gap between fractured surfaces to allow a "fracture" callus to form, then gradually distracted by approximately 1 mm per day as new bone forms between the fractured ends. Once the desired length is achieved, the two bone fragments are held rigidly to allow the final healing.

The introduction of biocompatible materials for internal fracture repair around the turn of the nineteenth century also paved the way for the first successful replacement of the hip joint nearly 50 years later, eventually leading to the development of prosthetic replacements for the knee, shoulder, ankle and other joints.

In this chapter, we introduce and provide an overview of a few of the more important aspects of the current technology used for the design and development of contemporary orthopedic implants.

12.2   BASIC PRINCIPLES

Designing an orthopedic implant requires considerations and poses limitations not encountered in the design of a conventional structure, such as a bridge. When designing a load-bearing structure, one must consider the following: (i) the mode of failure, (ii) the relationship between loads and the mode of failure, such as stress or deflection as a function of load, and (iii) the maximum stress or deflection considered to cause damage. In addition to these concerns, an orthopedic implant designer must consider biocompatibility, bone or tissue remodeling over time, and other issues related to biology, anatomy, and biomechanics. Many of these issues can be categorized as the optimization of strength and stiffness, maximization of fixation, and minimization of degradation. Another area of concern pertains to implant sterility and avoiding contamination, as bacterial infections are common problems with any implanted metal device or associated instrumentation.

12.2.1 Optimization for Strength and Stiffness

It is intuitively obvious that an implant or a device that fractures or plastically deforms has failed to serve its intended design goals. Therefore, the first goal is to design an orthopedic implant to be as strong as possible. Unfortunately, making an implant or device stronger often goes hand in hand with making the structural stiffness greater as well. As explained in the following examples, increasing the structural stiffness of an implant can produce an adverse mechanism of bone loss or degeneration, leading to the failure of the structure as a whole by loss of structural stability, commonly termed in orthopedics as *loosening*. Although the process of failure is typically gradual, loosening is nevertheless the most common mode of failure of orthopedic implants and, in extreme cases, may even result in device failure (such as implant fracture) as well. In order to understand the process of loosening, it is essential first to have an understanding of bone remodeling.

12.2.1.1 Bone Remodeling   Bones respond to external mechanical stimulus or rather, loads induced by activities such as walking, stair-climbing, carrying weights, or any type of exercise. Both the bone mineral density and the dimensions of bone change to adapt in response to the dynamic loading environment over time. This adaptation process is commonly referred to in the orthopedic literature as the law of bone remodeling, or Wolff's Law (Wolff, 1870). Prior to Wolff, Meyer and Culmann noted that the trabecular orientations in the proximal femur corresponded closely to the directions of principal stresses as a result of loading on the hip (Meyer, 1867; Cowin, 1986). The process of bone remodeling is one example of biological efficiency, using the minimum amount of material to produce the strongest structure necessary. Among other design goals, the majority of orthopedic implants are intended to substitute for or supplement the strength of an injured bone or damaged joint. However, the mechanical properties of the composite mechanical structure composed of bone and implant components are highly time dependent, because of the continuous process of bone and soft tissue remodeling.

12.2.1.2 Stress Shielding   The delicate process of bone remodeling is disrupted by the presence of a metal implant, which, unlike bone, has constant mechanical properties. For example, it has long been recognized that placing a metal stem inside the femur, such as a total hip replacement femoral stem, alters the distribution of stresses in the femur. That is, compared to the natural hip, with a metallic femoral component in the medullary canal, the stresses in the proximal femoral bone near the hip joint are decreased, whereas stresses in distal region, near the tip of the stem, are increased. In the proximal femur, these changes are said to produce stress shielding, since the metal stem shields the surrounding bone from some of the stresses to which it would naturally be subjected. The magnitude of stress shielding is directly related to the structural stiffness of the implant. Stiffer implant will produce a larger amount of stress shielding, and more flexible implant produce less stress shielding.

Nearly all orthopedic implants, whether intended for fracture repair or joint replacement, produce some degree of stress shielding. Stress shielding results in loss of bone. With time, this bone loss, or bone resorption, can lead to degradation of the integrity of the implant–bone composite structure. With fracture repair devices, such as plates, stress shielding is a transient problem, which may compromise the quality of the healed bone to some extent, but if the fracture heals within a reasonable time period, the device has served its function and may typically be removed. However, with joint replacements, which are intended to be permanent devices, stress shielding may eventually lead to mechanical loosening of the component. Loosening is in fact the most common mode of failure necessitating revision surgery of joint replacement implants (Malchau et al., 2002). Its incidence is orders of magnitude more frequent than device failure, such as implant fracture or gross deformation (Ebramzadeh et al., 1994). Moreover, many periprosthetic fractures, or fracture of the bone around the prosthetic implant, can also be attributed to gross loosening of implants, which results in eventual structural failure and collapse. It is important to recognize other causes of bone resorption aside from stress shielding, most notably, biological reaction to particulate wear; this issue is addressed briefly in Section 12.2.3.

As a result of these considerations, the design of an implant must be optimized to be (i) strong enough to avoid device failure under dynamic loads and (ii) structurally flexible enough so as to minimize stress shielding in the bone. In addition, a permanent implant such as a joint replacement implant must be stable enough at the interfaces to prevent fibrous tissue formation at the interfaces. Given the materials and alloys commonly used for implants, these two design goals are somewhat in conflict. For example, the modulus of elasticity (E, material stiffness) of cobalt-chromium alloy or stainless steel, typically used for femoral components, is approximately 200 GPa, substantially larger than that typical of cortical bone, generally 15–20 GPa. In contrast, the modulus of elasticity of polymethylmethacrylate (PMMA) cement, used for fixing a variety implants to bone, is about 2 GPa. These differences in material stiffness, combined with the shape of the stem, contribute to stress shielding.

The overall structural stiffness of any implant is a function of shape and material selection. For example, reducing the cross-sectional dimensions of a femoral stem or using a material with lower modulus of elasticity, such as titanium alloy ($E = 100$ GPa), decreases the amount of stress shielding in the proximal femur that an identically shaped cobalt–chromium alloy or stainless steel stem would produce. This is one reason for the popularity of titanium alloy for orthopedic implants. On the other hand, noncemented implants, which are designed to fill the medullary canal of the proximal femur, by design have larger dimensions and are therefore stiffer, producing larger amounts of stress shielding.

12.2.1.3 *Maximizing Strength*

When a long bone, such as a femur or tibia, breaks, a metal implant is often needed to carry the load that the broken bone is incapable of supporting during the healing process. These implants, generically termed *fracture repair devices*, need to be stronger and stiffer than the injured bone and often are removed months or even years later, after the injury has healed. The general approach in fracture repair is to anatomically reduce the fractured

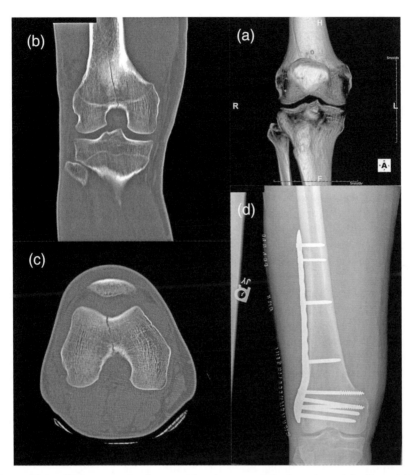

Figure 12.2 CT images of a femur with a fracture of the lateral condyle (knee joint). (a)–(c) show preoperative views of the injury. (d) The postoperative image includes fixation with a locked plate and screws. Note the repair method was intended to shield the bone from stress, to preserve the anatomy of the articulating surfaces (compliments of Robert Tonks, MD).

fragments, that is, to reorient the bones in their original natural alignment and to achieve good contact, stability, and a certain amount of compressive stresses at the fractured surfaces. Long bone fractures (Fig. 12.2) may be fixed with metal plates placed along the shaft, spanning the fracture longitudinally and screwed into the shaft below and above the fracture using multiple transverse metal screws through the cortical shaft. The plate not only serves to carry some of the load the bone would typically support but also shields the broken fragments from compressive forces that are believed to be necessary to stimulate the process of new bone formation. Therefore, different methods have been established to produce compression of the fractured surfaces. For example, in fractures fixed with a compression plate, this interfragmentary compression can be achieved by using a slotted hole with a chamfered edge for the screw head in the plate, which displaces the fragment while tightening the screw (Mow and Hayes, 1997).

While primary bone healing is often successfully achieved using interfragmentary screws or plates and screws, this method of repair is rarely used today for midshaft fractures, because with a stiff metal plate attached to the midshaft, the bone and plate construct is too rigid (high structural stiffness), hence inducing excessive stress shielding in the cortical bone. This may result in a weaker cortical shaft structure at the site of the healed fracture, which can be problematic, as the midshaft is typically subject to the high bending and torsional forces. In contrast, when a fracture heals with a more flexible mechanism, such as casting followed by a brace, there is less stress shielding and, hence, stronger new bone is built. In addition, implantation of a plate requires a large incision and bone exposure, as well as multiple drill holes for screws and disruption of the surrounding soft tissue, all of which are considered invasive and detrimental to healing.

In the initial postoperative period, implants used for fracture repair typically carry a large part of the load, or nearly the entire load, that is being transmitted from one fractured fragment to the other. As the fracture heals, the bone will carry a larger share of the load, until such time that healing is complete and ideally the implant can be removed. For a given fracture, the selection of the appropriate implant size, material, and design requires careful judgment not only to avoid implant fracture or failure before the fracture heals but also to minimize stress shielding.

It is important to recognize that plates and screws are still considered necessary in repair of fractures in or near a joint since perfectly aligned joint structures are essential to prevent arthritis in the long term. In this regard, locked plating has been developed. Unlike compression plating, in which tightening of the screws into cortical bone pushes the plate against bone producing high friction, in locked plating, not only the screws thread into cortical bone but the screw heads also thread into the plate, such that the screws become a rigid construct with the plate. In this way, the surgeon can avoid contact of the plate against bone altogether, minimizing injury to the periosteum, a thick, tough membrane surrounding the outside of bone that is essential in providing nutrition (Fig. 12.1). Although locked plating produces a larger amount of stress shielding, it is designed to maintain nearly perfect alignment of end segments and can also help protect osteoporotic bones by transmitting the loads to healthier and uninjured portions of the bone fragments via the screws, away from the fracture site.

Concerns with compression plating led to the popularity of intramedullary nails for the fixation of midshaft fractures in long bones. Typically, a tubular rod, or rather, nail, is inserted into the medullary canal and is fixed with one or two transverse interlocking screws at the proximal and distal ends of the long bone, if necessary, for stability or the prevention of limb shortening. The intramedullary nail is a great example of design optimization, as the structural strength of the tubular structure is typically much larger than that of the flat compression plate under bending and torsional loads (Fig. 12.3). This is clearly a desirable design goal since it minimizes the chances of device failure. Further, despite the increased

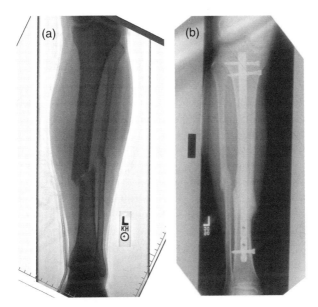

Figure 12.3 Midshaft tibia and fibula fractures. (a) Preoperative images of a fractured tibia and fibula. (b) Postoperative images show fracture repair (fixation) with an
intramedullary rod and callus formation, indicative of the early stages of healing. Note
that the fibula fracture was untreated, as the tibia, not the fibula, is responsible for carrying
most of the load in the lower leg.

strength, an intramedullary nail typically induces substantially less stress shielding
than a compression plate and screws. This is because a plate is screwed rigidly to
the femoral shaft near the fracture, producing a highly rigid and immobile structure,
whereas the intramedullary nail is generally less rigidly fixed within the canal and
is free to rotate to some extent. The reduced amount of stress shielding and the
large amount of motion allowed at the fracture surfaces are considered as major
advantages offered by intramedullary nails in providing faster and stronger healed
bones. However, as indicated, fractures close to joint surfaces typically require
periarticular plate (plate and screw) fixation to ensure optimal anatomical reduction.

The majority of intrameduallary nails are made of stainless steel. However, as
with joint replacement implants, titanium alloy versions have been designed to
further reduce stress shielding (McKellop et al., 1989).

In the preceding section, we used examples of fracture repair implants to high-
light the importance of maximizing implant strength and maximizing the strength
of the composite structure made of the bone and the implant. Clearly, adequate
strength is imperative to the function of every orthopedic implant, whether intended
to replace a joint or fix a fracture. The incidence of joint replacement device failure
is relatively low, although it is catastrophic for the patient and a difficult challenge
for the surgeon. However, since a fracture repair implant carries nearly the entire
load initially, it is subject to a greater risk of failure until healing is underway.

For this reason, fracture repair implants are generally at greater risk of failure and provide better examples in this regard compared to joint replacement implants.

12.2.2   Maximization of Implant Fixation to Host Bone

Repair of a fracture, whether with implants or using casting, involves providing sufficient mechanical stability only until healing takes over, after which bone gradually builds up structural strength and eventually the implant may be removed. It should be noted that a certain amount of motion between the fracture fragments is considered beneficial to the healing process by a cellular transduction mechanism that promotes bone formation and regeneration.

Unlike fracture repair devices, joint replacement implants, such as a hip or knee replacement, are designed to ideally last the patient's lifetime and, therefore, should generally be fixed to bone as strongly as possible. There are three main ways in which a joint replacement implant is designed to achieve stable and, ideally, permanent fixation within the host bone (Fig. 12.1). Early designs of joint replacements, popularized in the 1950s, were simply anchored against bone, such as a wedged stem placed inside the medullary canal of the femur. These early designs did not have any special surface coatings to promote bony attachment and relied entirely on mechanical anchorage for stability, yet many functioned well for several years (Amstutz, 1991).

A major advancement in fixation of joint replacement implants was made by Charnley who used PMMA, also referred to as *bone cement*, to fix both the femoral and the acetabular components of a total hip replacement to bone. Bone cement is supplied as a prepolymerized powder that is mixed with a liquid monomer in surgery for polymerization. During the few minutes it takes to set, cement turns into a low viscosity liquid and then a doughy material that can be used to fill the space between the implant and bone before it hardens. Fixation with bone cement provides adequate stability and good outcome and is still one of the most successful advances in joint replacement, used for total knee replacement, hip resurfacing implants and spine surgery. However, while highly successful, it was soon recognized that bone does not adhere to bone cement and that the strength of the bone–cement interface was purely by mechanical interdigitation. In fact, even as early as a few weeks following surgery, fibrous tissue can form between the cement and bone, albeit very thin, even with well-fixed implants (Fornasier and Cameron, 1976).

The next approach to obtain permanent fixation of implants was noncemented porous ingrowth fixation. Popularized in the early 1980s, it is designed to take advantage of the bone's ability to grow onto the surfaces or into the pores of a metal implant once sufficient mechanical stability is achieved, thereby providing porous ingrowth or ongrowth fixation. This approach requires initial press-fit fixation, which requires careful preparation of the cavity for the implant such that the implant surfaces have extensive areas of direct and stable contact with cortical bone. Studies have shown that the tangential motion (micromotion) between the bone and metal needs to be on the order of 50 μm or lower in order to achieve ingrowth or ongrowth; otherwise, a fibrous tissue membrane will form. The term *osseointegration* refers to bone ongrowth such that there is no fibrous tissue between

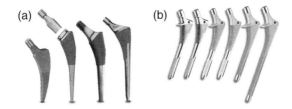

Figure 12.4 Examples of several current noncemented femoral stems. (a) A typical product line of modular noncemented femoral stems (VerSys® Hip System, Zimmer, Inc., Warsaw, IN). Four titanium alloy stems. For porous ingrowth, 3 on the left have titanium plasma spray roughened surface treatment; one (right) has porous metal tantalum surface treatment. The three stems on the left utilize titanium alloy with a titanium plasma spray roughened metal surface treatment for ingrowth. The last stem on the right is made of titanium alloy with a porous metal tantalum surface treatment for ingrowth. (b) Four noncemented femoral stems, varying largely in design, showing a variety of options available to surgeons.

the implant and bone. Osseointegration is a sign of a stable and biocompatible implant.

Representative examples of noncemented femoral stems for total hip replacement with surfaces intended for porous ingrowth are depicted in Figure 12.4. This type of technology is used in other joint replacements as well, such as knee replacements, shoulder replacements, and dental implants. The noncemented approach for implant fixation proved to be highly successful in particular with porous ingrowth acetabular cups (the socket component of total hip replacements), in part because the acetabular socket in the pelvis could be prepared accurately and precisely into a hemispherical shape, conforming to the back of the acetabular cup with extensive contact between the porous ingrowth metal surface and bone. A variety of porous ingrowth coatings and surfaces have been developed and applied, as we describe in the subsequent sections. Noncemented femoral fixation is viewed by most as preferable for younger and more active patients, as an osseointegrated implant is considered more durable than a cement interface. Today, noncemented fixation is extremely popular, particularly in the United States and Australia.

Implant design alone does not guarantee stable fixation and long-term success. Engineers must also design instrumentation, including rasps, reamers, and broaches used to prepare the bony cavity, such that good press-fit fixation and adequate direct contact with bone are achieved. Fixation is also a function of the surgeon's skills in using surgical instrumentation and proper selection of an implant design for a given patient, as well as surgical placement with proper alignment. Proper preoperative planning, using templates to select the best fit, size, and modular components are among the primary considerations.

12.2.3   Minimization of Degradation

A joint replacement, by design, transmits loads across the joint and provides mobility and function through the full range of motion (ROM). It is inevitable, therefore, that the surfaces of an artificial joint would deteriorate over time. In general, regardless of anatomical location, there are two types of deterioration that the surfaces of a joint replacement experience: wear of the articulating surfaces and fretting wear.

12.2.3.1 Wear of Articulating Surfaces

The first type of degradation occurs when a joint's articulating (or bearing) surfaces gradually wear, because of the intended function and motion of the joint. The best known examples of this type of joint wear is wear of the polyethylene acetabular cup as a result of articulating against the ball at the hip joint, although a similar type of wear is generated at the knee implant or other joint replacements.

The primary focus of the orthopedic research and development conducted in the past half century in this regard has been on development of new and modified materials and surface treatments to reduce joint wear and the volume of wear debris produced. The first bearing material combination for hip joints that resulted in substantial longevity in patients was a stainless steel or cobalt-chromium (Co-Cr) alloy ball articulating against an ultrahigh molecular weight polyethylene (UHMWPE) cup. This coupling of materials is known as *hard on soft*, as the softer material, polyethylene, is intended to wear more rapidly but is theoretically more easily replaced. One of the most significant advances in improving the wear resistance of joint replacement surfaces has been the introduction of cross-linked UHMWPE, particularly for the reduction of acetabular cup wear (McKellop et al., 1999). One of the technical challenges in this process was to minimize oxidation, which can lead to the degradation of mechanical properties. Different methods have been developed to cross-link polyethylene or, similarly, to reduce the number of free radicals. The most established method for cross-linking is γ-irradiation, which has long been used as a method of sterilization before packaging. Improvements in manufacturing technology have improved the surface hardness and resistance to damage of hard surfaces, such as femoral balls in total hip replacements and the femoral components in total knee replacements.

Hard-on-hard bearing combinations are typically made of CoCrMo alloy, alumina, or zirconia. Advances in design and manufacturing technology along with a better understanding of the potential failure mechanisms have enabled extremely highly polished surfaces, tighter tolerances, and larger diameter femoral balls, which reduce the risk of dislocation. These advances in wear reduction have led to the reintroduction of surface replacements for total hip replacement.

Unlike conventional total hip replacements, surface replacements retain the femoral neck and do not rely on anchorage of an intramedullary stem for fixation and are therefore considered by many to be less invasive and preferable for younger and more active patients. Surface replacements have large femoral balls, which necessitates the use of a larger diameter acetabular component. A hard-on-hard bearing combination reduces the required thickness of the acetabular component because of higher material strength.

12.2.3.2 Fretting Wear

A second mechanism for implant degradation is loosening or, rather, motion between fixation interfaces that are designed to remain attached. As indicated, the most common cause of failure for joint replacements is implant loosening. Loosening is a gradual process, with the rate determined by multiple factors.

As an implant loosens, micromotion occurs at the fixation interfaces resulting in fretting wear. Fretting is defined as unintentional small, oscillating, tangential displacement between the surfaces of two components of a mechanical structure (Hills and Nowell, 1994). Fretting wear is the removal of material from the mating surfaces owing to fretting (Waterhouse, 1982). Studies of implants retrieved at revision surgery or postmortem have shown that a significant amount of wear debris is generated not only at articulating surfaces but also at fixation and modular interfaces as well, as a result of fretting motion (McKellop et al., 1990; Collier et al., 1992; Barrack, 1994; Bobyn et al., 1994; Boggan et al., 1994; Mjoberg, 1994; Brown et al., 1995; Kawalec et al., 1995). Therefore, it is important to consider the wear properties and the potential to produce debris, regardless of whether or not the intended function of the implant surface is articulation.

There are many different failure mechanisms caused by fretting wear. In some instances, fretting can initiate cracks, thus decreasing the fatigue life of the implant or other structural components, such as acrylic cement (Hills and Nowell, 1994). With some metals in the body, fretting wear repeatedly removes the surface oxide layer, which may rapidly reform, leading to abrasive–corrosive wear, since the metal is directly exposed to the oxidizing environment of the body (Agarwala et al., 1982). Fretting corrosion also has been associated with increased crevice corrosion (Brown et al., 1995). Fretting and corrosion can also lead to implant fracture, commonly observed at high stress interfaces.

12.2.4 Sterilization of Implants and Instrumentation

Infection is one of the most confounding complications of any surgery and is of particular concern in orthopedic procedures where foreign materials are used to restore stability and function for the patient. *Staphylococcus aureus* bacterial infections are common when any unsterilized metal is being implanted in the body. For this reason, sterility is one of the main requirements not only for implants but also surgical instruments. Instrument systems must be designed to accommodate routine and repeated hospital grade sterilization. Autoclaving is the most common form of sterilization, involving high pressure steam in sequences of temperatures and time to terminally eliminate organisms (bacteria, viruses, fungi, and spores) that can cause infection. Autoclaving is particularly important for orthopedic systems that have many expensive and reusable instruments to assist the surgeon with the placement of the orthopedic implant (Fig. 12.5) (Matthews et al., 1994).

12.3 IMPLANT TECHNOLOGIES

In this section, we cover the examples of the technologies involved in implants for hip replacements and knee replacements, as well as technologies related to implants for spine surgery.

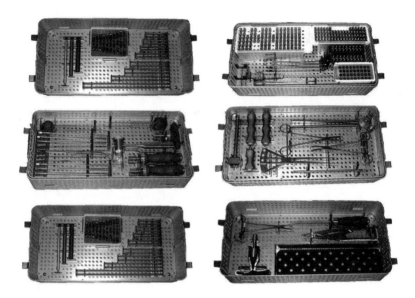

Figure 12.5   Sterilization trays for a spine surgery. Multiple trays and inserts for the system containing implants anodized with different colors to identify different pedicle screw diameters, lengths, and sizes. Also, instruments important for the surgical preparation, manipulation, and delivery of the implants are included in the sterilization trays.

12.3.1   Total Hip Replacement

12.3.1.1   Basic Components of a Hip Replacement

The hip joint consists of the femoral head at the proximal end of the femur, which is placed within the hemispherical shaped acetabular socket of the pelvis. That is, the natural hip joint is a ball-and-socket joint, which, by definition, has three degrees of freedom, meaning rotation in all three planes. Anatomically, these rotations are called *flexion/extension*, *abduction/adduction*, and *internal/external axial rotation*. Both the ball (femoral head) and the socket (acetabulum) are naturally lined by thick articular cartilage, providing low friction for all motions. The hip joint is encapsulated by a fibrous membrane, which contains synovial fluid, providing lubrication for the natural joint.

The first prosthesis that gained widespread popularity for a hip joint replacement was a hemiarthroplasty, so called since it replaced only one of the two components of the hip, that is, the femoral head. The Austin Moore hemiarthroplasty consisted of a metal ball that replaced the natural femoral head, attached to a long femoral stem that was placed in the femoral canal to stabilize the prosthesis (Fig. 12.1). This type of prosthesis is still in use today, but its application is limited to cases where only the femoral head is diseased or broken and the natural acetabular socket is a viable option for the metal head to articulate against.

Different diseases and conditions, such as different forms of advanced arthritis, make it necessary to replace both the femoral head and the acetabular socket lining altogether. The first total hip replacement, so called since is replaced both the femoral head and the acetabular socket, was the Charnley total hip replacement. The Charnley total hip replacement consisted of two components, the stainless steel femoral stem and the polyethylene acetabular cup. Similar to the Austin Moore prosthesis, the Charnley femoral component consisted of a ball attached to a stem that was placed in the femoral canal. However, unlike the Austin Moore, the Charnley femoral ball was only 22.25 mm, so as to allow room for the polyethylene acetabular socket. Another innovation of the Charnley total hip replacement was that both the femoral and acetabular components were fixed in position with PMMA, previously described in Section 12.2.

Modern total hip replacement, or arthroplasty, has been increasing in popularity since the 1960s and has a well-established, successful long-term clinical outcome. After more than five decades, Charnley's hip implant designs are still considered the gold standard, to which many other more recent designs have been compared (Herberts and Malchau, 2000; Malchau et al., 2002).

Despite many advances in material properties and surgical instrumentation and techniques, many basic principles introduced by Charnley are still in use today. However, issues related to implant fixation and polyethylene wear have been identified as key areas for improvement to provide even better long-term clinical outcomes, especially for younger more active patients, as the first generation of hip arthroplasties were typically limited to less active, nonoverweight patients who were at least 65 years old. Despite warnings and precautions, today, joint replacement implants are increasingly used in younger, heavier or excessively active patients.

Today's total hip replacement resembles the original Charnley design in many ways but addresses the challenges and demands posed on the implant. It also differs in several ways. While the femoral component of the Charnley design was one part, today's femoral component is invariably made of two or more parts: the femoral head is attached to the femoral stem via a morse taper and perhaps more parts are attached to the distal part of the stem to lengthen it or to increase the contact with bone. Modular components provide major advantages: they allow the ball and the stem to be made of different materials as deemed necessary by the designer or the size of the components can be mixed and matched for the patient during surgery. Similarly, the acetabular component is commonly made of two components. First, the liner is designed to articulate against the femoral head (ball) and is most commonly made of cross-linked polyethylene. This liner is typically mounted inside a metal shell that is designed to be fixed to bone, using specially designed surfaces and perhaps screws.

Despite the sophisticated technologies involved in developing and manufacturing a joint replacement prosthesis, the basic functional requirements are straightforward. First, it allows the necessary joint motion; in this case, the rotations of the ball within the cup, without producing excessive amounts of wear debris (ideally none). Second, it should maintain long-lasting or ideally permanent fixation to bone so as to provide structural stability of the implant.

The technical advances that have been made over the past few decades toward improving the performance and longevity of total hip replacements have addressed these two functional requirements, that is, minimizing wear and maximizing fixation. In this section, we cover some of these technological advances to meet the demands of a rapidly growing population of patients placing increasingly higher biomechanical demands on the implant.

First, cementing technology to achieve fixation of total hip replacement components is discussed. Most acetabular components used today are noncemented; therefore, the options for cementing will use only femoral component option examples. Next, metal surface technologies to achieve fixation of noncemented components are discussed. These surfaces are applied to both noncemented acetabular and noncemented femoral components. This is followed by a description of MOM total hip replacement technology, which is designed to minimize wear debris and to maximize the ball size.

12.3.1.2 Cemented Total Hip Replacements

One of the main engineering design goals for any orthopedic implant is to maximize fixation. In this regard, some of the most common cemented femoral stems have featured PMMA precoating, roughened oxide blasted surfaces, smooth glass-bead-blasted surfaces (matte), or alternatively, highly polished surfaces. Subtle differences in design variables such as surface finish, the presence or lack of a dorsal flange or collar, cement thickness, or stem size can have very profound effects on implant stability and long-term fixation (Ebramzadeh et al., 1994, 2003, 2004; Sangiorgio et al., 2004). These areas have been addressed in the design of contemporary cemented stem offerings with components to accommodate a variety of philosophies (Fig. 12.6).

Clinical studies have demonstrated that successful total hip arthroplasty outcomes can be achieved with any of a number of design philosophies as long as the appropriate implant design and surgical technique are combined together. The first design philosophy, shown on the left in Figure 12.6, is similar to the original Charnley, as well as many others commonly used in Europe and South America today, with well-established long-term clinical outcome. The matte surface finish is created by blasting the implant with a glass bead medium. The roughness of the surface morphology can be characterized as 0.4–1.0 μm average roughness, which is larger than polished stems and lesser than the roughened oxide blasted stems. A macrotextured surface is added to the proximal stem to enhance proximal stem–cement fixation. The dimples are sized to create interdigitation of cement and added resistance to interfacial shear stresses.

A second cemented stem design philosophy uses a combination of macrotexturing and circumferential PMMA precoating, resulting in a significantly stronger initial proximal cement/metal bond (Ahmed et al., 1984). The third type is the highly polished, noncollared cemented stem. It has a trapezoidal cross section to improve rotational stability and rounded corners to improve stress distribution at the stem/cement interface. Many investigators believe that the highly polished stem is the best cemented femoral stem design, since it can migrate distally within the

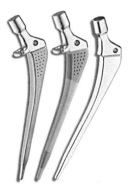

Figure 12.6 This is an example of the three main cemented stem design philosophies (Zimmer, Inc., Warsaw, IN). All three stems are modular, meaning that the ball is a separate component. The stem on the left includes a matte finish with macropores for cement interdigitation and a cement collar, to prevent distal axial migration of the stem. The stem in the middle includes the same features in addition to a precoat of polymethylmethacrylate, for maximum initial stem–cement adhesion. The stem on the right is collarless and highly polished, intended to debond from the cement mantle and migrate distally within the cement mantle, wedging into tight mechanical interlock.

cement mantle, achieving secure fixation by mechanically wedging into position (Ling, 1992; Verdonschot and Huiskes, 1997a,b; Verdonschot et al., 1998).

The cross-sectional shape of a femoral stem is essential for cemented components, as it influences the load distribution throughout the prosthesis, cement, and bone (Crowninshield et al., 1980a,b; Ebramzadeh, 1995; Ebramzadeh et al., 2003). The broad lateral face is intended to provide uniform distribution of lateral tensile stresses on the cement, reducing the potential for debonding at the cement/stem interface. The broad medial face is intended to distribute medial compressive stresses uniformly on the cement, reducing cement strain and the potential for fracturing the cement mantle (Pellicci et al., 1979; Mjoberg, 1994). The medial edges of the proximal stem are rounded, to reduce stress concentrations that could induce fracture of the cement (Robinson et al., 1989; Kelman et al., 1994). The anterior and posterior surfaces of the proximal implant taper, thereby reducing the shear stresses at the stem/cement interface compared to the nontapered implant.

12.3.1.3 Porous Ingrowth Surfaces   Initially introduced for dental applications, biological ingrowth, known as *osseointegration*, has proven to be an effective long-term orthopedic fixation method (Branemark et al., 1969; Albrektsson and Albrektsson, 1987; Albrektsson and Jacobsson, 1987). Examples of porous ingrowth surface options include sintering of CoCrMo beads, a compressed mesh of metal strands referred to as *fibermetal* or *fibermesh*, and tantulum-based trabecular metal (TM). Most of these treatments can be applied to titanium alloy, Co–Cr, or stainless steel implants. TM is a unique porous surface that has a

(a)   (b)   (c)

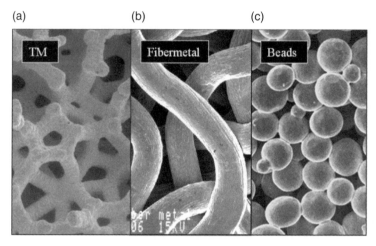

Figure 12.7 Examples of porous ingrowth surface options under high magnification. (a) Trabecular metal (TM), (b) fiber metal or mesh, and (c) sintered beads. All are currently used surface treatments intended to achieve porous ingrowth. These types of surfaces are used for fixation of hip replacement femoral and acetabular components as well as knee replacements and other joint replacement prostheses.

cancellous-bone-like structure with a very high coefficient of friction against bone, intended to enhance initial fixation and fully interconnected pores of similar size and shape to promote bone ingrowth (Fig. 12.7).

All of these porous surfaces are time-tested technologies and are offered to accommodate surgeon preference. The size, shape, and circumferential placement of the porous area are designed to provide the greatest area for ingrowth without compromising implant strength or performance. For example, because of the inherent notch sensitivity of titanium alloy, the extent of the porous coating is limited for titanium substrate stems.

Today, the vast majority of acetabular cups used for total hip replacement are noncemented metal shells, with replaceable inner liners. Regardless of whether the femoral stem is cemented or noncemented, noncemented cups have demonstrated excellent fixation strength and long-term clinical success (Woolson and Maloney, 1992; Malchau et al., 2002). An example of a contemporary porous acetabular cup design is shown in Figure 12.8, which utilizes TM as a porous ingrowth surface, with the option of a cross-linked polyethylene or a highly polished metal liner as a bearing surface against which the ball of the femoral stem articulates.

The acetabular components of a modern total hip replacement typically utilize a metal substrate with a porous surface on the pelvic bone side as well as a provision to accept modular bearing inserts on the inner diameter. The modularity of the system provides the surgeon with multiple options for bearing surfaces including UHMWPE, highly polished metal, and ceramics. In addition, at the time of any subsequent revision surgeries, if the porous ingrowth shell is well fixed, the surgeon has the option to retain it and replace the liner with either an identical unworn liner

Figure 12.8   A modular trabecular metal acetabular cup system (Zimmer, Inc., Warsaw, IN). The same acetabular porous ingrowth cup, with or without holes for screws, may be used with a variety of polyethylene, ceramic, or highly polished metal liners (cups). This allows the liner, in some material combinations, to be replaced in subsequent revision surgeries with a new, unworn component, without needing to remove a potentially well fixed, porous ingrowth shell. This also allows a surgeon to switch to a larger ball diameter, by replacing the femoral ball and the acetabular liner, without having to compromise a potentially well-fixed acetabular cup with bony ingrowth.

or one of a different material or inner diameter, to match the new femoral stem and ball used for the revision surgery.

12.3.1.4 Metal-on-Metal Hip Bearing Technology   Current MOM bearings offer several potential advantages over the more commonly used metal-on-polyethylene bearings for contemporary total hip replacements. Laboratory studies and clinical data have suggested that wear rates are significantly lower, which, combined with the high strength and stiffness allow the engineer to design larger bearing diameters with thinner acetabular cups. Larger diameter joints (typically femoral ball/surface replacement >36 mm) reduce the risk of dislocation over conventional femoral ball sizes (28–36 mm). Thinner acetabular cup shells also minimize the amount of bone removed from the pelvis to prepare for impaction, thus maintaining the strength of the underlying bone. An example of a modern large diameter MOM total hip replacement bearing couple and the femoral and acetabular components of a representative resurfacing prosthesis are shown in Figure 12.9 (Amstutz et al., 2011).

The performance advantages of MOM bearing surfaces come at a substantial cost in manufacturing. To achieve extremely low wear, MOM bearings rely on fluid film and boundary layer lubrication. This requires both very highly polished surfaces and small diametral clearances (100–200 μm) between the head and the shell. This in turn imposes a requirement for micrometer level tolerances for the articular surfaces

Figure 12.9 Large diameter metal-on-metal hip replacement options. (a) A large diameter metal-on-metal total hip replacement with a femoral stem (A-Class Hip System, Wright Medical, TN). (b) Metal-on-metal surface replacement. (c) Implanted metal-on-metal surface replacement (Conserve Plus Hip Resurfacing, Wright Medical, Memphis, TN).

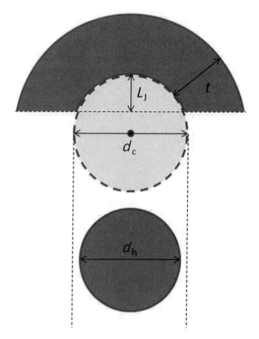

Figure 12.10 Schematic of cup and ball bearing design parameters. d_c, inner cup diameter; d_h, diameter of head; L_j, jump distance; t, cup thickness.

of both components. The discussion of the technology involved in this section has been organized in terms of design, testing, manufacturing, and quality control.

The design objectives for MOM bearing components are to minimize wear while maximizing the ROM and increasing stability, that is, the ability to resist dislocation. From the standpoint of design, stability is primarily a function of the jump distance, which is the distance that the femoral head would have to translate relative to the cup before dislocation would occur. For example, in Figure 12.10, if the cup design covers a full hemisphere (180°) then the jump distance is equal to the spherical radius. Typically, the cup design covers less than 180°, since lower coverage angles will increase the ROM, although they will also decrease the jump distance slightly. Jump distance is more meaningfully measured with the cup at an anatomical inclination. A lower coverage angle may also make the bearing more susceptible to edge loading if the cup is implanted with a high inclination (abduction or anteversion) angle.

The main design factors affecting wear in a MOM bearing are the diametral clearance ($d_c - d_h$), the material characteristics, surface finish, and sphericity. As shown in Figure 12.10, if the clearance is too small (or negative, i.e., ball diameter larger than socket diameter), this can lead to equatorial contact, which increases frictional torque and leads to high wear and potential loosening of the acetabular component. An optimum clearance results in polar contact between the cup and the ball, which puts minimal frictional torque on the cup and minimizes the rate of wear. If, on the other hand, the clearance is too high, this can lead to smaller polar contact area, higher polar stresses, a loss of lubrication, and again, excessive wear. Therefore, the optimum clearance for a given size and design of bearing has to be determined empirically through functional testing, wear evaluation, and strength considerations.

Sphericity is a measurement of how closely a given ball or cup surface resembles a perfect sphere and is defined as the ratio of the surface area of an ideal sphere with the same volume divided by the actual surface area of the ball or cup. This ratio is always less than 1 because of imperfections in the actual surface. From a design standpoint, the engineer must take into account what is achievable in manufacturing given the state of the art in tooling and materials. Excessive deviation from a sphere, low sphericity, reduces the effective clearance, leading to an increase in asperity contact and increased wear.

Surface finish is also critical to the performance of a MOM bearing, and the most commonly used measure to quantify surface roughness is average roughness, Ra is the average of the absolute values of vertical deviations of the roughness profile from the mean line. A smaller Ra indicates a smoother surface. If surfaces are smooth enough, the asperities are small enough and the clearance is optimized, a state of fluid film lubrication may be achieved in which there is essentially no contact between asperities on the two surfaces.

The film thickness is a function of many factors, including lubricant viscosity, clearance, and relative velocity of the surfaces. The effect of velocity explains in part why larger diameter MOM bearings wear at a lower rate than small diameter bearings. The higher relative speed between surfaces entrains more lubricant into the gap. In clinical use, lubrication conditions may seldom be ideal, with start-and-stop activities and constant direction reversals. However, for a given film thickness, reducing the surface roughness of both components will minimize metal to metal contact and reduce wear. As with sphericity, the engineer must balance the benefit of improved surface finish against the increased manufacturing cost of tighter tolerances.

The mechanical and physical properties of the bearing material have a major effect on the performance. A high modulus of elasticity is required to withstand joint forces without excessive deformation that would cause the cup to "pinch" and interfere with motion. The material must also have high surface hardness in order to maintain a highly polished surface finish and avoid damage. Finally, the alloy must have a high corrosion resistance in order to be biocompatible.

Any new implant must go through a process to verify that the design requirements have been met. This process begins even before prototypes are fabricated, with the use of finite element analysis (FEA) to verify that stresses under load do not exceed the yield strength or fatigue limits of the material. Most design verification consists of mechanical and functional testing in the laboratory. Fatigue testing under various standardized loading conditions is conducted. For example, the implant system must be able to withstand a minimum of 10 million cycles of loading under conditions that represent a clinical worst case, such as extreme activity by heavy patients. Functional testing of a new bearing design includes wear testing, which is conducted on joint wear simulators that produce representative motion and loading on the hip during walking or other activities. Testing is typically conducted according to ASTM F1714, or ISO 14242, which describe loading and environmental conditions for testing and specify a gravimetric method of assessing wear rates. The lubricant is typically a mixture of saline and bovine serum proteins, which simulates the lubrication conditions provided by synovial fluid.

Achieving the tolerances required for well-functioning MOM bearings, especially with materials such as cobalt-chromium alloys that are difficult to machine, poses major challenges to the manufacturing engineer. In many ways, the success of these systems is directly attributable to improvements in manufacturing technology. High precision CNC machining tools and advances in grinding media technology have enabled repeatable production of components with micrometer level tolerances in clearance and sphericity, along with mirrorlike surface finishes.

The manufacture of a cast MOM bearing system begins with a very old technology. The lost wax technique of investment casting was developed thousands of years ago, and its basic principle is unchanged. Replicas of the parts to be fabricated are injection molded from low ash wax and are attached with molten wax to a sprue and runner assembly to form a "tree." The tree is coated with refractory slurry using robotic manipulators and then fired in an oxidizing oven to remove the wax and preheat the mold. The cavity is then filled with molten cobalt-chromium alloy that has been melted in an induction furnace.

Before further processing, cast components are usually treated by hot isostatic pressing to reduce the internal porosity, followed by solution annealing to promote carbide dissolution and establish a more uniform microstructure. Next, the nonarticular features of the acetabular shell or femoral component are machined, and any surface modifications are applied to the bone interface. For example, at this stage, sintered beads or arc plasma coatings are applied.

Machining of the articular surface is the last fabrication operation because any subsequent thermal or mechanical process could cause the part to be out of specification. Modern multiaxis CNC lathes with extremely rigid construction are essential for producing a spherical surface with the necessary tolerances. Cutting hard materials such as cobalt-chromium alloy also requires specialized grinding media and coolants.

12.3.2 Technology in Total Knee Replacement

Modern total knee replacement has been performed for about 40 years and is highly successful in restoring function and providing pain relief for end-stage osteoarthritis. The knee is highly dependent on soft tissue support from the ligaments for stability and resistance to dislocation. The types of components used to treat knee cartilage degeneration are typically selected based on the condition of the ligaments and the level of constraint required. If the ligaments are very lax, a knee is prone to dislocation and a more constrictive knee replacement device is necessary for the patient to ambulate without falling.

There have been numerous advances in the design and materials used for total knee replacements with varying degrees of success and prevalence. For example, modular components are used today to achieve proper sizing and soft tissue tension and general balance, customized for each patient, regardless of height, weight, age, or activity level. In addition to improvements in the designs, there have also been advances in surgical techniques and instrumentation, which include advanced technologies for component alignment and soft tissue balancing, as well as soft tissue preservation and minimally invasive techniques.

12.3.2.1 *Basic Components of a Total Knee Replacement*   The most commonly used total knee replacement incorporates three components: a tibial component, a femoral component, and a patellar component; collectively, these components resurface and replace all of the articular cartilage surfaces of the knee. One less invasive alternative to a total knee replacement is a unicondylar replacement. The knee is a bicondylar joint, and most frequently, the medial condyle is extensively damaged, while the lateral condyle appears relatively healthy. Hence, unicondylar designs partially resurface the femoral condyle and tibia only in the area of damaged cartilage, retaining the stabilizing ligaments and soft tissue that would have been removed with a total knee replacement (Fig. 12.11).

12.3.2.2 *Modularity and Constraint*   Modularity and constraint, or rather, the degrees of freedom and extent of motions allowed, are essential to a comprehensive total knee replacement system (Fig. 12.12). Modularity of a product line allows surgeons to customize components based on patient, size, bone quality, gender, and the condition of the patient's muscles and ligaments. This is extremely important for revision cases, where bone quality may be severely compromised and the muscles and ligaments may not be capable of providing as much support (constraint) as they did in the initial (primary) operation. The system chart in Figure 12.12 depicts the range of components available in a total knee replacement product line.

The differing levels of constraint in the tibiofemoral interface of the components is provided by varying the geometry and the design features of interaction between the metal femoral component and the polyethylene bearing (alternatively called the *articular surface*) insert. Moving from the left to the right side of Figure 12.12, it becomes apparent that the articular surface inserts employ progressively more dishing (concavity) and the addition of post-like features. The increased dishing and

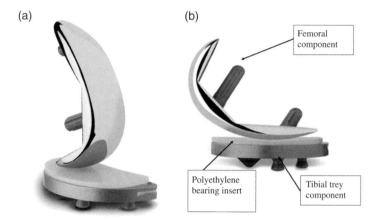

Figure 12.11 Two views of a unicondylar knee replacement system, intended as a less invasive treatment alternative to a total knee replacement, as one condyle is left intact.

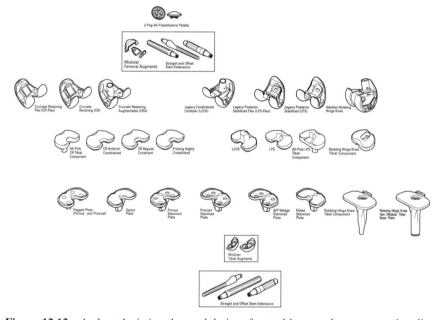

Figure 12.12 A chart depicting the modularity of a total knee replacement product line. This example is taken from NexGen® Complete Knee Solution (Zimmer, Inc., Warsaw, IN)

the mating interactions of the femoral components with the post features increase the resistance to translational and rotational movements, as well as prescribing certain motion patterns through cam and post interactions. The specific geometries and design features employed in the progressive levels of constraint provided in the system are necessary to appropriately accommodate either patient-specific disease states of ligament viability or individual surgeon preference for retention or substitution of important ligament structures of the knee. Four ligaments stabilize each knee: the two collateral ligaments prevent dislocation to the left or right and the two cruciate ligaments prevent dislocation to the front or back. The majority of contemporary (bicondylar) total knee designs sacrifice the anterior cruciate ligament and either retain or substitute for the posterior cruciate ligament.

Designs that retain the posterior cruciate ligament are commonly referred to as *cruciate retaining* (CR) devices and those that substitute it with mechanical constraint are often referred to as *posterior stabilized* (PS) devices. Examples of these two types of components are depicted in Figure 12.13.

There are slight variations in these two types of designs (CR and PS) based on patient/surgeon preference of high flexion accommodating designs (Flex components) and/or the ability to add stem extensions and build up augmentation to the components. The constrained condylar components are an example of a product line of constrained knee replacements, which, in addition to substituting for a missing or insufficient PCL with post and cam features, employ a tight fit between the metal femoral component box and the prominent metal-reinforced polyethylene post. The slip-fit between the components provides additional rotational and rocking moment constraint for collateral ligament insufficiency (preventing pigeon-toed, knock-kneed, or bow-legged deformities and instability). Additional operative

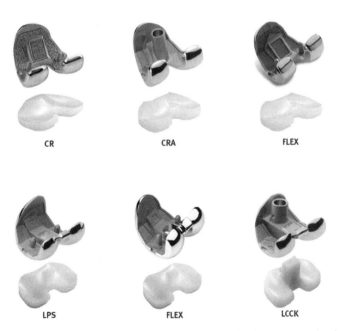

CR   CRA   FLEX

LPS   FLEX   LCCK

Figure 12.13 Representation of increasing levels of total knee constraint in modular total knee replacement (Zimmer, Inc., Warsaw, IN).

Figure 12.14 Depiction of the most constrained option in total knee replacement, a rotating hinge design (Zimmer, Inc., Warsaw, IN).

options are also provided through the use of multiple thicknesses of articular surface inserts. The multiple thicknesses allow the surgeon to obtain optimal soft tissue tensions for a given patient based on ligament and bone condition intraoperatively. The final level of constraint employed in this type of system is referred to as a *rotating hinge design*, in that the femoral component rotates about a fixed pivot or hinge point and hence, allows no anterior or posterior translation at the tibiofemoral interface. This type of design is depicted in Figure 12.14.

Constrained total knee replacement designs, such as the rotating hinge design, are typically utilized for revision surgery of primary implants because of a desire to provide additional tibiofemoral constraint through either the hinge mechanism or the slip fit of femoral component over the reinforced polyethylene post. These components also have the ability to accept stem extensions and build up augmentation, which is often necessary because of compromised bone or ligaments in surgical revision of prior total knee components.

12.3.2.3 Computer-Aided Alignment

In a recent position statement put out by the American Association of Hip and Knee Surgeons (AAHKS) on computer-assisted orthopedic surgery (CAOS), some of the key principles of both total hip and total knee surgeries are (i) restoration of joint and limb alignment; (ii) correction of joint and limb deformity; (iii) restoration of joint motion; (iv) maintenance or restoration joint stability; (v) proper sizing, positioning, and fixation of the joint replacement implants; and (vi) balancing of soft tissues.

CAOS is a relatively recent addition to the surgeon's toolbox of instrumentation and technique methods. The basic principles of this method are the ability to visualize either the total hip or the knee replacement components with respect to the patient's specific bones and soft tissues through the use of a special camera during surgery, which is then coordinated with computer programs to aid the surgeon in the operating room. The visualization of the patient's joint is recreated in the computer, which has routines and methods predefined to provide information to the surgeon to aid in primarily limb alignment, sizing, and positioning. This in turn leads to potentially better stability and joint motion because of properly maintaining adequate soft tissue balancing. Both the camera and computer are fed information during surgery to establish proper bony resections and component placement by the user digitizing anatomical landmarks of the joint and limb (Figure 12.15).

Potential CAOS benefits are improved accuracy in component placement (position and rotation), fewer surgical complications (fat emboli and blood loss), and enhanced visualization for minimally invasive surgery (MIS). In contrast to standard open conventional incision procedures, are somewhat controversial in hip and knee surgeries but have the potential to shorten hospital stays, allow patients a quicker return to activities of daily living, provide less pain and larger ROM in the first six months, and leave a smaller scar. In terms of surgical technique issues, the ability of the surgeon to properly balance the soft tissues through the complete ROM of the joint remains one of the key challenges to surgeons relative to patient perceptions of joint stability and proper motion patterns after joint replacement.

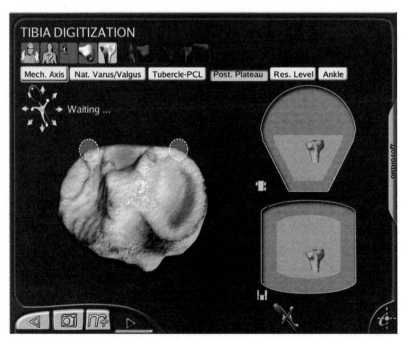

Figure 12.15   An example of a computer-aided alignment screenshot seen by the surgeon during an operation, where the surgeon is prompted to identify bony landmarks on the tibia for a knee replacement surgery.

12.3.3 Technology in Spine Surgery

The spine is perhaps the most complicated of orthopedic structures, consisting of both hard and soft tissue structures, including the intervertebral disks and ligaments. Each intervertebral disk consists of a nucleus pulposus and the surrounding annulus fibrosis. The posterior aspect of the spine forms a protective channel for the spinal cord and nerve roots that exit from each level of the spine. The posterior column also has numerous attachment points for ligaments and tendons. Some of the motions of the spine are accommodated by the deformation of the disk, while others are facilitated by the facet joints in the posterior column of the spine. The spine has several ligaments that provide additional stability by limiting motion. Together, the components of the spine provide motion to the core of the human body, while protecting the vital spinal cord. The spinal cord runs down the center of the spine and, if injured, may be irreparable. Such injury can leave a patient paralyzed. Therefore, it is important to protect the cord at all costs during surgery.

The spinal column is generally divided into three anatomical regions: the cervical spine with 7 vertebrae, the thoracic spine (including the rib cage) with 12 vertebrae, and the lumbar spine with 5 vertebrae. Each of these three regions has very unique biomechanical properties and therefore different requirements for surgery or implant design. For example, the cervical region (neck) requires the greatest ROM in all three planes compared to the other two regions of the spine, particularly the thoracic region. On the other hand, the lumbar spine provides a lower ROM, but the ability to support much higher dynamic loads, involving the mass of the upper body, and anything carried or supported by the upper body.

12.3.3.1 Spine Fusion Technology
The most established surgical treatment for severe disk disorders and spine deformities is immobilization of the spine or spine fusion. Typically, immobilization is achieved by the removal of the disk(s) to allow two or more adjacent vertebral bodies to fuse together, hence the term *fusion*. The gap created by the removal of the disk is typically filled by a bone graft or bone substitute, such as a cage. In order for the bones to fuse, much like a fractured bone, the spine must be stabilized. This stability is provided by the addition of instrumentation, such as rods, screws, hooks, and/or plates (Fig. 12.16).

Spine fusion was initially introduced in the 1950s, by Drs. Harrington and Luque, to treat spinal deformities (Kurtz and Edidin, 2006). Today, fusion is still the most common treatment for the correction of scoliosis deformities and considered to be the gold standard for many other spine disorders. This procedure is used successfully for treating herniated disks owing to traumatic injury, degenerative disk disease in the middle-aged population, and stenosis in elderly patients, for example.

One of the most common applications of fusion is for treatment of radicular leg and low back pain. This procedure is typically performed on the lumbar spine using posterior instrumentation and may also be performed in conjunction with anterior column stabilization achieved via anterior lumbar interbody fusion (ALIF), posterior lumbar interbody fusion (PLIF or TLIF), or lateral lumbar interbody fusion (XLIF) (Kurtz and Edidin, 2006; Goodrich and Volcan, 2008). Specifically,

(a)   (b)

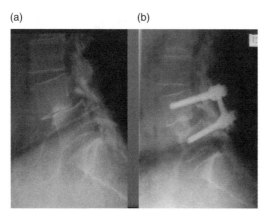

Figure 12.16   (a) Preoperative radiograph illustrating loss of disc space at L4–L5 and severe lumbar instability. (b) Postoperative radiograph illustrating reconstruction with an XLIF interbody cage and pedicle-screw-based fusion instrumentation.

pedicle screws are used to anchor rods, placed bilaterally, on the posterior spine, to restrict all motion until the fusion is complete (Fig. 12.16). Although this procedure is by and large successful, if healing does not occur as anticipated, over time, the hardware can loosen, fracture, or deform, resulting in the need to reoperate (Rutherford et al., 2007).

Another example of a clinical application for fusion is cervical disk degeneration. Cervical disk degeneration is a condition associated with the normal aging process, frequently exacerbated by injuries and trauma, thus affecting a wide age group of patients. Symptoms of patients with severe cervical spine ailments include axial neck pain, pain radiating into the arms or regions innervated by the exiting nerve roots (radicular arm pain), and in severe cases, compression of the spinal cord which can result in functional deficits.

Surgical techniques for cervical spine fusion, like that of the lumbar spine, are generally grouped by surgical approach including anterior, posterior, or combined (anterior plus posterior). Anterior surgery is the most prevalent for the cervical spine, particularly in the United States. An anterior approach typically involves removing the disk for decompression of the pathology, followed by insertion of an interbody fusion device, and stabilization with an anterior plate and screws (Fig. 12.17). The interbody fusion device could be autograft harvested from a different bone from the patient, allograft bone harvested from a donor, or synthetic materials that provide structural support until the bone healing occurs. There are many designs and features to anterior plates that provide varying degrees of loading to the graft material, such as the material, number and diameter of the screws, mode of attachment, and other features that facilitate dynamic loading of the bone graft (Haid et al., 2002).

12.3.3.2 *Total Disk Replacement and Dynamic Stabilization*   Until recently, fusion was viewed as the only surgical option for advanced degenerative

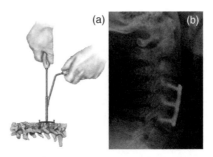

Figure 12.17 Example of an anterior cervical fusion: (a) surgical instrumentation and approach and (b) postoperative radiograph depicting two-level fusion of the cervical spine.

disk disease and instability in the spine and, even today, most view it as the gold standard. In a fusion procedure, two or more spine vertebrae are fused together, rendering the joint immobile. While fusion alleviates pain, there are disadvantages: the disks adjacent to the fused levels may be at risk of accelerated degeneration because of higher motion and stresses (Hilibrand et al., 1997; Harrop et al., 2008). Moreover, fusion is an invasive procedure, requiring a long recovery time (Kurtz and Edidin, 2006). Finally, fusion is a motion-restricting procedure, reducing the mobility of the spine.

In the past two decades, interest in motion-sparing devices as an alternative to fusion for the spine has increased, producing a multibillion dollar market internationally. These include joint replacement devices designed to replace portions of or the entire disk and dynamic stabilization devices intended to provide semirigid fixation without requiring fusion or the removal of the disk, and many other devices (Kurtz and Edidin, 2006).

Owing to the success of joint replacement for the hip, knee, and even shoulder joints, total disk replacement implants have been introduced for the cervical and lumbar spine (Fig. 12.18). Most can be categorized as articulating (ball-and-socket devices or ball-on-trough, for example) or nonarticulating compliant materials intended to mimic the mechanical properties of the disk. Most of the articulating disk replacements employ the same material combinations that have been clinically successful for the hip and knee joint replacements. For example, typical bearing materials include metal-on-UHMWPE and MOM (CoCrMo).

The main advantage of a total disk replacement compared to fusion is to retain or restore motion, as in the natural healthy spine. Another potential advantage of disk replacement may be a reduced risk of accelerated adjacent segment degeneration. Following a fusion, ROM may increase at the levels adjacent to the fusion to substitute for the motion loss at the fused level. This increase in motion (or stresses) commonly results in accelerated degeneration at these adjacent levels. Therefore, allowing motion at the treated level is considered advantageous (Pimenta et al., 2007).

The clinical results from prospective, randomized studies between total disk replacement and fusion in both the lumbar spine and the cervical spine have

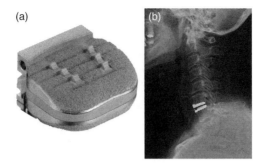

Figure 12.18 Example of a cervical total disc replacement system. (a) Image of the top surface (bone ingrowth side) of a cervical disc replacement with fixation "V-teeth" for short-term fixation and textured titanium with calcium phosphate coating for long-term fixation. The articulating surfaces are CoCrMo with a polyethylene disc insert. (b) Postoperative radiograph depicting successful implantation of the disc in the cervical spine.

demonstrated positive results with both technologies (Guyer et al., 2009; Murrey et al., 2009). Longer term follow-up studies have also been encouraging. However, complications have been reported and the strategies for addressing complications in the lumbar spine are more complicated than those for the cervical spine (Patel et al., 2008). Since the majority of lumbar total disk replacements are implanted through an anterior approach and the great vessels (aorta and vena cava) bifurcate at the L4–L5 level, revision for the lumbar spine involves much greater risk for the patient. In general, greater clinical success has been reported for cervical disk replacements than for lumbar disk arthroplasty (Phillips and Garfin, 2005; Pimenta et al., 2007). This is likely due to the higher stresses on an implant in the lumbar spine, as compared to cervical spine, predisposing it to risk of expulsion, loosening, and excessive wear.

A recent technological advance in lumbar spine surgery has been the development of a new class of devices known as *posterior dynamic stabilization implants*. Generally, these devices are designed to save the intervertebral disk, while at the same time, restricting some motion, particularly extension or hyperextension. The majority of these devices can be categorized as interspinous-process spacers, or pedicle-based distracters mostly with dynamic rods, or other shock-absorbing devices. Depending on the implant and system design, they may be implanted through conventional open approaches where the anatomy is visually identified before insertion or more minimally invasive approaches. Each of these technologies have associated trade-offs, but the intention is to enable earlier surgical intervention with less bone loss and potentially less damage to the surrounding soft tissue.

12.3.3.3 *Minimally Invasive Approaches to Spine Surgery*

Minimally invasive approaches have been desirable and of increasing interest to both surgeons and patients based on the perceived advantages of faster recovery, less blood loss, and less damage to surrounding structures (McAfee 2010). However,

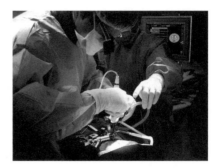

Figure 12.19   Intraoperative photo of minimally disruptive spine surgery, depicting XLIF technique (NuVasive, Inc., San Diego, CA).

earlier attempts many minimally invasive techniques were fraught with challenges of reproducibility and effective clinical outcomes; this has been changing in the last five years or so (McAfee 2010). Recent advances such as kyphoplasty for vertebral body compression fractures and placement of percutaneous pedicle screws have been important developments. Both techniques involve increased use of fluoroscopy or intraoperative radiography to provide anatomical information concerning the bony structures to the surgeon in order to reduce the larger incisions and retraction of muscle and other soft tissue structures to visualize the targeted anatomy.

A novel minimally disruptive technique for a lateral approach to the lumbar spine fusion introduced a split-blade retractor with controllable, customizable aperture (Ozgur et al., 2006). Safety and reproducibility increased with use of surgeon-driven neurophysiology based on stimulated electromyography (EMG) to identify nerves and safe path through the psoas muscle (Fig. 12.19). Implant design is important as the strongest ligament in the lumbar spine, the anterior longitudinal ligament, is preserved and the implant is placed on the strongest region of the bone.

12.4   SUMMARY

The field of orthopedic technology is constantly evolving. When total joint replacements were first popularized 50 years ago, the average patient was over 65 years old, not overweight, not employed in manual labor, and did not have an overly active lifestyle. Today, many patients are between 40 and 65 years old, overweight, if not obese, and fully expected to assume active lifestyles following joint replacement surgery (Ebramzadeh et al., 2007). For spine surgery, nearly 40% of the patients undergoing cervical or lumbar fusion are under the age of 45 and 80% are under the age of 65. Consequently, the main concern in orthopedic implant design, regardless of anatomical location, has been to maximize the longevity of the device and its fixation to meet the increasing patient expectations and demands.

An orthopedic implant is first and foremost a device that serves a mechanical function within the human body. As such, the implant must meet the structural strength requirements under repetitive loads and be biocompatible. At the same time, an ideal implant allows an optimal level of load transmission to the host bone, so as to minimize stress shielding and subsequent bone resorption. Furthermore, the implanted material must be sterile so as not to introduce an infection and the materials must not elicit a foreign body reaction.

The second and perhaps, most challenging requirement for an orthopedic implant is to achieve and maintain stable fixation. For fracture repair devices, stability is only required until the fracture has healed; however, for a joint replacement device, or spine fusion devices, the clinical need may be nearly as long as the patient's lifespan. For example, an adolescent with corrective scoliosis surgery living into the 90s would require 80 years of clinical performance from the fusion hardware.

A third requirement, essential to the design of any orthopedic implant, is to minimize material degradation. In this regard, it is important to consider not only the articulating or bearing surfaces but also the nonarticulating fixation surfaces and interfaces that might undergo fretting or micromotion.

To meet these requirements, an orthopedic implant designer must use the latest engineering technology. However, modern technology alone cannot substitute for the decades of clinical experience. Therefore, the recent engineering graduate, armed with the arsenal of the latest computer-aided design (CAD), FEA, and CNC technology and internet search engines, must nevertheless consider lessons learned in orthopedics over the past century.

PART V

IMAGING AND IMAGE-GUIDED TECHNIQUES

CHAPTER 13

Endoscopy

GREGORY NIGHSWONGER

KARL STORZ Endoscopy-America, Inc., El Segundo, CA

13.1 INTRODUCTION

Endoscopy has played a pivotal role in the development of modern minimally invasive surgical (MIS) methods. The term *endoscopy* refers to the use of small telescope devices to look inside the body. *Endoscope* applies generally to the optical devices (telescopes) used for endoscopic procedures, which are most often given more specific names based on their application areas, such as gastroscopes for gastrointestinal use, bronchoscopes for use in the lungs, arthroscopes for examining and treating the joints, and so on. *Laparoscopy* is a procedure in which the abdominal cavity is inspected with a rigid endoscope and has long been used in medicine to view abdominal and pelvic organs and provide necessary treatment (Fig. 13.1).

This area of development is deeply rooted in some of the earliest explorations in the practice of human medicine. Early physicians sought a way to peer inside the human body, unlocking the secrets to understanding the form and function of organs and learning how disease originates and progresses, how injuries affect the body's systems and structure and, most importantly, to offer surgical methods that promote rapid healing.

Today, considerable effort continues to be focused on discovering ways to more effectively examine the body's interior and enhance methods for treating injury and illness while offering less pain, shorter recovery times, and reduced scarring.

Endoscopic and laparoscopic technologies now provide minimally invasive solutions for a broad range of surgical specialties, from gastroenterology, obstetrics and gynecology, orthopedics and otorhinolaryngology, to thoracic and urological surgery, as well as many bariatric procedures. According to recent reports, for

Medical Devices: Surgical and Image-Guided Technologies, First Edition.
Edited by Martin Culjat, Rahul Singh, and Hua Lee.
© 2013 John Wiley & Sons, Inc. Published 2013 by John Wiley & Sons, Inc.

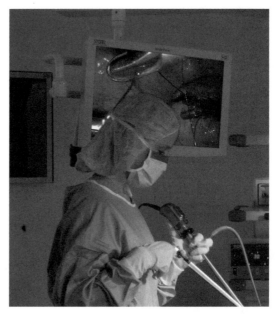

Figure 13.1 Endoscopic surgery enables surgeons to view inside the body through tiny incisions. More recently, video endoscopy systems such as the one shown allow surgeons to operate while viewing high definition images on a monitor.

example, general and pelvic endoscopic/laparoscopic surgical procedures, such as gastric bypass, endometrial ablation, laparoscopically assisted vaginal hysterectomy (LAVH), appendectomy, and prostatectomy, reached a total of more than 2.8 million in 2008 in the United States. The total US market for products marketed for general and pelvic endoscopic surgery exceeded $3.3 billion in 2008, and growth is expected to continue in the coming years, approaching $5.2 billion by 2013 (Life Science Intelligence, 2010).

Modern endoscopy represents a combination of diverse modern technologies: new materials, software, electronics, and mechanics. These technologies work in concert to offer surgeons crisper, sharper views of anatomical structures and pathologies, while enabling them to use smaller, more precise instruments with a greater degree of control.

13.2 ANCIENT ORIGINS

It has been suggested that the development of what could be considered true modern endoscopy required four hurdles to be overcome.

- Creating an opening or expanding a natural opening into the patient's body for examination and treatment
- Safely providing adequate light inside the patient's interior space

- Providing a means for transmitting a clear and magnified image back to the eye
- Expanding the surgeon's vision with higher magnification and resolution (Nezhat, 2008)

It can be argued that mankind's earliest healers and surgeons recognized the importance of being able to peer within the human body to examine and possibly treat conditions affecting their patients. Early attempts to use a patient's natural orifice to examine the internal cavities date as far back as 1600 B.C. Physicians at that time in Egypt are believed to have performed urologic procedures to evacuate a blocked bladder while minimizing trauma, as well as having performed certain gynecological procedures. Earlier records from India, some dating to 2800 B.C., describe a range of medical procedures that even include early descriptions of a rudimentary speculum used to examine the inner ear. Hippocrates (460–377 B.C.), recognized as the "Father of Medicine," described early catheters used to treat overfilled bladders, as well as a rudimentary resectoscope that was remarkably similar in principle to contemporary devices.

While early physicians made their first attempts to peer inside the body using available natural light, providing higher levels of illumination required centuries of development. A key threshold was passed in 1806 when a young obstetrician named Phillip Bozzini introduced an instrument that combined a means of examining the inside of the patient's body with a source of light. His "Lichtleiter" incorporated angled mirrors capable of projecting an image of internal organs and structures for viewing by the physician. A single candle served as a source of illumination as a double set of tubes could be inserted into the orifice that was being studied. Although there is no evidence the device was ever used on human subjects, it prompted other inventors to attempt subsequent improvements on the Lichtleiter concept.

In 1853, Antonin J. Desormeaux introduced a novel version of a cystoscope that bore little resemblance to Bozzini's device, aside from using an elongated funnel. A lamp that burned alcohol and turpentine provided from a reservoir was incorporated in the device's handle, while a concave mirror was used to reflect light into the body. Desormeaux used his device to perform the first urethrotomy and excision of a urethral papilloma, an achievement for which he is often referred to as being among the "fathers of endoscopy."

German urologist Maximilian Nitze built upon the foundations of several of his predecessors to achieve a major shift in endoscopic concepts. Nitze and Viennese instrument maker Joseph Leiter are generally credited with being the first to apply microscope optics of the time to the endoscope and, thus, enhance the physician's ability to view inside the patient's body. They eventually introduced a small version of Edison's electric light bulb to the endoscope. This pioneering work of Nitze and Leiter became the foundation of much of the subsequent progress by other researchers.

By the dawn of the new century, the principal challenges of developing endoscopic technologies were gradually being overcome, yet the field of vision offered by endoscopes remained limited, which limited their use for more complex procedures. Nevertheless, the foundation of modern endoscopy had become well established.

13.3 MODERN ENDOSCOPY

By the middle of the twentieth century, efforts to advance the development of endoscopy were focused on increasing the magnification and field of view, as well as offering more powerful sources of illumination.

In 1945, Karl Storz founded a surgical instrument company in Tuttlingen, Germany, with his principal objective being to find effective solutions to these challenges. Although he began by producing instruments for ear, nose, and throat specialists, his intention was to develop instruments that would enhance the ability of physicians and surgeons to look inside the human body. He designed and built his first endoscope as early as 1953, using a conventional optical system. And, in 1956, Storz developed the first extracorporeal flash and a system for transmitting light using a quartz rod. The result of the more effective illumination of this latter device was that surgeons were subsequently able to perform endoscopic photography with unprecedented quality. These achievements were followed by key developments that would have significant influence on modern endoscopy, including the introduction of a cold light sources and rod-lens endoscope technology.

13.3.1 Creating Cold Light

Among the challenges faced after World War II was that the technologies available at the time were quite limited. One significant impediment was that illumination needed for examining the interior of a patient's body was provided by small electric lamps. Alternatively, some attempts were being made during this period to reflect light from a source located outside the body through the endoscopic tube.

Karl Storz set out to introduce very bright but cold light directly into the body cavities through the instrument. The result would be to establish a superior source of illumination that could enhance the capabilities of physicians and surgeons through improved transmission of images.

In 1960, Storz recognized that by using a remote light source, a fiber-optic cable could be used to transmit light through an endoscope to the examination site. This discovery made possible the first cold light endoscopy (Fig. 13.2). The system no longer relied on the use of a distally mounted, heat-emitting incandescent bulb that could be easily broken. The transmission of large amounts of light into the body made possible the capture of endoscopic images of unprecedented quality and with minimal patient trauma. This key development provided the basis for significant growth in endoscopy in the following years.

13.3.2 Introduction of Rod-Lens Technology

Dr. Harold H. Hopkins had registered a patent for a rod-lens system with the British Patent Office in 1959, while a professor of physics at the University of Reading

Figure 13.2   The introduction of the cold light source in 1960 had a significant impact on endoscopy.

in the United Kingdom. Although his invention attracted considerable scientific attention, British and American companies took little notice. But a recommendation was made to Karl Storz to meet with Hopkins, an event that took place in 1965 (Reuter et al., 1999). The result of their meeting was commercial development of the rod-lens system (Fig. 13.3).

The invention of the Hopkins rod-lens system has been said to mark the most important breakthrough in optics since the development of the conventional lens system by Max Nitze in 1879. At the time, conventional lens designs entailed creating a tube of air in which a series of thin lens of glass was placed. The Hopkins rod-lens system was essentially a tube of glass in which a series of thin lenses of air was created.

According to Hopkins, the total light transmission using his new system was 80 times greater than could be achieved with previous systems. Essentially, this was achieved by replacing the air spaces between the objective lens and eyepiece lens of the telescope by using coated rod glass lenses. The superior properties of rod-lens optics allowed the diameter of the endoscope to be made smaller, yet the device delivered images with remarkable brilliance and contrast, as well as accurate representation. Combined with the Karl Storz fiber-optic light transmission technology, this represented a breakthrough in modern endoscopy, as the devices were more durable, provided excellent illumination and transmitted images that were sharp and clear (Smith et al., 2006).

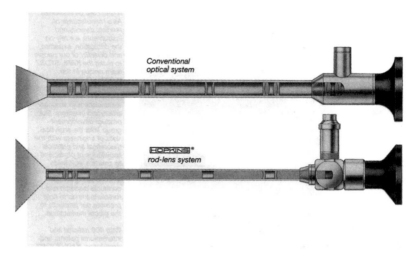

Figure 13.3 Comparison of a conventional rigid endoscope with a Hopkins rod-lens endoscope. In each example, the solid gray areas with curved lines are glass lens locations and the areas between are air spaces.

It should be noted that the development of the coherent fiber-optic bundle system used for illumination of the rod-lens endoscope also provided the basis for flexible endoscopes, which are based on the use of optical fiber bundles. The method used was to wrap a small quartz bundle of numerous fibers around a drum, then making a single cut across all fibers to leave them the identical length at each end. By thus ensuring that the individual fibers at one end were identical in position to the cut counterpart at the other end, the coherent fiber-optic bundle created could transmit images, as well as light (Fig. 13.4) (Miller and Cohen, 2008).

The typical rod-lens endoscope consists of

- the proximal end, which interfaces with either the user's eye or a video camera by using an optical coupler;
- the distal end, which is the furthest end from the user's eye or video interface;
- a light post that allows connection of a fiber-optic light guide leading to the illumination source;
- an outer tube, which houses the rod-lens system and illumination fibers, as well as any working channel to allow insertion of instruments.

In defining the characteristics of a given endoscope, there are a number of principle design concerns, including

- working length or the actual length of the outer tube that can be applied during use;
- insertion diameter, described as the external diameter of the endoscope's working length that is inserted into the patient's anatomy;

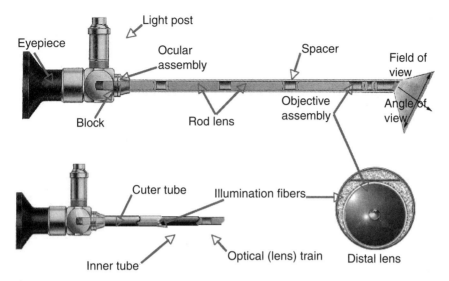

Figure 13.4 Illustration of the HOPKINS rod-lens system illustrating lens placement and illumination fibers as well as other key components.

- instrument axis;
- optical axis;
- angle of view;
- optical field of view.

While differences in functionality and design considerations may exist, similar fundamental components can be identified for most other types of scopes, that is, semirigid and flexible endoscopes, as well as video endoscopes.

Typical flexible endoscope designs provide for bundles of optical fibers used both to transmit light and images. The bundles are encased in a flexible tube, the distal portion of which can be bent at various angles, facilitating insertion into the patient for diagnostic or therapeutic procedures. The proximal end of the tube is connected to the endoscope body that incorporates the ocular and optical components and functions as a handle. The flexible tube portion of the endoscope is coated with a material selected to promote atraumatic insertion into the body while providing sufficient durability and protecting the scope's internal components from body fluids and other liquids, such as those used for sterilization. Both functional channels (used for transmitting light and images or for insertion of working instruments) are located within the tube, and mechanical systems (used for bending the endoscope and for control of air/fluid flow) are located within the endoscope body.

13.4 PRINCIPLES OF MODERN ENDOSCOPY

Like all medical technologies, the development of new generations of surgical endoscopes entails the utilization of the design and engineering methods used for most medical devices. The process is also guided in part by the same regulatory oversight and need to comply with specific standards.

Beyond this, however, modern endoscopic technologies also combine four essential and yet diverse fields of technology and engineering: optics, mechanics, electronics, and software. If complete endoscopic systems are to function as intended, the individual components representing each of these fields must work in optimum harmony.

13.4.1 Optics

A modern endoscope must generate as brilliant an image as possible of otherwise concealed body cavities. Decisive factors in this consideration are light intensity, depth of focus, magnification, contrast, and resolution.

As stated earlier, the basis of an optimal image transmission in endoscopic systems was the development and introduction of the rod-lens system by Hopkins. This achievement allowed a highly realistic image of the surface and structure of internal organs to be produced. This lens system has been subject to continual improvement and reflects a precise harmonization of all parameters for perfectly matched optics. This is exemplified in a number of further developments of the basic technology, including video endoscopes, fiber-optic endoscopes, 3D imaging systems, high magnification contact endoscopes, and others.

13.4.2 Mechanics

Today, industrial manufacturing processes generally mean mechanical series production. The exceptionally high demands placed on mechanical quality levels for medical/surgical devices, however, continue to reflect the precision and workmanship represented by the skills of the master instrument maker. The development and manufacture of endoscopy products represent a blending of highly developed functional quality, as well as ergonomic consideration of the finished products.

13.4.3 Electronics

The inherent advantages of endoscopic techniques lie not only in providing a means for examining the inside the human body for diagnostic purposes but also in being able to endoscopically offer effective treatment to patients with minimal trauma. Currently, systems such as those used for tissue disintegration, lithotripsy, high frequency surgery, insufflation, and irrigation are among the standard range of products that have extended the range of endoscopic surgical intervention.

Innovative application of state-of-the-art electronics and micromechanics enable therapeutic units to provide a maximum of safety and operational convenience while offering clinical effectiveness. Furthermore, the ability of the appliances to be networked to information systems allows an integrative systems solution to be created for superior efficiency and benefit to the patient, surgeon, and operating room (OR) personnel alike.

Therapeutic devices and their delivery systems for use inside the human body are designed using the latest computer-supported development and simulation facilities, in accordance with national and international standards and guidelines for medical products. They continue to be subjected to numerous quality assurance measures throughout the manufacturing process and undergo a 100% final inspection before delivery to the customer to ensure unsurpassed quality of electronic systems and system components.

13.4.4 Software

Generally speaking, modern electronics are designed to yield satisfying results in combination with dedicated software. Software is used to improve image quality in video systems, reduce optical defects inherent to certain parts of the system, such as the Moiré pattern in fiberscopes, and more. Software also represents an enabling device control with unsurpassed precision and intuitive operation provided by a user-friendly menu control.

In modern ORs, the use of software is critical to increasing the overall utility of contemporary endoscopy and related systems. In the endoscopic OR, peripheral devices can be integrated via defined interfaces, allowing all relevant devices to be operated and controlled from one central point. Even speech control of surgical devices and systems from within the sterile field is possible today. Complex tasks can be simplified and optimized through the straightforward use of predefined, stored settings.

Aside from the endoscope itself, design of endoscopic instrumentation must also address ergonomic and haptic considerations. Compared with open surgery, endoscopic procedures must reflect sound ergonomic principles to avoid fatigue during lengthy procedures. While the ability to feel, grasp, and manipulate internal organs with the fingers was an important aspect of open surgery, this type of feedback is now transmitted through the handles of often long metal instruments.

There are also rapid, secure capabilities for image and video documentation and transfer, as well as integration of sophisticated multimedia applications for audio and video communication, which provide the basis for transmitting surgical procedures from the OR to lecture halls or for obtaining specialist consultation, all over virtually limitless distances.

13.5 THE IMAGING CHAIN

In the context of modern endoscopic surgery, which now often includes the use of video capabilities in many specialties, the scope itself is but one link in an "imaging chain" comprising the light source, telescope, camera head, camera control unit (CCU), monitor, and image management system. With increased reliance of video endoscopy systems and the introduction of high definition (HD) technology, there are two essential qualities to virtually all imaging chains: each component is essential and the imaging chain is only as effective as its weakest component.

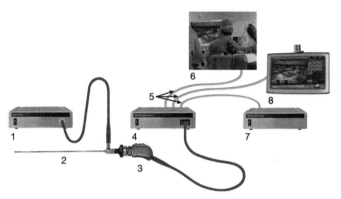

Figure 13.5   Traditional imaging chain components: (1) light source, (2) endoscope, (3) camera head and cable, (4) camera CCU, (5) cables, (6) video monitor, and (7) and (8) image management systems.

In fact, the impact of HD video capabilities on current endoscopic surgery techniques has been substantial enough that HD capability should be considered an integral part of virtually any discussion of today's video endoscopy applications. Recent studies have demonstrated, for example, that use of HD video systems during laparoscopic procedures can improve suturing and knot tying, which demand precise depth perception (Hagiike et al., 2007). Its critical influence on each aspect of the imaging chain must also be considered (Fig. 13.5).

13.5.1   Light Source (1)

Light remains the essence of endoscopic imaging. Endoscopic light sources must combine sufficient brightness and contrast, as well as the proper lighting characteristics to enable surgeons to distinguish healthy tissue from suspect tissue requiring treatment. During therapeutic procedures, the light source must adequately illuminate the operative field to allow the surgeon to clearly visualize anatomical structures and control the delicate movements of surgical instruments.

The use of HD technology places similar, if not greater, demands on light sources. To acquire crisp, clear endoscopic images, optimum illumination must be provided to the operative field. For HD-enabled systems, the cameras have lower sensitivity because of the smaller pixel size, and therefore, a powerful 300W Xenon light source is generally recommended.

13.5.2   Telescope (2)

The telescope generally captures images through a distal-mounted objective lens. The image is then relayed through a rod-lens system to a proximal-mounted ocular lens that magnifies it for the surgeon. In classic optics, optical resolution is limited by diffraction—point objects are converted into spots, and the size of those spots is determined by the optical aperture, image magnification, and optical wavelength.

Larger telescopes with lower magnification generate smaller spots, resulting in higher resolution. The telescope is a key component since the lens design and production quality are critical factors in achieving optimum resolution and optical quality throughout the entire imaging chain of either standard-definition (SD) or HD endoscopy systems.

13.5.3 Camera Head (3)

As with SD endoscopy systems, the HD camera acquires image data from the telescope. Image quality, however, will depend on the camera acquisition standard applied to a given HD system. Using HDTV technology as an example, use of the 720p HD standard with 16 : 9 aspect ratio would yield considerably lower image quality in comparison with the 1080i HD standard with 16 : 9 aspect ratio.

For HD endoscopy, 1080p is the highest standard available for acquiring and displaying images. Progressive scanning offers twice the temporal resolution when compared with interlaced scanning, making it well suited for visualization of fast moving objects without motion artifacts or blurring, as well as for capturing still images for documentation purposes.

13.5.4 Camera CCU (4)

The camera CCU is a virtual "hub" connecting various elements of the imaging chain, capturing and processing video signals from the camera head for display on the monitor and for transfer to recording and printing devices.

In addition to processing digital HD images, the camera CCU should be capable of either down-converting HD signals to SD or up-converting SD signals to HD. This allows SD printers and capture units to continue being used with the HD system. In this way, compatibility of the HD system with both SD and HD components extends the capabilities of facilities that still rely on SD components for some applications.

13.5.5 Video Cables (5)

Video Cables that carry digital image data between the camera head, camera CCU, and monitoring and recording devices must offer sufficient bandwidth for video transmission of HD data. For example, a 1080p system provides approximately 2 million pixel resolution at 60 frames per second, requiring twice the bandwidth as a 1080i system at 30 frames per second.

Optical fiber provides an optimum video cable solution for transmitting HD signals over long distances. This provides the ability to transmit other HD signals from imaging sources such as picture archiving and communication systems, or PACS.

13.5.6 Monitor (6)

Virtually all monitors used in endoscopic surgery today are thin-film transistor (TFT) liquid crystal display (LCD) type. Use of TFT technology produces a display that provides advantages of low power consumption (as little as one-third the energy required for cathode ray tube displays), light weight, and space savings. And, in the HD endoscopy environment, a wide-screen video monitor enables surgeons to experience more natural vision and operation than is possible with standard monitors. More importantly, visualization in this larger format is more in tune with human anatomy.

To gain the full benefit of HD imaging and maximize performance, the monitor resolution must be properly matched to the camera head acquisition resolution. Full HD flat-panel monitors display image data as a progressive scan. This means that a 1080i HD signal must be deinterlaced to match the monitor format, and a 720p HD signal may be up-converted for an HD monitor with 1200 vertical lines. On the other hand, a 1080p HD signal requires no conversions for a 1920×1200 HD monitor when small black bars at top and bottom are acceptable.

Although not every surgical specialty gains the same benefits from HD endoscopy, there are clear advantages for laparoscopic surgeons. With images filling a $16:9$ wide-screen monitor, laterally placed laparoscopic instruments actually appear in the field of view earlier than when viewing on a monitor with a $4:3$ aspect ratio. This generally enables surgeons to reach the operational site faster and improves identification and recognition of anatomical structures. The result is less fatigue for surgeons and a higher margin of safety during procedures.

13.5.7 Image Management Systems (7)

The ability to capture and control images is another key benefit of HD endoscopy systems. Archiving clinical images and video satisfies documentation needs while also providing a valuable resource for sharing clinical data with peers and for education and training purposes.

The use of optical fibers for HD image transmission effectively extends the reach of the image management system, allowing distribution of HD image data to a broader range of imaging sources and destinations. In addition to delivering desired video performance, the HD infrastructure should support both existing and anticipated imaging modalities.

13.6 ENDOSCOPES FOR TODAY

The application of the rod-lens technology is guided in part by the expanding range of surgical specialties in which endoscopy is being used, the needs of specific procedures, and by collaboration with key thought leaders within the medical community, among other factors. Among the considerations that are shaping new endoscope designs are flexible versus rigid or semirigid construction, outer diameter and tube length, angle of view, use of new coatings and materials, incorporation of emerging video capabilities, and use of other innovative technologies. This section describes a number of these considerations and innovative design solutions.

13.6.1 Rigid Endoscopes — Designs to Enhance Functionality

While the overall concept and design of rigid endoscopes is generally similar regardless of application, there are important differences that will vary, depending on the specialty for which the device is being developed. Thus, for many laparoscopic applications, the laparoscope may have a larger diameter, up to 10 mm or even 14 mm, and will offer a variety of viewing angles, from 0° to 70°. The 0°, called a *straight-viewing* or *forward-viewing* lens is used generally for minimally invasive abdominal operations (Fig. 13.6). Angled laparoscopes, on the other hand, enhance visualization by enabling the user to "look around" corners by rotating the laparoscope (Heniford and Mathews, 2001).

Other specialties, such as neurosurgery, require endoscopes with very different characteristics — generally smaller diameters and methods for more precise control and handling, for example. Although these rigid endoscopes have vastly different physical dimensions, they share a common design.

13.6.1.1 More Maneuverable Neuroendoscopes A surgeon's selection of a neuroendoscope (rigid, semirigid, or steerable flexible design) is based on the specific indications and clinical requirements of a given procedure. Among the factors involved are

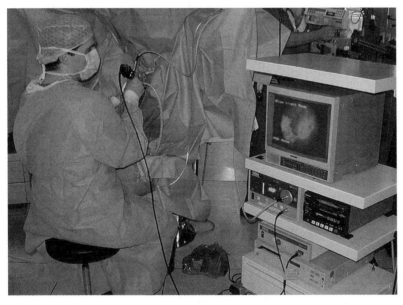

Figure 13.6 A laparoscopic surgeon operates using a rigid scope with a forward-viewing lens. With a digital camera attached, the surgeon works with a video monitor to identify cancerous tissue.

- high resolution image versus endoscope mobility;
- Anticipated surgical procedures, such as fenestration, resection, and cutting/ coagulation, for required microinstruments;
- The assist/support system, with considerable understanding of the limitation associated with each individual form of neuroendoscopic surgery (Oi, 2009).

In performing neurosurgery, surgeons benefit from having endoscopes and instruments that are small enough to permit use in all patients (including newborns and patients with small ventricles), yet allow ample illumination for good visualization of all instruments at all times. It must also be light enough to allow surgeons to hold the endoscope easily, particularly during free-hand maneuvers, and yet able to withstand handling during cleaning and sterilization.

This issue has been addressed by the design characteristics of the Oi HandyPro rigid-shaft neuroendoscope that combines high resolution imaging with bright illumination, mobile manipulation, and a lightweight body with fine surgical instruments.

The device is used with a 2 mm, 20 cm Hopkins rod-lens telescope with a $0°$ angle of view, placed in the lower two-thirds of the single lumen of an oval-shaped outer sheath. Three inlets in the upper third of the sheath can be used for irrigation/suction and operating instruments. Semiflexible microinstruments, 1.3 mm diameter, can be used, including grasping forceps, scissors, and unipolar and bipolar coagulation electrodes. While the diameters of monopolar and bipolar coagulation devices have been reduced to 1.3 mm, coagulation capacity is unchanged.

Weighing only 550 g (not including a camera head and fiber-optic light cable), the Oi HandyPro combines good image quality with a small diameter and ease of control during procedures. Use of a pistol grip allows the neuroendoscope to be held safely in the surgeon's nondominant hand, and a unified configuration enables the surgeon to maintain proper orientation during use.

Using the nondominant hand to maintain a firm grip and exercising precise handling of the device enable surgeons to guide and control semiflexible microinstruments more easily through the inlet/outlet orifice. The ability to use microinstruments with such ease is a significant benefit to surgeons (Oi, 2009).

Another advantage of the device is the ability to employ free-hand maneuvers to gain mobility during procedures. This enables surgeons to perform more than one task using a single burr hole.

13.6.2 Less Traumatic Ureterorenoscopes

One of the important goals in designing endoscopes for use in urologic applications is to minimize trauma to patients.

One approach to attaining this goal was taken in developing a ureterorenoscope that incorporates a distinctive shaft design with a barely perceptible transitional "step," from 7 to 9 Fr. This transitional step provides easier and less traumatic navigation of the device toward the operating field, helping to promote reduced operating times and more positive surgical results during procedures such

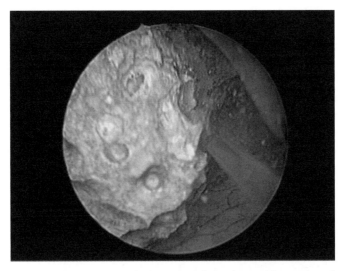

Figure 13.7 The ureterorenoscope includes both irrigation and instrument channels that allow use of laser lithotripsy to remove stones under continuous irrigation that helps maintain a clear field.

as lithotripsy. The design also provides a large working channel that allows passage of instruments and probes up to 3 Fr.

The ureterorenoscope also incorporates laterally placed instrument and irrigation ports that help simplify manipulation of auxiliary instruments, while maintaining optimal weight balance. The presence of the second channel, designed for irrigation, aids in maintaining an optimal view of the operating field during laser lithotripsy by allowing continuous irrigation (Fig. 13.7).

13.6.3 Advances in Flexible Endoscope Design

Flexible endoscopy has demonstrated particular usefulness in the practice of urology, providing advantages in kidney stone management and other applications, and in OB/GYN procedures, among other specialties. Flexible endoscopes combine three fundamental systems: an optical and illumination system that uses fiber-optic bundles and lenses at the proximal and distal ends, a deflection mechanism to aid maneuverability of the endoscope's distal tip, and a working channel passing through the length of the endoscope that allows insertion of instruments while permitting adequate irrigation of the surgical field. Designs must provide sufficient controllable flexibility to allow use in exploring the restricted confines of various anatomical structures, such as the urinary tract and kidneys, and reproductive organs. They must also exhibit durability for long-term use and subsequent repeated sterilization processes. Flexible endoscope designs can also be configured to take advantage of current video technologies, such as HD systems, to expand the range of capabilities (Fig. 13.8).

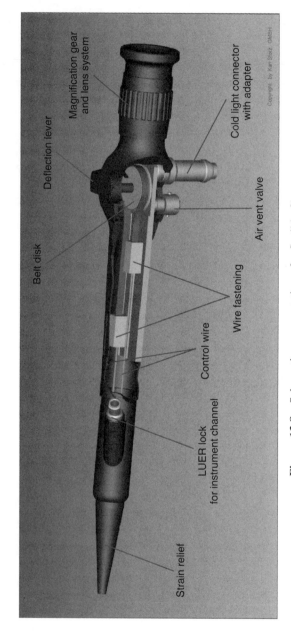

Figure 13.8 Schematic representation of a flexible fiberscope.

Magnification gear and lens system

Deflection lever

Cold light connector with adapter

Belt disk

Air vent valve

Wire fastening

Control wire

LUER lock for instrument channel

Strain relief

Copyright by Karl Storz GMBH

292

13.6.3.1 Tip Deflection Mechanisms   The flexible ureteroscope has been shown to be an essential tool in urologic endoscopy, with its ability to extend the reach of the urologist throughout the ureter and intrarenal collecting system (Bagley and Grasso, 2009). Flexibility of the endoscope allows its use in the treatment of calculi in the upper urinary tract, filling defects in the upper urinary tract, diagnostic procedures, treatment of gross unilateral hematuria, and upper tract neoplasms. Additional applications include endopyelotomy, ureterotomy, intrarenal incision, antegrade ureteroscopy, biliary tract use, pediatric ureteroscopy, and pediatric endoscopy.

Conventional flexible ureteroscopes typically feature a mechanism that enables distal portions of the shaft to be deflected up or down for enhanced maneuverability and control within the urine collection system of the kidneys. Types of deflection may be either active or passive in nature. An active deflection system is manually operated with a lever near the eyepiece that actuates control wires leading to a mechanism in the distal end that provides controlled movement of the tip in a single plane of approximately 180°. Passive deflection is provided by the flexible ureteroscope having a more flexible shaft segment near the portion that provides active deflection. Passive deflection is achieved by bending the ureteroscope tip against the renal pelvis, which moves the deflection point more proximally on the ureteroscope and provides a greater degree of deflection than active deflection alone. Gaining access to the lower pole calyx, a short cup-shaped tube in the lower portion of the kidney, generally entails using a combination of passive and active deflection to maneuver in this challenging area of the anatomy. Even when the lower calyx can be reached using this technique, use of instruments inserted through the working channel of the ureteroscope while active deflection is being used can remain difficult.

The Flex-X2 flexible ureteroscope was designed to offer a modified mechanism for active tip deflection to provide an "exaggerated deflection" (Fig. 13.9). The Flex-X2 provides continuous-controlled dual deflection of the tip, allowing for tip deflection of 270° in each direction, and the radius of deflection is broader (more exaggerated), combining aspects of both active tip deflection and passive shaft deflection. In addition to facilitating access to the lower pole of the kidney, the broader deflection also makes it easier to place and use working instruments into the lower calyx (Bagley and Grasso, 2009).

13.6.3.2 Protection from Laser Damage   The distal tip of conventional flexible ureteroscopes is often prone to inadvertent damage caused by delivery of laser energy to the interior of the working channel or from arcing of energy during laser lithotripsy. To protect the working channel, special coatings can be applied to the surfaces of ureteroscopes. The Flex-X2 has an innovative tip design that has a proprietary laser-resistant ceramic coating at the distal end of the working channel. The addition of the ceramic material protects the interior surface of the instrument's tip from thermal or electrocautery damage while allowing work to be done closer to the tissue being treated. This can reduce unnecessary downtime and minimize endoscope repairs from laser damage and promotes instrument longevity.

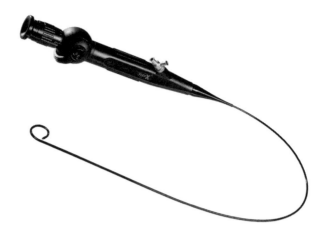

Figure 13.9   The KARL STORZ Flex-X2 flexible ureteroscope was designed to provide both functionality and durability during urologic procedures.

13.6.4   Broader Functionality with New Technologies

A range of new technologies, from device coatings that offer enhanced properties to novel ways of using existing devices and instruments to perform more advanced procedures, are shaping the design goals of new generations of endoscopes and related instruments. While some new designs are being driven by growing interest in entirely new approaches to performing surgery (such as procedures that further minimize or even eliminate surgical scars), others are utilizing technologies that are found more and more often in the homes of consumers (namely, HD video systems). Still others incorporate novel mechanisms for enhancing or extending the utility offered by new versions of proven devices, making their use more efficient, or they may be combined with other types of medical products to make them useful for new methods of diagnosing or treating disease.

13.6.4.1   Benefits of Variable Direction of View   Until recently, surgeons performing laparoscopic procedures using conventional rigid telescopes had to choose in advance as to which telescopes and which direction of view they would need to use. Although the surgeon might require telescopes with 0°, 30°, or 45° viewing angles, for example, they were limited to using only one selected viewing angle throughout surgery unless they chose to exchange telescopes intraoperatively to obtain a different viewing direction. This lengthened procedure times and required additional steps that are inconvenient to the surgeon. The solution to this challenge was to develop a laparoscope that could provide the advantages of a variable angle of view while maintaining operational characteristics of a rigid endoscope (Fig. 13.10).

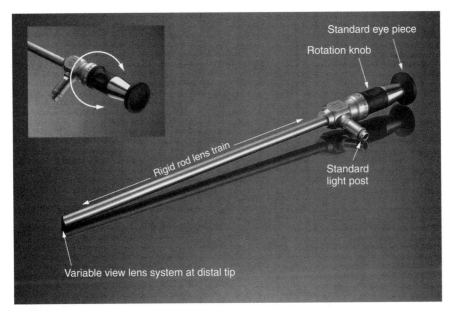

Figure 13.10   The EndoCAMeleon uses a "swing prism" component to provide a variable direction of view to surgeons.

The design of the EndoCAMeleon laparoscope combines the handling and procedural comfort previously found in a conventional 10-mm 0° HOPKINS rod-lens telescope with the advantages of a laparoscope capable of offering a variable viewing direction. The design uses a special "swing prism" component at the distal tip to enable users to quickly and easily adjust the viewing direction between 0° and 120° as required by the operative conditions and without limiting the working field. While the image orientation is maintained as in any conventional rigid telescope, the direction of view is selected by using a simple rotating wheel to select the desired angle. The device also incorporates a standard eyepiece to allow connection to any camera head, including HD video systems.

The new design has reduced the need to replace intraoperative telescopes during surgery, as surgeons are able to retain an optimal view of the entire surgical field throughout the procedure. The EndoCAMeleon achieves the goal of providing optimal visualization of regions that cannot be viewed with traditional laparoscopes without equipment changes.

Applications of this innovative and universal laparoscope include all laparoscopic and/or endoscopic procedures in surgery, urology, and gynecology. In addition, a variation on the initial design has been developed to accommodate the specific needs of more challenging interventional procedures, such as those in the field of bariatric surgery where variation of the original EndoCAMeleon has been developed with a longer telescope design.

13.6.4.2 Photodynamic Diagnosis Systems to Aid Tumor Detection

A critical difficulty in diagnosing bladder cancer is the broad variation in appearance associated with this type of tumor, which can make it difficult to distinguish cancerous tissue from healthy tissue. Similarly, during cystoscopic tumor resection using conventional white light for illumination, it can be challenging to successfully remove the tumor in its entirety while safeguarding the bordering healthy tissue. Ensuring complete removal of cancerous tissue is of course critical to minimizing the potential for tumor recurrence.

Photodynamic diagnosis (PDD) combines several technologies to aid in detection of bladder cancer when used as an adjunct to conventional cystoscopy. In PDD, a special photosensitizing drug (Cysview, developed by the Photocure ASA, a pharmaceutical company of Norway) is administered to the patient before the procedure. This drug accumulates in cancerous cells and can cause them to fluoresce with a reddish color when exposed to light in a specific blue wavelength. During cystoscopy, this fluorescence effect can aid identification of cancerous tissue for diagnostic or therapeutic purposes.

The KARL STORZ D-Light system for PDD is used as an adjunct to conventional white light cystoscopy in combination with Cysview for detection of nonmuscle invasive papillary cancer of the bladder. The PDD system includes (Fig. 13.11)

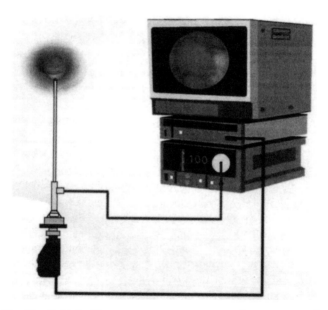

Figure 13.11 The D-Light C system encompasses several components that provide a functional PDD solution for recognizing certain bladder tumors when used as an adjunct to white light cystoscopy.

- The special D-Light C light source, which produces both white light as well as a blue light in a defined spectrum for PDD applications;
- Rigid PDD endoscopes, which differ from standard endoscopes in that they have an optimized light transmission system and a special filter fitted into the eyepiece;
- Light cables;
- A specially designed camera system, including a PDD-specific camera head and CCU for observation using a video monitor.

In its white light setting, the system functions similar to a conventional cystoscopy device. When generating blue light in its PDD setting, with Cysview having been administered to the patient and with observation using the cystoscope incorporating the special filters, tumorous areas fluoresce with a red color. This allows cancerous tissue to be more easily distinguished from healthy tissue in the bladder with a high level of contrast (Fig. 13.11). Visualization with pure white light alone does not permit such differentiation.

Special rigid cystoscopes with an optimized light transmission system have been developed specifically for use with the PDD system and are produced with $0°$, $12°$, $30°$, or $70°$ viewing angles. The instruments also have a filtering component that, under the photodynamic blue light, aid endoscopic diagnosis and allow transurethral or laser resection of tumors to be performed with an emphasis on achieving complete tumor removal while conserving healthy tissue (Fig. 13.12).

The D-Light C system used for photodynamic blue light cystoscopy is closely related to a previous system developed by KARL STORZ that, rather than being combined with a special photosensitizing drug, was based on the detection of tissue autofluorescence (AF). The D-Light AF system was used with special flexible bronchoscopes to identify and locate abnormal bronchial tissue for biopsy and histological evaluation (FDA, 2010). Future applications for PDD and AF technologies could extend to fields such as otorhinolaryngology, neurosurgery, laparoscopy, and gynecology.

13.6.4.3 *Natural Orifice and Single Portal Surgery*   In the fields of surgery and gastroenterology, research is constantly striving to create better therapy options,

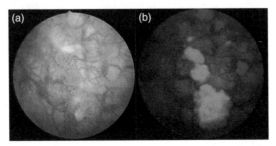

Figure 13.12 Using the D-Light C system's special filters, tumorous areas can be identified. (a) A cystoscopic view with conventional white light; (b) The same view but with PDD blue light illumination. The light-colored areas in the PDD view are fluorescing cancer cells.

so that patients can be treated even more gently and effectively. New approaches such as NOTES (natural orifice transluminal endoscopic surgery), single-port surgery, and similar techniques are focused on using either surgical entry via natural orifices (transoral, transanal, transvaginal, etc.) or only a single surgical site.

In this relatively young area of development, there are many approaches under consideration. While a pure form of NOTES has yet to become a reality, there is an array of hybrid procedures that combine transumbilical access with use of a natural body opening. These approaches can use both rigid and flexible instruments, or rigid instruments only.

NOTES-related procedures were first performed on animal models in 2004, followed by procedures on human patients in Europe and the United States in 2007. In June 2007, a series of hybrid NOTES cholecystectomies were performed that left no visible scars on the patients. These procedures entailed use of extra-long rigid laparoscopes and instruments using an approach combining transumbilical and transvaginal methods (Zornig and Mofid, 2009).

While these first NOTES cholecystectomies used what could be considered conventional laparoscopy equipment, the instruments also represented an innovative application of the existing technologies. Specifically, the devices were extended-length instruments designed originally for specialized applications, such as an extra-long, 42-cm 45° laparoscope intended originally to be used in procedures involving bariatric patients. The longer laparoscopes, for example, were designed to provide bright, clear images despite their added length.

Because of the reduced space available when performing transvaginal and/or single-port operations, instruments that differ in their length in order to avoid conflicts when manipulations and movements are performed are needed. Such designs now include a 5-mm HOPKINS endoscope with an extended length (50 cm working length) to support such techniques. Beyond this, a number of specific solutions have been developed to address the new challenges of NOTES, single-port surgery, and other approaches.

S-Portal (Single Port) components represent a new concept the encompasses a broad range of surgical techniques based on using a small, concentrated area (i.e., single site) for surgical access. During this approach, triangulation of instruments remains one of the foundations of laparoscopy with the preferred location of the laparoscope being in the vertical plane at the bisector of the angle. During S-Portal procedures, this can be achieved by using curved instruments, as well as the special extra-long telescope or the EndoCAMeleon, which reduces the risk of instrument collisions. Special flexible trocar sleeves have been developed to allow the introduction of curved instruments, even in combination with other manufacturers' single-port systems.

The X-Cone is the first device to provide reusable, stable access for transumbilical laparoscopy. Designed to offer high instrument mobility, it also ensures stable instrument guidance and an introduction technique that is more ergonomic for the surgeon. The design includes three working channels to permit the introduction of instruments up to 12.5 mm in size (e.g., clip applicator, stapler, etc.) (Fig. 13.13).

Figure 13.13 The X-Cone system with specially designed instruments provides functional solutions for innovative surgical methods that results in minimal visible scarring.

Despite its simple design, the X-Cone also provides high stability and a secure hold on the abdominal tissue during use in single-port procedures. The centerpiece of the design is the novel X-Cone seal that centrally stabilizes the telescope while still providing a full lateral range of motion for the working instruments used during the procedure. This provides advantages over fixing the instruments at a pivot point. The device's slender design makes it especially desirable for reconstructive procedures.

A minilaparotomy with a 2.5-cm incision is performed at the umbilicus, followed by digital exploration. The atraumatic X-Cone halves are inserted in a manner similar to retractors and joined together with a simple pivoting motion to form a sealing cone. The X-Cone seal is snapped on, which protects the assembly from loosening. Again, special curved instruments provide greater working space for the surgeon's hands while offering flexible use of the instruments. For example, the 5-mm telescope can be switched to one of the working channels, and instruments up to 12.5 mm in size can be inserted through the central port.

13.6.5 Enhancing Video Capabilities

The addition of video capabilities to endoscopic procedures was a milestone event that began in the 1960s with relatively crude attempts to create a practical interface between the telescope's eyepiece and the bulky video cameras available at the time.

Advances in camera technology have gradually led to more practical solutions to recording video images of surgical views. Such recordings have proven valuable not only in documenting procedures but also in making it practical to share surgical knowledge through education and training enhanced by use of surgical videos. It was many years, however, before video technology gained acceptance as a practical tool that could enhance the surgeon's capabilities when performing endoscopic procedures. Part of the challenge was to convince surgeons that it was possible to precisely manipulate endoscopic instruments while watching a video monitor rather than looking through an eyepiece. It was largely through the efforts of key opinion leaders within the surgical field and educational programs supported by a number of medical organizations throughout the 1980s taht the use of video endoscopy gradually became accepted as practical and effective MIS technique (Nezhat and Page, 2011).

An equally challenging hurdle was for video technology itself to be improved (e.g., cameras made smaller, lighter, and more sensitive). A significant advance was the introduction of three-chip cameras in 1989 to replace single-chip technology and the all-digital video platform in 2002. Single-chip cameras use one CCD (charge-coupled device) chip to sense, or capture, the primary colors (red, green and blue) in generating the endoscopic image. The three-chip camera uses a prism to allow three separate chips to sense and process each color. Three-chip technology offers certain advantages, including higher image resolution and producing endoscopic images offering more accurate color reproduction. The launch of the all-digital video platform helped to ensure that consistent image quality was maintained throughout the optical chain and provided a key step toward the development of HD endoscopic imaging solutions.

Not long after the introduction of the all-digital video platform, people started to become familiar with HD television through personal experience as consumers—perhaps in their own home. The challenge was to combine HD technology with the latest camera and imaging systems used for endoscopy to enhance the capabilities of surgeons performing minimally invasive procedures. As suggested earlier in this chapter, the addition of high resolution HD technology to key links in the imaging chain would mean greater clarity and visibility for the surgeons and enhance their capabilities.

With the addition of HD capabilities, the goal was to enhance the utility of video technology to surgeons working directly from the monitor (instead of looking directly through the endoscope eyepiece) and to offer video recordings of exceptional quality for training, documentation, and other uses. These efforts have included development of new HD camera heads that are connected to the endoscope eyepiece and video endoscopes that actually use the camera in place of an eyepiece, as well as advanced HD CCUs and wide-screen monitors to ensure that surgeons gain the benefits of high resolution HD throughout the imaging chain.

Systems were designed to offer FULL HD, defined as generating 1920 × 1080 resolution and capable of acquiring and displaying surgical images in a 16 : 9 aspect ratio to provide the widest, most natural endoscopic views possible. Progressive scanning at 60 frames per second enable surgeons to capture fast moving objects and produce sharper still images with less distortion, especially during rapid instrument movement.

13.7 ENDOSCOPY'S FUTURE

In looking forward, the course to be followed by endoscopy research in coming years is likely to be shaped largely by the capabilities being offered by digital technologies. Using HD video solutions and new generations of endoscopy equipment designed to interface easily with digital systems creates a foundation for expanded capabilities.

Capturing and storing digital information, including images as well as critical patient information and data, is already expanding the abilities of surgeons and others to benefit from telesurgery and telementoring programs, enhancing education and training, and making it easier and more efficient to share knowledge through live conferencing, podcasts, and more. At the same time, digital endoscopy solutions interface more efficiently with OR integration systems, enabling surgeons to more easily control related surgical systems from within the sterile field and to access key information and data about their patients.

In considering the future course of endoscopy, one might be tempted to reflect back on its earliest goals. Despite continued innovation and development of technology, we are continuing to pursue four familiar goals:

1. As in NOTES and associated procedures, less traumatic openings are required or better use of natural openings into a patient's body are being sought.
2. Through the developing of PDD technologies, new ways of using illumination made available inside the patient's body are being explored.
3. Through the use of HD technology and new optical mechanisms, more effective methods are being developed for transmitting clear and magnified images back to the eye.
4. Similarly, these same technologies and others are being used to expand surgeon's field of vision.

CHAPTER 14

Medical Ultrasound Devices

RAHUL SINGH and MARTIN CULJAT

Departments of Bioengineering and Surgery, University of California, Los Angeles, CA

14.1 INTRODUCTION

Ultrasound waves are pressure waves that propagate through a medium with a frequency greater than 20 kHz. The audible range for humans is 20 Hz to 20 kHz, so ultrasound falls immediately above the range of human hearing. The first major application of ultrasound came in the early twentieth century as a tool for detecting the range of icebergs and submarines, motivated by the Titanic disaster and the advent of submarine warfare in World War I. The first system, developed in 1916, was a simple hydrophone, or underwater listening device that hung over the side of a ship; this is the basis for the high tech sonar systems that naval vessels and fishing fleets use today. Ultrasonic nondestructive testing (NDT), another application of ultrasound, has been used since the late 1920s to detect defects, corrosion, and fatigue in metals and other solids, to determine when parts or structures need to be repaired or replaced. Ultrasonic NDT has been applied to various fields such as the automotive, military, petrochemical, and semiconductor industries. Structural health monitoring is closely related to the NDT field and features networks of ultrasound sensors embedding within structures for longer term monitoring.

Diagnostic medical ultrasound, or medical sonography, was first researched following World War II, using surplus naval sonar equipment (Dubose, 2000). In the mid-1950s, simple ultrasound systems were used to generate images of the eyes and the cerebrospinal fluid in the brain (Hedrick et al., 1995). Research and development work in the 1970s led to more advanced imaging systems, in which sector scanners and arrays were constructed that could display static two-dimensional grayscale images on a screen for ultrasound technicians to study. These static images were soon replaced by real-time images, and color Doppler images became commonplace in the 1980s, allowing tissues and fluids to be viewed in motion. Medical

Medical Devices: Surgical and Image-Guided Technologies, First Edition.
Edited by Martin Culjat, Rahul Singh, and Hua Lee.
© 2013 John Wiley & Sons, Inc. Published 2013 by John Wiley & Sons, Inc.

ultrasound has also been adapted for many nonimaging uses, such as monitoring, measuring vital signs, hyperthermia, therapeutic heating of tissues, and high intensity focused ultrasound (HIFU) ablation of tissues in the body. This chapter provides an introduction to ultrasound physics, ultrasound transducer design and fabrication, and applications of medical ultrasound.

14.2 BASIC PRINCIPLES OF ULTRASOUND

14.2.1 Basic Acoustic Physics

Ultrasound waves travel at a fixed speed of sound through an acoustically conductive medium, such as air or water. These waves travel with oscillating zones of high pressure (compression) and low pressure (rarefaction), with one high pressure–low pressure zone corresponding to a single oscillation, or wavelength. The number of oscillations per second is the frequency, and the frequency f is related to the wavelength λ and speed of sound c by

$$c = f\lambda \tag{14.1}$$

The speed of sound within a homogenous and isotropic medium depends on the bulk modulus β and density ρ of that medium and is described by the relation

$$c = \sqrt{\frac{\beta}{\rho}} \tag{14.2}$$

where the bulk modulus is the stress–strain ratio under isotropic compression. In general, ultrasound waves travel faster in materials that are stiffer and less compressible, and hence the speed of sound is usually highest in solids and lower in liquids and gases.

In the basic form of ultrasound imaging, pressure waves are produced and directed from an ultrasound transducer to a target that is to be imaged or detected. The pressure waves reflect off of the target, producing echoes. These echoes return and are received by the transducer, and are converted to electrical signals (Fig. 14.1). The target can be a submarine in the case of sonar, an interface within a bridge in the case of NDT, or a tissue interface in medical sonography. In pulse-echo imaging, also often referred to as A-mode imaging, this technique is used to detect the distance, or range, of objects. If the speed of sound c of the medium is known, the range (distance) of the object R can be determined by measuring the round-trip time t required for the pulse to travel from the transducer to the object and to reflect back to the transducer. In sonar, the medium is typically salt water, and in medical ultrasound, the medium may be a combination of acoustic scanning gel and soft tissues that lie in the path of the structure to be imaged. The range can be calculated using the range equation

$$R = \frac{ct}{2} \tag{14.3}$$

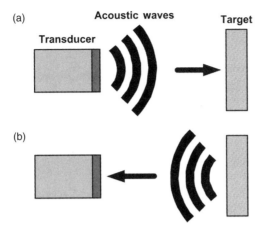

Figure 14.1   Pulse-echo imaging. (a) Propagation of ultrasound waves from transducer to target and (b) reflection from target back to transducer. Catheterization of arteries with catheter and guide wire.

where the denominator accounts for the round trip of the echo. This technique can also be used to measure the thickness or relative distance between two objects within the field of view of the transducer.

When pressure waves travel toward an object, a portion of the energy is reflected, a portion is transmitted into the object, and a portion is absorbed or scattered. The amplitude of the signal detected by the transducer is proportional to the pressure amplitude of the returning echo and is, for simple cases, related to the reflectivity of the target. The reflectivity of the target depends on the relative difference in the characteristic acoustic impedance of the target to that of the characteristic acoustic impedance of the medium. The characteristic acoustic impedance Z is the acoustic analog of the index of refraction in optics and is defined for an isotropic and homogeneous medium as the product of the density and speed of sound, or

$$Z = \rho c \tag{14.4}$$

The reflectivity is based on the fraction of the pressure wave that is reflected at an interface and can be described by the pressure reflection coefficient Γ as

$$\Gamma = \frac{Z_2 - Z_1}{Z_2 + Z_1} \tag{14.5}$$

where Z_1 and Z_2 are the acoustic impedances of the medium and the target, respectively, or more generally, the first and second materials comprising a given interface. Assuming no acoustic losses, the fraction of pressure transmitted through the interface is $1 - \Gamma$. The amount of energy reflected at the interface, known as the *intensity reflection coefficient*, is Γ^2.

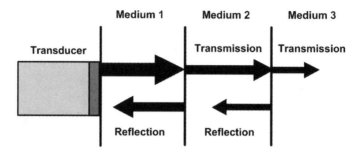

Figure 14.2   Transmission through multiple interfaces. A fraction of the acoustic energy is reflected at each interface.

Since a portion of the ultrasound energy is transmitted through an interface, more than one interface along a given line of sight can be detected (Fig. 14.2). At each interface, a fraction of the remaining energy propagates back toward the transducer, such that information about the internal structure of an object can be determined. The larger the impedance difference at each interface, the larger is the reflected echo, with air-filled organs and bone interfaces most reflective. Acoustic scanning gels must be used to couple acoustic energy between a transducer and the body in order to minimize highly reflective air gaps in the acoustic path. Table 14.1 provides various acoustic properties of materials, including acoustic impedance (Culjat et al., 2010; Onda Corporation, 2011).

Table 14.1   Acoustic Properties of Materials

Material	Speed of Sound (m/s)	Density (kg/m^3)	Attenuation (dB/cm MHz)	Acoustic Impedance (MRayl)
Air	330	1.2	—	0.0004
Water	1480	1000	0.0022	1.48
Blood	1584	1060	0.2	1.68
Bone, Cortical	3476	1975	6.9	7.38
Fat	1478	950	0.48	1.40
Muscle	1547	1050	1.09	1.62
Tendon	1670	1100	4.7	1.84
Soft tissue, Average	1561	1043	0.54	1.63
PZT	4350	7500	—	33.0
PVDF	2300	1790	—	4.2
Epoxy	2640	1080	4–8 @ 2 MHz	2.85
Silicone rubber	1050	1180	2.5 @ 0.8 MHz	1.24
Tungsten	5200	1940	—	101.0

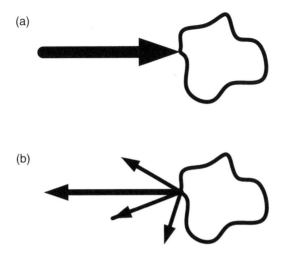

Figure 14.3 Illustration of (a) incident beam and (b) resulting diffuse reflection.

14.2.2 Reflection and Refraction

Also affecting relative strength or amplitude of the echoes is the topography of the target and the incident angle of the beam when it interacts with the target. Mirror-like (specular) targets result in near-complete reflection of the signal, whereas objects with considerable topography on the scale of the acoustic wavelength can cause diffuse reflection (Fig. 14.3), in which portions of the beam are redirected away from the transducer.

Refraction can reduce signal strength when the acoustic beam is not perpendicular to a target or the interfaces within the target. When the beam is directed off-angle towards an object, the approximate angle of transmission can be calculated using Snell's Law:

$$\frac{\sin \phi_1}{\sin \phi_2} = \frac{c_1}{c_2} \tag{14.6}$$

where ϕ_1 and ϕ_2 are the incident and transmitted angles, respectively, and c_1 and c_2 are the speed of sound within the incident and transmitted media, respectively (Fig. 14.4). Surfaces of objects, especially biological tissues, are rarely specular or parallel, so diffuse reflection and refraction are both a common occurrence with ultrasound waves in medicine.

14.2.3 Attenuation

Attenuation is the primary acoustic loss mechanism within a medium and consists of scattering and absorption. Scattering is the result of pressure waves interacting

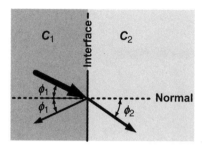

Figure 14.4 Snell's law of refraction ($c_2 > c_1$). Acoustic energy is reflected at the incident angle (ϕ_1) and refracted at an angle (ϕ_2) dictated by Snell's law.

with particles within a medium that are smaller than the acoustic wavelength and can redirect and randomly reflect portions of the beam and weaken the signal. Absorption is the process in which acoustic energy is transformed to another form of energy, such as heat. Heat generation due to absorption is primarily affected by the viscosity and the relaxation time of the medium.

Attenuation increases exponentially with distance and can be expressed as

$$A = A_0 e^{-(a_s + \alpha)z} \tag{14.7}$$

where A is the peak amplitude of the beam at a distance z into the medium, A_0 is the original amplitude, and a_s and α are the scattering and absorption coefficients, respectively. Although frequency is not included in Equation 14.7, the absorption coefficient is known to increase linearly with frequency in soft tissues. Therefore, neglecting scattering, attenuation would increase exponentially with frequency as well as distance in soft tissue. The absorption coefficient is generally related to the square of the frequency in liquids, while in solids, absorption has a negligible dependence on frequency. The effect of attenuation influences the frequency ranges used in sonar, NDT, and medical sonography.

14.2.4 Piezoelectricity

The active element in an ultrasound transducer is the piezoelectric material. In 1880, French physicists Pierre and Jacques Curie discovered that if a pressure was applied to a quartz crystal, an electric charge was created in the crystal. This phenomenon was termed the *piezoelectric effect* and is now employed in many common household products, for example, to create sparks in cigarette lighters and gas burners and to power the LEDs in children's flashing shoes. The year after the Curies' discovery, Franco-Luxembourgish physicist Gabriel Lippmann deduced that the application of an electric field to a piezoelectric material would conversely result in a deformation of the material (Fig. 14.5). This was referred to as the *reverse piezoelectric effect*.

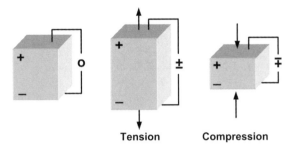

Tension   **Compression**

Figure 14.5   Reverse piezoelectric effect in a piezoelectric material.

When a sinusoidal electric input signal is provided across electrodes on either side of a piezoelectric material, the material alternates between tension and compression at the same frequency as the electric input signal. This tension and compression, typically not more than a few micrometers displacement, causes acoustic pressure waves to propagate from both ends of the material. It is this conversion of electrical energy into mechanical energy that results in the generation of acoustic waves. When a high frequency (>20 kHz) input signal is provided, the acoustic waves that are emitted are ultrasound waves.

For imaging and monitoring applications, the same piezoelectric materials are used to generate and receive ultrasound signals. When the high frequency pressure waves propagate outwards and reflect off of a target, such as a submarine in the ocean or a liver in the human body, the reflected energy can be captured by the material. This is possible because the returning waves again compress the piezoelectric material, and the resulting voltage change can be detected, processed, and displayed. Therefore, the piezoelectric material behaves both as an *actuator* when acoustic waves are emitted using the reverse piezoelectric effect and as a *sensor* when receiving waves by taking advantage of the piezoelectric effect. Because the device behaves as both a sensor and an actuator, it is termed a *transducer*, a generic term for a device that coverts one form of energy into another.

It was later discovered that these effects extend to a number of materials, including a variety of crystalline materials, ceramics, and polymers. Common, naturally occurring piezoelectric crystals include quartz, Rochelle salt, and topaz. However, man-made ceramic piezoelectric materials are more frequently used today in sonar and medical ultrasound transducers because of their superior electromechanical coupling efficiencies. Some common piezoelectric ceramics are barium titanate ($BaTiO_3$), the first piezoelectric ceramic; lithium niobate ($LiNbO_3$); and lead zirconatetitanate (PZT), which is the most widely used these days. Polyvinylidene fluoride (PVDF) is the most common piezoelectric polymer and while not as efficient as ceramics, has advantages such as the ability to be deposited as a thin film,

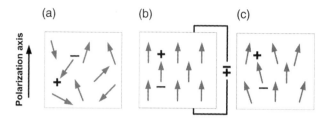

Figure 14.6 Dipole arrangement in piezoelectric material (a) before poling, (b) during poling, and (c) after removal of electric field.

a low acoustic impedance for efficient acoustic matching to water and soft tissues, and low cost. Piezoelectric materials such as PZT can also be grown as a single crystal for higher efficiency or deposited as a thin film using a sol–gel process, although with a lower efficiency than conventional PZT. Piezoelectric composites, or piezocomposites, are also used when a lower acoustic impedance is desired. The lower impedance is achieved by embedding PZT rods within a low density polymer.

Piezoelectric ceramics are made by mixing fine metal oxide powders and subsequently heating and shaping through various steps, until the powder sinters and a dense crystalline structure is formed. Electrodes are deposited atop either end of the ceramic following sintering. In order to maximize the piezoelectric efficiency of the material, the ceramic must first be polarized, or poled, at a high DC electric field and high temperature to align internal dipoles within the crystalline structure (Fig. 14.6). Each piezoelectric material has an associated poling field and temperature threshold, or Curie temperature; if the material is exposed to an electric field larger than its poling field and a temperature higher than its Curie temperature, its dipoles will orient themselves with the electric field. Once the temperature is reduced and the field is slowly removed, the dipoles relax slightly, but remain aligned. With this permanent alignment, the material will experience greater expansion on application of an AC field (to generate acoustic waves) or become more sensitive when compressed (in reception of acoustic waves). If the material is unpoled, many of the dipoles in the material effectively cancel each other out, therefore decreasing the electromechanical coupling efficiency. If a piezoelectric material is accidentally exposed to a high electric field or temperature, the material can depole.

14.2.5 Ultrasound Systems

The two essential components of any ultrasound system are the transducer and the electronic transmitter. The transmitter produces the driving electrical signal that is applied to the transducer, which converts the electrical signal into mechanical pressure waves. For imaging systems, an analog receiver is also required. In this case, the acoustical echoes are detected by either the same transducer or a second transducer and are converted to electrical signals that are then passed to the receiver.

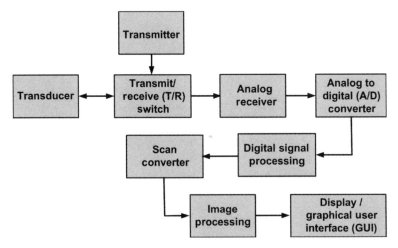

Figure 14.7   Block diagram of generic ultrasound imaging system.

The primary purpose of the analog receiver is to amplify the desired signals and filter out the noise or unwanted signals. From the analog receiver, the signals are then typically converted to a digital signal, on which sophisticated signal processing algorithms can be applied. Once digitized, the signals are then further processed to produce an image, often referred to as *scan conversion*. Image processing is then performed and finally the image is presented to the operator. In addition, the system provides controls for customizing the display of the ultrasound image and setting parameters for the ultrasound scan. A block diagram of a generic imaging system is given in Figure 14.7.

As described above, A-mode, or amplitude-mode, is the most basic form of ultrasound imaging. In this case, a short acoustical pulse is emitted from a transducer, and a series of received echoes are converted to an electrical signal by the transducer along a 1D line scan image (Fig. 14.8). The vertical axis of the line scan corresponds to the amplitude of the returning echoes, and the horizontal axis to the depth into the object. The amplitudes of the received echoes are plotted against time, or the time axis can be converted into distance using Equation 14.3, assuming a constant speed of sound. A-mode is commonly used in NDT to measure the distance of defects in solid materials and in ophthalmology to measure tissue thicknesses in the eye.

B-mode, or brightness-mode, imaging is used to facilitate the interpretation of ultrasound data and to provide a means to present it in a 2D format. In B-mode imaging, the relative intensity of each echo in a scan line is mapped to a gray scale value, where high intensity points appear as white pixels on a screen, low intensity as black, and the others appear in various shades of gray. A 2D image is then produced by collecting a series of scan lines across the imaging plane and reconstructing a B-mode image after accounting for the position and angle at which each scan line was taken. This reconstruction is performed by the scan converter.

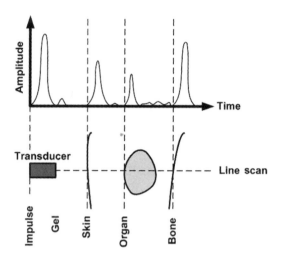

Figure 14.8   A-mode image, corresponding to a single line scan from a transducer. The four large echoes in the image (top) correspond to the electrical impulse from the transducer and ultrasound echoes from the skin surface, organ surface, and bone surface, respectively (bottom). A B-mode image can be created by taking a set of line scans by either translating the transducer across the skin surface or rotating the transducer and mapping the intensities of the line scans into a grayscale image.

The simplest way to obtain a 2D B-mode image is to linearly translate a single-element transducer across a sample, to capture scan lines at fixed intervals, and to assign each scan line as a row of pixels. It is also common to rotate a transducer through a series of angles and obtain a scan line at each angle. This produces an image with a pie-shaped field of view that corresponds to the full angle of rotation.

In medical imaging, translation or rotation of a single-element transducer across a sample is time consuming, requires precise positioning of the transducer, and is not typically done in the clinical setting. A more viable approach is to use an ultrasound probe that can automate the scanning process. Two common embodiments are sector scanning transducers, which mimic the mechanical rotation of a single-element transducer, and array transducers, which can mimic the mechanical translation of a single-element transducer. These transducers are described in more detail below.

14.2.6   Resolution and Bandwidth

Of particular interest in most imaging modalities are the spatial and temporal resolutions. Spatial resolution describes the ability of a system to detect and display two adjacent objects as separate entities, and temporal resolution is the ability to

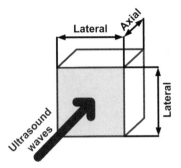

Figure 14.9   Three-dimensional voxel illustrating axial and lateral resolutions.

resolve two distinct objects in a given amount of time. Temporal resolution is limited by the speed of sound in the medium of travel and affects the frame rate that can be used in ultrasound systems.

Spatial resolution includes axial resolution and lateral resolution (Fig. 14.9). Axial resolution is the degree to which two adjacent objects can be distinguished along the axis of the ultrasound beam, and with pulsed ultrasound transducers, depends on the pulse width. The theoretical axial resolution limit of an ultrasound system, established by the Rayleigh criterion, corresponds to one half of the pulse width. Lateral resolution describes the ability to resolve two objects that are perpendicular to the ultrasound beam. Lateral resolution is primarily dependent on beam width, as a smaller beam results in a smaller lateral spot size. Beam width is strongly dependent on frequency, aperture size, and the material properties of the medium.

The axial resolution is strongly dependent on the bandwidth of the transducer. Most transducers have a resonant behavior over a range of frequencies rather than a single frequency; this range of frequencies is called the *bandwidth* (Fig. 14.10). A transducer with a narrow bandwidth tends to continue to oscillate, or ring, following an impulse excitation. Narrow band transducers are useful for continuous wave applications operating at a single frequency, as well as for therapeutic applications where the goal is to deliver a large amount of acoustic energy to a target. However, the bandwidth is also inversely proportional to the pulse width, and the extended ringing resulting from narrow band transducers effectively broadens the pulse width. Broad bandwidth, or broadband, transducers are more desirable for most imaging applications because of their narrower pulse width, which results in better axial resolution. In the reception of acoustic energy, broadband transducers are less sensitive but can more easily resolve adjacent echoes because of the decreased ringing. A bandwidth of at least 50% is typically desired for medical imaging applications.

Spatial (axial and lateral) resolution, attenuation, and the desired penetration depth all play key roles in determining the frequency ranges used in sonar, NDT, and medical sonography. Since frequency is inversely proportional to the acoustic wavelength (Eq. 14.1), and the wavelength influences the resolution, frequency

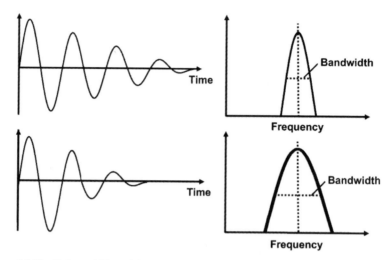

Figure 14.10 Pulse width and bandwidth. A pulse with a longer ringdown has a wider pulse width and narrower bandwidth. A pulse with a shorter ringdown has a smaller pulse width and broader bandwidth.

strongly influences the resolution. However, attenuation increases with both distance and frequency. Since sonar requires waves to travel over a long distance, lower frequency kilohertz waves must be used, resulting in a relatively poor spatial resolution. On the other hand, in medical ultrasound, megahertz frequencies may be used since the waves are only required to travel centimeters into the body, resulting in the ability to resolve much finer features. NDT generally falls between the two. Acoustic microscopy generally operates with frequencies greater than 1 MHz, sometimes above 100 MHz and, therefore, has a much shorter working distance than typical NDT.

14.2.7 Beam Characteristics

The ultrasound beam emitted from a transducer is complex and has a direct impact on the achievable spatial resolution and the imaging depth (Fig. 14.11). With ultrasound waves, and all waves in general, diffraction causes divergence of the beam. The angle of divergence or the rate at which the beam spreads depends on the size of the aperture of the source; the smaller the aperture, the faster the beam diverges. The divergence of the beam causes the acoustic intensity to decay with distance, with a decay factor related to $1/R^2$, where R is the range from the transducer face. This decay can be compensated for with time–gain amplifiers, in which the gain of the signal increases with time. Divergence can also be reduced by focusing the transducer or with beam forming in the case of an array.

Nonuniformities are also present in an ultrasound beam. In accordance with Huygens' principle, a diverging but uniform beam exists in the far field but a nonuniform, nondiverging beam exists in the near field. The near-field depth (NFD),

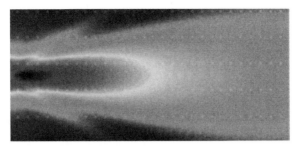

Figure 14.11 Acoustic beam profile, with diverging beam and side lobes. *Source*: Courtesy of www.biosono.com.

or the near-field/far-field boundary for a nonfocused single-element transducer is estimated to be

$$\text{NFD} = \frac{D^2}{4\lambda} \tag{14.8}$$

where D is the element diameter and λ the wavelength in the medium. In the near field, the intensity of the pressure field alternates with distance, thus making imaging in the near field challenging. In addition to variations along the axis of the beam, the beam is nonuniform at the extents of the main lobe of the beam in the near field. The nonuniformity of the beam produces side lobes, or low intensity beams adjacent to but angled away from the main beam (Fig. 14.11). These side lobes create unwanted artifacts in an image by producing false echoes.

To maintain lateral resolution, it is preferable to operate within the near field or just after the transition to the far field, before significant divergence of the beam. However, the intensity variations in the near field also create additional imaging artifacts. A broadband transducer, one that is able to produce a very short pulse in time, produces many frequencies rather than just one and can considerably reduce these variations across the intensity pattern.

The discussions so far have primarily dealt with flat, unfocused ultrasound transducers. Focusing is desired in applications such as medical sonography, acoustic microscopy, or high intensity focused ultrasound (HIFU), where improved spatial resolution or energy density is desired in a specific region of the acoustic field. In medical sonography and acoustic microscopy, a focused acoustic beam allows objects at a fixed distance to be more readily distinguished from adjacent objects. In HIFU, focusing allows ultrasound to travel harmlessly through tissues outside of the focal zone of the transducer, while heating and ablating tissues such as tumors within the focal zone (Fig. 14.12). Mechanical focusing of an ultrasound transducer can be achieved either by using a focused piezoelectric material or by using an acoustic lens (Fig. 14.13). Focusing will narrow the beam waist in the near field, with the beam diverging again past the near-field/far-field transition zone.

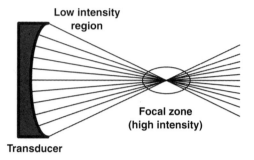

Figure 14.12 High intensity focused ultrasound (HIFU), with low intensity regions outside the focal zone.

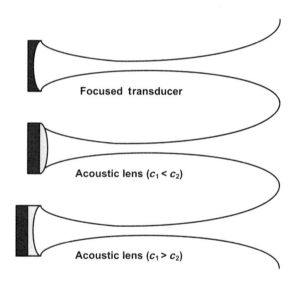

Figure 14.13 Focusing using spherically focused transducer and acoustic lenses. A convex lens is typically used in medical ultrasonography. A concave lens is used when the speed of sound of the lens (c_1) is greater than that of the medium (c_2).

14.3 ULTRASOUND TRANSDUCER DESIGN

The typical ultrasound transducer is composed of a number of layers and materials, with their dimensions, compositions, and material properties carefully selected. The most critical components are the piezoelectric material, which generates and receives the ultrasound energy; the matching layers, which maximize transmission of acoustic energy through the transducer to the body; and the backing layer, which

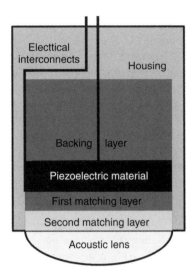

Figure 14.14 Typical medical transducer design and packaging, with imaging direction downward. Relative acoustic impedance is represented by shading, with darker corresponding to higher impedance.

mechanically damps the device. An acoustic lens is also often used to focus a transducer. Together, these layers influence the acoustic performance of the transducer. These layers are illustrated for a single-element transducer in Figure 14.14 and described below.

14.3.1 Piezoelectric Material

When designing an ultrasound transducer, the piezoelectric material properties, dimensions, and packaging arrangement must be carefully selected. The material properties dictate the electromechanical coupling efficiency, acoustic impedance, speed of sound, and dielectric properties of the piezoelectric material. The thickness and speed of sound (and therefore the wavelength λ) influence the resonant frequency; the resonant frequency is inversely proportional to the thickness, with a thickness of $\lambda/2$ corresponding to the fundamental resonance and additional resonances, or harmonics, occurring at odd multiples of $\lambda/2$ (i.e., $3\lambda/2$, $5\lambda/2$, ...). At these thickness multiples, there is maximum constructive interference of acoustic waves within the piezoelectric material, at resonance, leading to a maximum transfer of energy from the electrical to acoustic domain at that frequency. The resonant frequency is typically the most critical design parameter, as it influences the resolution, penetration depth, and various acoustic beam characteristics. The lateral dimensions of a piezoelectric material also have a considerable impact on transducer performance. The lateral dimensions can shift the resonant frequency of the device from the resonance predicted by the thickness of the material, because of modal coupling (McKeighen, 1998).

For reference, using Equation 14.1, the half-wave thickness of PZT, with a speed of sound of 4350 m/s, is 435 μm at 5 MHz. Therefore, the thickness of piezoelectric materials in medical ultrasound transducers is typically on the order of one half a millimeter; the materials are thinner for higher frequency applications such as intravascular ultrasound (IVUS) and thicker for lower frequency applications and sonar systems.

14.3.2 Backing Layers and Damping

With the application of an electrical pulse, a piezoelectric transducer will enter into free oscillation, or ringing, in the absence of damping or other loss mechanisms. This continued expansion and contraction of a piezoelectric material effectively elongates the pulse and subsequently reduces the axial resolution. Damping can be used to reduce ringing by causing resistance to the mechanical movement and converting the mechanical oscillation energy into heat energy.

A common approach to damping is to add a backing layer to reduce internal reverberations that cause the ringing, by absorbing the acoustic energy propagating backward from the piezoelectric material. The ideal backing layer should have an acoustic impedance that is close to that of the piezoelectric material to allow it absorb, rather than reflect, the backward-propagating acoustic energy. It should also be acoustically lossy so that acoustic energy is attenuated and not reflected back into the piezoelectric material. One technique is to apply epoxy loaded with fine grain particles of a high impedance material, such as tungsten or alumina. These dense materials are mixed with epoxy, increasing the acoustic impedance of the epoxy to that of PZT. The high viscosity of epoxy and large corresponding absorption coefficient causes a conversion of the backward-propagating energy into heat. The fine grains, on the order of the acoustic wavelength, can further increase the attenuation of the backing layer and scatter the acoustic energy. The result of the backing layer is a transducer with a broader bandwidth and a narrowed acoustic pulse width. However, addition of a backing layer also reduces the efficiency of the device by reducing the overall transmitted energy, so no backing layers (i.e., air backing) are used for applications requiring high acoustic power or when pulse width is not important.

14.3.3 Matching Layers

Acoustic matching layers are materials that are applied to the front (transmitting) surface of a transducer and serve as an acoustic transformer between the high acoustic impedance piezoelectric material, such as PZT, and the lower acoustic impedance target, such as soft tissue (Table 14.1). Because of its position between these two layers, it is important to select a matching layer, or a series of matching layers, that optimizes the transmission of energy through it. Matching layers should have low attenuation in order to minimize losses through the layer and, when a single layer is used, should have an acoustic impedance of

$$Z_m = \left(Z_p Z_t^2\right)^{1/3} \tag{14.9}$$

where Z_p and Z_t are the acoustic impedances of the piezoelectric and the target (tissue), and Z_m is the acoustic impedance of the matching layer (Desilets et al., 1978). Multiple matching layers are often used in medical ultrasound transducers to further increase the transmission of energy between the piezoelectric material and the body and to reduce the acoustic pulse width (Mooney and Wilson, 1994). If two matching layers are used, the desired acoustic impedances become

$$Z_{m1} = \left(Z_p^4 Z_t^3\right)^{1/3} \tag{14.10}$$

and

$$Z_{m2} = \left(Z_p Z_t^6\right)^{1/7} \tag{14.11}$$

where Z_{m1} and Z_{m2} are the first and second matching layers (McKeighen, 1989). Common matching layer materials are epoxy resins, sometimes loaded with dense metal powders such as tungsten or alumina (McKeighen, 1989). The high density of the metal powders is used to increase the acoustic impedance to the desired range, and the thinness of the layers minimizes losses because of attenuation.

Careful selection of the thickness of a matching layer is important. A matching layer should have a thickness that is an odd multiple of a quarter wavelength in that material, or

$$\Delta x = \frac{\lambda}{4}, \frac{3\lambda}{4}, \frac{5\lambda}{4}, \ldots \tag{14.12}$$

where Δx is the layer thickness and λ the acoustic wavelength within the layer. A quarter-wave thickness takes advantage of constructive interference and reduces internal reverberations within the layer, therefore reducing the pulse ringdown and subsequently reducing the pulse width. During ultrasound transducer fabrication, matching layers are carefully polished down to closely match the desired quarter-wave thickness. For an epoxy matching layer, with a speed of sound of 2640 m/s, the quarter-wave thickness at 5 MHz is 132 μm (Eq. 14.1).

14.3.4 Mechanical Focusing

As mentioned above, ultrasound transducers can be mechanically focused using either a focused piezoelectric material or an acoustic lens. Piezoelectric polymers, such as PVDF, can be deposited onto a concave substrate to create a focused ultrasound transducer with a fixed radius of curvature. Piezoelectric ceramics can also be formed with a concave spherical shape by casting the material, machining, or press-focusing the material into a spherical shape with a steel ball bearing at a high temperature (Lockwood et al., 1994).

Acoustic lenses in medical sonography have historically been made from RTV (room temperature vulcanizing) silicone rubbers because of their biocompatibility, appropriate speed of sound, and low attenuation (Yamashita et al., 2008). Over the past four decades, considerable research effort has been spent adjusting the material properties of silicone rubbers to more closely match the speed of sound, density, and acoustic impedance of the lens to soft tissue. Various formulations of silicone rubber lenses have been created by doping with inorganic or heavy metal powders (Yamashita et al., 2008). Acoustic lenses are also often made using epoxy. Because of Snell's law (Eq. 14.6), acoustic lenses are convex for a lens material with a lower speed of sound than tissue (such as silicone rubber), and concave if the speed of sound is higher than that of tissue (such as epoxy) (Fig. 14.13).

14.3.5 Electrical Matching

To maximize the energy transfer between the transmitter/receiver and the transducer, the electrical impedance of the transducer and the electronics should be matched. A poor electrical impedance match results in ringing in the electrical domain between the transducer and the electronics. As with acoustic ringing, electrical ringing is detrimental to the bandwidth and pulse width of a device. For most transducers, a matching circuit or matching network must be designed and placed in series between the transducer and the electronics, transforming the complex electrical impedance presented by the transducer to that of the electronics.

The dominant component of an ultrasound transducer structure is capacitive at the center frequency. A simple example of a matching network is an inductor and resistor in series with the transducer. The inductor value is chosen to shunt the capacitive load at resonance (i.e., the total imaginary impedance is zero), and the resistor is chosen so that the total real impedance is approximately 50 Ω, a common electrical impedance for many electronics.

14.3.6 Sector Scanners

Sector scanners were first developed in the 1970s as a means to provide real-time 2D image data. Sector scanners feature a single piezoelectric element that rotates partially within a probe and forms an image with a pie-shaped field of view, or *sector*, that corresponds to the angle of rotation (Figs 14.15 and 14.16). A single scan line is obtained at each angle by sending a pulse and receiving the reflected information from different depths at that angle, as in A-mode imaging. Scan lines are taken at each angle, and a 2D image is reconstructed by combining the scan lines together, as in B-mode operation. The number of scan lines depends on the sampling rate of the scanner.

Sector scanners contain a mechanical drive shaft that rotates the piezoelectric element near the tip of the transducer. The element is typically suspended within a coupling medium, such as oil or glycerin, to ensure uniform coupling between

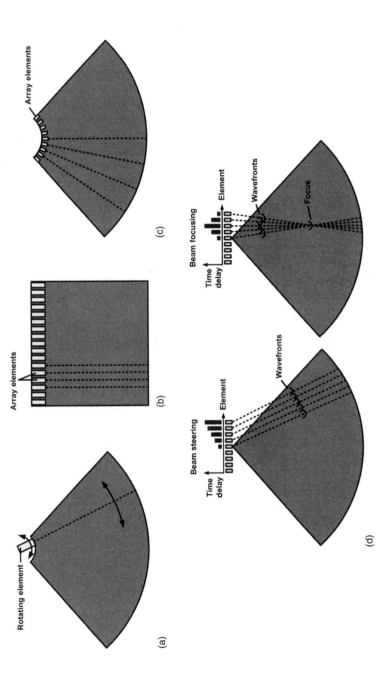

Figure 14.15 Sector image obtained from a sector scanner (a), rectangular image obtained from curved linear array (b), sector image obtained from a linear phased array (d). The sector scanner mimics the rotation of a single-element transducer, and the linear and curved linear arrays scan by sequentially selecting groups of elements. Phased arrays can perform beam steering (left) and focusing (right), by electronically delaying the electrical impulse into the elements of the array, consequently modulating the timing of the resulting ultrasonic wavefronts.

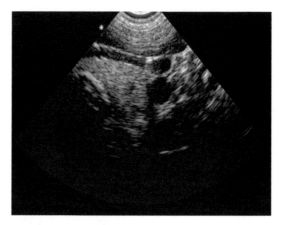

Figure 14.16 Sector image of the carotid artery, jugular vein, and thyroid gland, captured from a sector scanner.

the element and an outer casing or lens as the element rotates. A backing layer is often used, matching layers may be used, and the piezoelectric element may be spherically focused. Today, sector scanners are primarily used for applications with small working spaces, such as ophthalmology, transrectal ultrasound, and intravascular ultrasound.

14.3.7 Array Transducers

Ultrasound array transducers were developed shortly after sector scanners were commercialized and are the most common transducers used in medical sonography today. The majority of array transducers are composed of 64–256 elements, with most having either flat or convex shapes depending on the desired application. The three most common types of arrays are linear arrays, curved linear arrays, and linear phased arrays.

Linear arrays generate a rectangular shape, because of their linear arrangement of elements, and curved linear arrays generate a sector image, because of their convex arrangement of elements (Figs 14.15, 14.17, and 14.18). In their most basic form, linear arrays mimic linear translation of a single-element transducer, with individual scan lines obtained by electronically selecting individual elements along the array in a sequential pattern. However, in practice, scanning with both linear and curved linear arrays is performed by selecting groups of elements in a sequential pattern, increasing the aperture of the transducer (Fig. 14.15).

Phased arrays utilize pulse sequencing of each element in the array to electronically focus and steer a single beam using the entire array. When different elements in the array are fired with different timing patterns, the wave fronts propagating from each of the elements combine to focus the beam or steer the beam in different regions (Figs 14.15 and 14.19). By varying the timing across an array, the wave fronts from each element constructively interfere to steer the beam in the direction

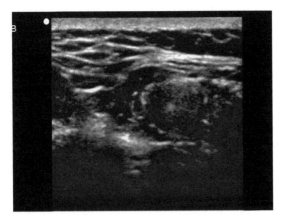

Figure 14.17   Rectangular image of the soft tissue adjacent to the ribs, obtained from a linear array probe.

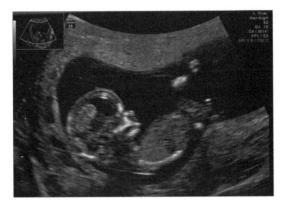

Figure 14.18   Sector image of the fetus at 12 weeks, obtained from a curved linear array probe.

opposite from the element which is first fired. If the outer elements are fired symmetrically before the inner elements, the beam can be focused, with the timing of the pattern influencing the focal depth. These principles can be combined to steer and focus the beam simultaneously.

Linear phased array transducers have a flat shape and a smaller aperture than linear and curved linear arrays. Because they have a small aperture and generate a sector image, phased arrays are ideally suited to applications such as cardiac imaging, where the beam must be fed through a small gap between the ribs to image a large heart at a considerable distance. Phased arrays are not well suited to imaging of superficial anatomy, such as superficial blood vessels, because of the small image area at close range. Phased arrays are also considerably more complex than linear arrays, both in their mechanical design and their operation.

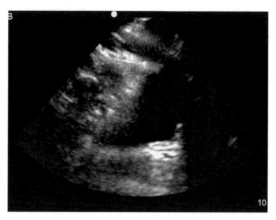

Figure 14.19 Sector image of the bladder, obtained from a linear phased array probe.

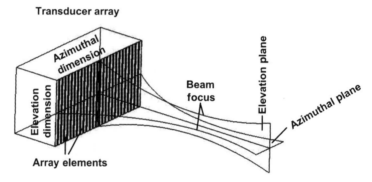

Figure 14.20 Focused beam from an ultrasound array.

In practice, many linear and curved linear arrays also have limited electronic steering and focusing capabilities. However, with all 1D arrays, electronic focusing is only possible in the azimuthal dimension, or long axis of the transducer (Fig. 14.20). Focusing of the beam in the elevation dimension, or short axis of the transducer, is fixed and is usually performed using a convex acoustic lens. The focus of the beam can be set by controlling the curvature and thickness of the lens, the aperture in the elevation dimension, and the speed of sound of the lens material (Yamashita et al., 2008).

Recently, 2D arrays and mechanically rotating linear arrays have been developed and are now available for clinical use (Fig. 14.21). 2D arrays can electronically focus and steer the beam in both the azimuthal and the elevation dimensions. Mechanically rotating linear arrays are hybrids between arrays and sector scanners, in which the entire array assembly is mechanically rotated. Therefore, array imaging is performed in one dimension, and sector scanning in the other. Both types of

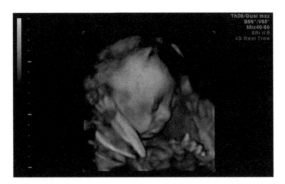

Figure 14.21 Four-dimensional ultrasound image of a fetus at 28 weeks, using a 2D array transducer.

probes provide real-time 3D imagery and are commonly referred to as *4D ultrasound* transducers, with the fourth dimension being time. Mechanically rotating arrays are well suited to imaging of static organs, while 2D arrays are preferred for imaging of organs that are in motion, such as the heart.

Another form of an ultrasound array is an annular array, which features a concentric set of circular elements. Annular arrays have been developed as alternatives to single-element transducers. These transducers can be dynamically focused, and also provide some advantages over linear arrays, but are not widely used in medical imaging because of the requirement to mechanically rotate or translate the transducer to form an image.

14.3.8 Transducer Array Fabrication

Fabrication and packaging of medical ultrasound transducers, particularly linear and phased arrays, is complex and requires high precision. The elements of the array must have equivalent compositions, dimensions, and physical orientations in order to ensure consistent sensitivity throughout the array. Each layer used in the process must be carefully polished to achieve a flatness and thickness with a tolerance smaller than a fraction of a wavelength. The combination of these layers is often referred to as the *acoustic stack* and is fundamentally the same in composition as that of a single-element transducer (Fig. 14.14), discussed earlier.

The desired center-to-center spacing, or pitch, between elements in a phased array is typically less than $\lambda/2$ in width, in order to minimize unwanted grating lobes in the ultrasound beam, where λ is the wavelength of the medium in which the transducer is operating. For linear arrays, elements should have a pitch ranging between 0.75λ and 2λ (Shung, 1996). Therefore, at 5 MHz and assuming a speed of sound in water of 1480 m/s, the pitch must be between 600 and 200 μm for a linear array and less than 150 μm for a phased array transducer (Fig. 14.22). For higher frequency applications, the pitch can decrease further and the kerf, or space between elements, becomes large compared to the element width. Dicing saws are

Figure 14.22 Piezoelectric slab diced into a 128-element linear array.

usually used to create the kerf, and it is challenging to create a kerf much smaller than 50 μm. High frequency (>20 MHz) ultrasound arrays are a topic of active research in the medical ultrasound community today (Shung et al., 2008).

A process flow illustrating the transducer array fabrication steps used commonly is given in Figure 14.23. A piezoelectric ceramics first sintered, cut, and carefully polished to the appropriate thickness so as to operate at the desired resonance frequency (1). Isolation channels are diced into the ceramic (2) to allow separate metal electrodes to be evaporated on the front (top) and rear (bottom) surfaces of the array (3). The ceramic is then poled (4), and two matching layers are added (5,6) to the front of the acoustic stack. A dual-layer flexible printed circuit (FPC) is then soldered to the signal and ground electrodes (7), with the isolation channel conveniently allowing both sides of the FPC to be bonded to the rear (nonimaging) side of the stack. A lossy backing layer is then either deposited or bonded directly to the rear side of the stack (8).

With the fundamental components in place, the acoustic stack is then diced using a dicing saw to create the elements of the array (9). The kerf is dictated by the width of the dicing saw blade. The kerf is filled with a lossy material, such as epoxy (10), to minimize acoustical cross-talk between elements. Following kerf fill, the stack may be coated with a chemically resistant film (11) that ensures that the array is not damaged from frequent exposure to disinfectants and coupling gels. Finally, an acoustic lens is cast and bonded to the front of the acoustic stack (12), or poured and cast directly onto the stack.

Following fabrication of the acoustic stack, the transducer assembly process continues by soldering the FPC to a rigid circuit board, which is then connected to microcoaxial cabling (Fig. 14.24). The cabling is connected on the opposite end to a connector housing (Fig. 14.24), which features electrical impedance matching circuitry and a pin connector with an array of pins that allows for electrical input to each element in the array from an ultrasound system.

The most labor-intensive portion of the transducer fabrication process, and the source of the largest component of the cost of the transducer, is the cabling (Brunner, 2006). The microcoaxial cables can have between 48 and 256 microcoaxial cables for linear arrays and up to 512 microcoaxial cables for newer 2D arrays (Fig. 14.25). Each of the microcoaxial wires in the cable bundle must be individually stripped, and their ground and signal wires carefully hand soldered to the

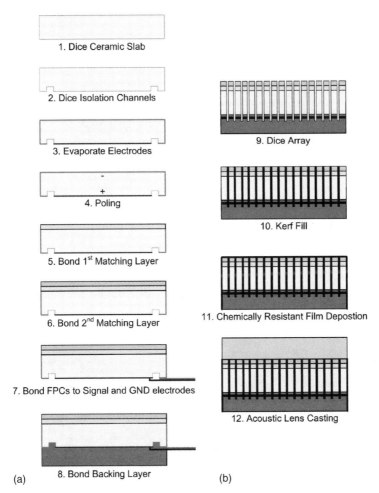

Figure 14.23 Ultrasound transducer array fabrication process flow. (a) Steps 1–8 are shown in the elevation dimension and (b) steps 9–12 are shown in the lateral dimension. The front (imaging direction) is at top of the slabs and the rear is at the bottom.

correct position on both the transducer-end and connector-end circuit boards. The cable itself is also expensive, as a result of a large quantity of small diameter microcoaxial wires that must allow for high flexibility, durability, and low loss. Owing to the high cost and complexity of the overall fabrication process, medical ultrasound transducers often cost more than $10,000, with some high end probes, such as transesophageal (TEE) probes, exceeding $45,000.

14.3.9 Regulatory Considerations

Ultrasound devices are regulated by the Food and Drug Administration (FDA) in the United States and the International Electrotechnical Commission (IEC) in the

Figure 14.24 Disassembled ultrasound transducer, showing the microcoaxial cable bundles, circuit board, flexible printed circuit connectors, and lens. The piezoelectric array, with matching layers and backing layers, are hidden beneath the lens and circuit boards.

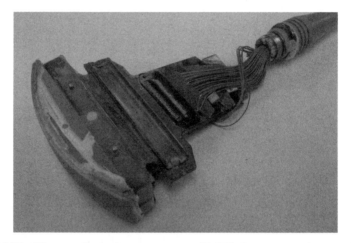

Figure 14.25 Ultrasound transducer connector with 128 pins, next to micro-coaxial cable bundle for 128 element.

European Union. In order to market an ultrasound device, it must satisfy several criteria related to acoustic output levels; thermal levels; mechanical, electrical, and electromagnetic safety; clinical measurement accuracy and system sensitivity; material biocompatibility; probe sterility; and certain software and labeling guidelines. For safe diagnostic use, the global maximum derated spatial peak temporal-average intensity (I_{SPTA}) should be less than or equal to 720 mW/cm^2, and either the global maximum mechanical index (MI) should be less than or equal to 1.9 or the global maximum derated spatial peak pulse-average intensity (I_{SPPA}) should be less than or equal to 190 W/cm^2. Most diagnostic ultrasound devices that satisfy the safety criteria can be cleared through an expedited 510(k) process, based on equivalence to preexisting diagnostic ultrasound devices. Therapeutic devices that exceed the FDA-approved acoustic output levels, such as HIFU systems, must pass a more rigorous review process. For additional detail on ultrasound device safety and acoustic testing, including intensity levels and the mechanical index, the reader is referred to the FDA diagnostic ultrasound guidance document (FDA, 2008), the American Institute of Ultrasound in Medicine (AIUM)/National Electrical Manufacturers Association (NEMA) standards documents (NEMA, 2009, 2010), and the many IEC international standards documents (www.iec.ch).

14.4 APPLICATIONS OF MEDICAL ULTRASOUND

There are a number of manufacturers of ultrasound imaging systems and transducers, including GE (USA), Philips/ATL (Netherlands), Hitachi/Aloka (Japan), Toshiba (Japan), Siemens/Acuson (Germany), Esaote (Italy), Samsung Medison (South Korea), Sonosite (USA), Mindray (China), and many others. Console-based imaging systems (Fig. 14.26) are most commonly used to control the operation of the transducers and allow the sonographer or radiologist to perform tasks such as steering and focusing of the beam, adjusting the gain and dynamic range, taking Doppler velocity measurements and color flow Doppler images, labeling points on the images, and recording patient data. Most systems include an assortment of ultrasound probes that the sonographer or radiologist can select from depending on the application (Fig. 14.27). These probes typically have between 128 and 256 linear elements and have center frequencies between 3 and 8 MHz, depending on the desired penetration and spatial resolution.

The variations in acoustic properties that occur along tissue boundaries make ultrasound imaging possible in the human body. Fluid-filled organs and organs with uniform internal structure appear the most clearly, while bones and air-filled organs reflect acoustic energy and are difficult to penetrate through because of large impedance mismatches. The bulk of medical ultrasound procedures are for scanning of internal organs, nodules, vasculature, and other tissues through the abdominal, thoracic, or pelvic regions of the body and utilize curved linear arrays. Because the scans originate from outside the body, they are known as *extracorporeal* procedures. Perhaps, the best known ultrasound application is obstetric sonography (Fig. 14.18), in which fetal development is observed through the abdomen during pregnancy. Medical transducers are also commonly used now for image guidance for a variety of procedures and have been developed more recently for interventional procedures, allowing imaging of internal structures from directly within the body. Of course, ultrasound is not limited to diagnostic uses and has found application for therapy, ablation, and monitoring. Because most discussions in the chapter have so far been focused on traditional extracorporeal probes, the remainder of this chapter is dedicated to less-traditional procedures and novel uses of ultrasound.

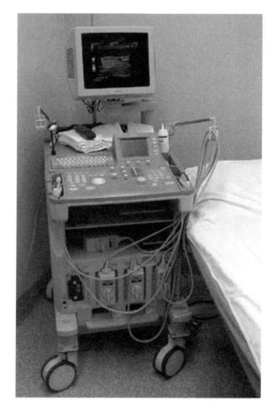

Figure 14.26 Medical ultrasonography system.

14.4.1 Image Guidance Applications

Because ultrasound is a real-time diagnostic tool that allows the operator to see through tissues, ultrasound is ideally suited as an image guidance tool. Ultrasound guidance is commonly employed during biopsies of the thyroid, liver, breast, and other tissues, with the operator using one hand to hold the probe and the other to insert the needle (Fig. 14.28). Because a metal needle has a large acoustic impedance compared to soft tissues, it is possible to view the needle as it travels through various tissue layers to a suspected mass or the desired tissue sample location. Some ultrasound probes that are used for image guidance have a needle guide attached to the center of the longitudinal axis of the transducer, allowing a physician to carefully insert the needle and monitor its depth of insertion.

Many central venous catheterization procedures also utilize ultrasound image guidance, with the operator using ultrasound to insert a needle into blood vessels in the neck, shoulder, arm, or groin and to subsequently pass a catheter deep into the vasculature or heart for administration of intravenous antibiotics, chemotherapy agents, or other drugs. Serious complications can result, the most common being

Figure 14.27   Various diagnostic medical ultrasound transducers.

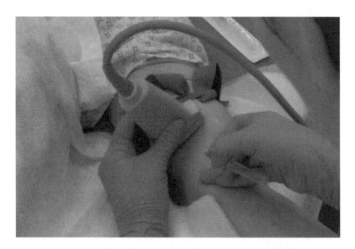

Figure 14.28   Ultrasound-guided needle inseration.

inadvertent puncture of arteries, hematoma, and pneumothorax. Ultrasound guidance has been shown to improve the success rate of peripherally inserted central catheter (PICC) procedures from 75% using the traditional antecubital anatomical method to more than 95% (Stokowski et al., 2009; Cardella et al., 1993). During these procedures, ultrasound can also be used to determine anatomical variation, vein patency, and vein course and can reduce the incidence of venous inflammation, or phlebitis, since trauma to the vessel wall can be reduced by visualization of the vein.

14.4.2 Intravascular and Intracardiac Applications

Ultrasound transducers have recently been integrated into catheters themselves, allowing for imaging of the vasculature and the internal structure and function of the chambers of the heart. The majority of these small-catheter-based transducers fall into three categories—linear phased arrays, radial phased arrays, and rotational single-element transducers.

Linear phased array transducers are integrated along the side of a catheter, near the tip (Fig. 14.29), and can obtain pie-shaped longitudinal image slices along the blood vessel or cardiac wall. These are most similar to traditional phased array probes. Radial phased arrays have an array of elements around the circumference of the catheter, also near the tip (Fig. 14.29), and obtain 2D axial image slices. Rotational single-element transducers, similar to sector scanners, feature a single piezoelectric element that is rotated 360° using a mechanical drive shaft, also providing 2D axial image slices (Fig. 14.30).

The catheters housing the ultrasound transducers are as small as 8 French (F), or 2.7 mm in diameter, severely complicating the fabrication process and putting constraints on the number of piezoelectric elements, the arrangement of the elements, and the number of wires that can be fed through the shaft of the catheter. 3D cylindrical imagery can be obtained by manually or automatically rotating linear arrays or by translating circular arrays or rotational transducers through a vessel with a constant pullback velocity.

Higher frequency operation, between 7 and 40 MHz, can be used in such catheter-based applications because of the limited scanning depth required to image vascular and cardiac walls and results in very high resolution images.

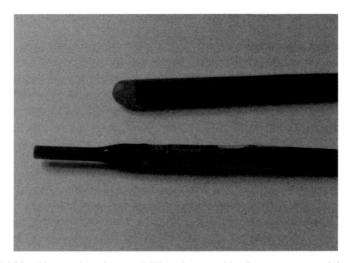

Figure 14.29 Linear phased array ICE catheter, with elements arranged in the long dimension of the catheter (top), and radial phased array IVUS catheter, with elements arranged around the circumference of the catheter (bottom).

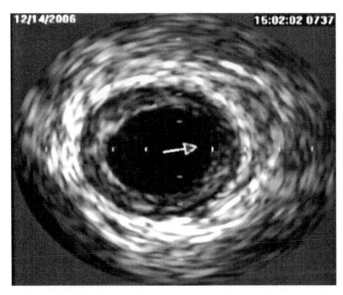

Figure 14.30   IVUS image of coronary artery, with plaque toward center of lumen. The image was obtained with a 360° rotational single-element IVUS probe.

IVUS typically refers to ultrasound imaging of the blood vessels and is used to characterize plaque buildup, stenosis, and elasticity within the vessels and to guide the placement of stents. Intracardiac ultrasound, or intracardiac echocardiography (ICE), has numerous uses, including imaging of the internal cardiac structure, evaluation of blood flow through the heart, and guidance or monitoring of interventional procedures that are performed within the heart. IVUS and ICE catheters are inserted into the vasculature and the heart using procedures similar to those used for insertion of central venous catheters, commonly using vessels in the arms, legs, or neck as access points.

14.4.3 Intraoral and Endocavity Applications

Another new method for imaging the heart is transesophageal echocardiography (TEE). In more traditional transthoracic (extracorporeal) echocardiography, a hand-held phased array probe is placed against the chest and imagery of the heart can be obtained through gaps between the ribs. TEE probes, on the other hand, have been developed recently for insertion directly through the mouth into the esophagus (Fig. 14.31). Because of the position of the esophagus directly behind the heart, TEE can be used to generate images of heart without requiring the beam to navigate around muscles, bones, and the lungs. Therefore, clear images of the heart can be generated using TEE that cannot be obtained with transthoracic echocardiography. TEE is especially useful for imaging of the rear structures of the heart such as the left atrium. TEE transducers have a partially rotating array, often 32 elements, on the tip of a long flexible cable. The position and orientation of the array and tip can

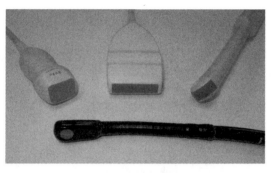

Figure 14.31 Medical ultrasound transducer, including phased array (top left), linear array (top center), curved linear endocavity array (top right), and transesophageal echo (TEE) probe (bottom).

be controlled using a mechanical wire-based system controlled by knobs, which allows the operator to bend and rotate articulations in the cable as it is inserted into the esophagus. Knobs are also used to rotate the array in different imaging planes, allowing for access to the heart in multiple orientations.

Endobrachial ultrasound (EBUS) transducers are also inserted into the mouth using a transducer that is integrated into flexible articulated cable. However, EBUS is passed into the larynx and the lungs for the diagnosis of lung cancer and infections, and for viewing of enlarged lymph nodes. EBUS probes have integrated aspiration needles to obtain tissue and fluid samples from the lungs and surrounding lymph nodes.

Endocavity transducers are rod-shaped probes that are inserted into the rectum or vagina to obtain high quality imagery of internal organs and structures (Fig. 14.31). These probes feature phased arrays or sector scanners that are used either in a forward-looking configuration to generate an image through the tip of the probe, or in a side-looking configuration along the side shaft of the probe. Transrectal endocavity probes are used to view the prostate, bladder, and seminal vesicles from directly within the rectum (Fig. 14.32). Transvaginal endocavity probes can generate images of the ovaries and uterus from within the vagina. Both can have needle attachments that allow ultrasound-guided biopsies, such as prostate or endometrial biopsies, to be performed from within body cavities. The procedures are commonly referred to as transrectal ultrasound (TRUS) and transvaginal ultrasound (TVUS). Catheter-based IVUS probes have been used for transurethral ultrasound (TUUS) imaging of the urethra and prostate, but dedicated TUUS catheters are still under development (Pavlica et al., 2003; Holmes et al., 2002).

14.4.4 Surgical Applications

Laparoscopic ultrasound (LUS) is used during minimally invasive surgery (MIS). In MIS, small keyhole-sized cuts are made in the body to allow passage of an endoscope and laparoscopic instruments that are used to manipulate, cut, and cauterize

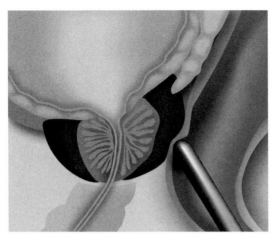

Figure 14.32   Transrectal ultrasound. The probe is inserted through the rectum to image the prostate.

tissue, and tie sutures. MIS was developed to reduce trauma to the patient, while providing for high precision procedures to be performed throughout the body, and is frequently performed in the abdominal and pelvic cavities. LUS probes can be inserted into the same ports used for insertion of laparoscopic instruments and can be placed against tissues to view their internal structures, such as blood vessels and nerve fibers. Most LUS probes have a flexible joint toward the tip and also allow for rotation of the instrument, in order to facilitate positioning of the LUS probe during the procedure. Recently, LUS probes have also been developed for use in robotic MIS. In this case, the robotic instruments can be used to hold an ultrasound probe, therefore allowing the surgeon, rather than an assistant, as is the case with LUS during robotic surgery, to manipulate the ultrasound probe.

14.4.5   Ophthalmic Ultrasound

Ultrasound transducers are available for ophthalmologic imaging, with two types of ultrasound probes available today: A-scan and B-scan transducers. A-scan ultrasound uses single-element transducers to calculate the thickness or distance between tissues in the eye in order to determine the health of the eye. The echo strength reflecting from each of the interfaces can be measured, with this technique used to measure the refractive power from the cornea to aid in the implantation of artificial lenses. A-scan probes can also be used to view and characterize masses within the eye. B-scan transducers use the sector scanning principle and are used to generate 2D grayscale images of intraocular structures of the eye, including the lens, vitreous tissue, retina, choroid, and sclera. In both A-scan and B-scan ultrasound, the probe is pressed against the open eye (or sometimes the closed eye), with acoustic scanning gel placed between the probe and the eye.

14.4.6 Doppler and Doppler Applications

For simplicity, this chapter has focused on grayscale imaging. However, most ultrasound array probes are also used for Doppler measurements, in which pulsed wave or continuous wave energy is transmitted, and the phase shift of the signal resulting from moving tissues or fluids is measured. In Doppler color flow imaging, a color overlay can be superimposed over a 2D grayscale image, showing blood flow in the heart or a blood vessel. Color indicates the direction and speed of motion, where red indicates movement away (positive frequency shift) and blue indicates movement toward (negative frequency shift) the probe. There are many types of Doppler measurements and imaging techniques, with some also used to measure blood flow velocities and their frequency spectra. Doppler is used for evaluation of blood flow through the arteries and veins throughout the body, in the fetus, in the heart, or even through the skull. Doppler measurements are often used to locate and observe narrowing or occlusion of blood vessels, which may lead to stroke or pulmonary embolism, or to assess function of the heart valves and calculate cardiac output. The Doppler shift occurs in the audible range and can, therefore, also be broadcast through speakers.

Fetal heart monitoring, which utilizes the Doppler principle, was first developed in the 1950s and commercialized in the 1960s, predating most traditional diagnostic ultrasound techniques. During labor, an ultrasound probe is either strapped onto the abdomen or placed transvaginally, and a pulsed wave signal is transmitted toward the fetus and the reflected phase shift from the beating heart is detected by the monitor. Fetal heart monitors have a number of piezoelectric elements arranged within the monitor but are considerably simpler in construction than ultrasound arrays (Fig. 14.33), as they are used for monitoring rather than imaging. Fetal heart monitors were originally developed as a tool to detect asphyxia during labor (Parer and King, 2000). They are now common in hospitals and birth centers worldwide and have been shown to reduce labor costs by allowing fewer nurses or midwives to manage a larger number patients (Kennedy et al., 2008).

14.4.7 Therapeutic Applications

The majority of the applications discussed above describe diagnostic uses of ultrasound. There are also many therapeutic applications of ultrasound, in which ultrasound is used to treat a specific disease or condition. Therapeutic ultrasound is most commonly used to heat tissues via absorption of acoustic waves. Heating can be accelerated by increasing the intensity, frequency (to increase absorption), or pulse repetition frequency, with continuous wave operation and air backing of transducers maximizing energy generation. Because tissues in the body have different absorption coefficients, it is possible to selectively heat certain tissues while leaving others unaffected. Use of focused ultrasound can further increase heating within a localized region, while minimizing heating of adjacent structures (Fig. 14.13).

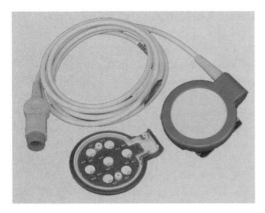

Figure 14.33 External fetal heart monitor (top) and disassembled external fetal heart monitor, with seven piezoelectric elements incorporated onto a circuit board (bottom). *Source*: Courtesy of Tenacore Holdings, Inc.

Many procedures also depend on the nonthermal effects of therapeutic ultrasound, which include cavitation and acoustic streaming. Cavitation is the expansion, contraction, and bursting of microbubbles in tissues, often causing acoustic shock waves that disrupt or damage cells. Acoustic streaming is a localized flow of liquid around the bubbles, causing shear stresses at the cell surfaces and varying the transport of ions and molecules into cells (Martinez, 2010). Both thermal and nonthermal effects are often desired, and in many cases, they cannot be distinguished (Martinez, 2010).

Therapeutic ultrasound can generally be split between low and high intensity applications. Low intensity therapeutic ultrasound is performed using both unfocused and focused transducers, most frequently for physical therapy of conditions such as osteoarthritis, carpal tunnel syndrome, and tennis elbow. Tissues are usually warmed to approximately $38°C$ for 10 min at intensities (I_{SPTA}) in the 1 W/cm^2 range (Bailey et al., 2003). Another low intensity application is hyperthermia, in which tumors are heated during radiation therapy to approximately $43°C$ for 1–2 h at roughly 10 W/cm^2 (Bailey et al., 2003). High intensity applications use focused transducers or arrays and include HIFU, shockwave lithotripsy of kidney stones, and ultrasound-enhanced drug delivery into tissues. During HIFU, tissues are typically heated to temperatures higher than $70°C$ for 1–3 s, at intensities in the 1000 W/cm^2 range (Bailey et al., 2003). Frequencies between 700 kHz and 3.3 MHz are most common in therapeutic ultrasound, and with maximum energy absorption occurring at depths between 2 and 5 cm (Martinez, 2010).

Some additional therapeutic uses of ultrasound are more vibratory in nature. These include the harmonic scalpel, for cutting and coagulation of tissues during surgery, tools used by dental hygienists for teeth cleaning, and devices for emulsification of the lens in the eye during cataract surgery.

14.5   THE FUTURE OF MEDICAL ULTRASOUND

The large diagnostic ultrasound manufacturers continue to develop high end ultrasound probes and systems with incredible resolution (Fig. 14.18). At the same time, there has been a considerable increase in the availability of lower-end ultrasound probes and systems that are now priced low enough for hospitals and clinics in developing countries. The majority of these systems have come from the profusion of ultrasound probe and system manufacturers in China. High end manufacturers now offer 4D ultrasound systems (Fig. 14.19), and further improvements in ultrasound array technology is expected with high density microfabricated arrays, such as piezoelectric micromachined ultrasound transducers (PMUTs) (Mina et al., 2007) and capacitive micromachined ultrasound transducers (CMUTs) (Jingkuang, 2010), and from improved piezoelectric materials, cabling, and packaging. High frequency (>20 MHz) ultrasound transducers are an active area of research and may enable high resolution imaging for applications such as dermatology, ophthalmology, and IVUS (Shung et al., 2008).

Among the most significant recent advances in medical sonography is the advent of portable ultrasound systems. These systems were first pioneered by Sonosite in collaboration with the US Department of Defense for military use and have since become commonplace in environments where large console-based systems are not practical because of cost or size constraints. Physicians are already benefiting from this technology in emergency settings, point-of-care offices, rural areas, and third world countries. Even smaller cell-phone-sized ultrasound systems have recently been introduced, such as the GE Vscan system. With the size and cost of ultrasound systems continually decreasing, it is conceivable that ultrasound units may ultimately be marketed for personal home use. However, ethical and safety concerns remain.

Additional applications of ultrasound continue to mature. Ultrasound elastography is a promising diagnostic technique that mimics manual palpation by imaging tissues, such as tumors and plaques based their stiffness, and has recently been introduced clinically. Application of HIFU beyond treatment of uterine fibroids has been challenging, particularly due to respiration and other patient movements. However, advanced imaging techniques have enabled its approval in other countries for ablation of cancers of the prostate, liver, kidney, pancreas, and bone. Clinical trials are now underway that are evaluating MR-guided HIFU as a tool for transcranial ablation of brain tumors (Hynynen and Clement, 2007). Low intensity focused ultrasound (LIFU) neuromodulation has recently shown promise as a tool to noninvasively suppress seizures, reduce neuropathic pain, and reversibly open the blood–brain barrier to facilitate drug delivery into the brain (Culjat et al., 2013). Fusion of ultrasound imagery with MRI and other modalities has found success in applications such as neurosurgery and prostate biopsies (Natarajan et al., 2011). Diagnostic and therapeutic ultrasound devices may soon be controlled robotically for trauma care (Yoo et al., 2010) or may be made as small as a patch that can be worn on the body (Culjat et al., 2009). Ultrasound has been a mature field for almost half a century, but advances in the fields of materials science, microfabrication, electronics, signal and image processing, and image fusion will continue to drive the field and allow clinicians to use ultrasound for purposes never imagined.

■■■■■■■ CHAPTER 15

Medical X-ray Imaging

MARK RODEN

Medical Vision Systems, LLC., Playa Del Rey, CA

15.1 INTRODUCTION

The most commonly acquired medical images in a modern hospital rely on highly energetic photons known as *X-rays*. X-rays interact with individual atoms in their path, a physical phenomenon that has been harnessed for over a century since their discovery by Röntgen in 1895 (Röntgen, 1895). As of 1999, 55% of all medical images acquired in the United States were X-ray radiographs (Maitino et al., 2003). X-ray imaging provides the physician with noninvasive exploratory techniques that readily distinguish tissues by their relative densities to X-ray absorption. Such techniques include standard two-dimensional radiographs, mammograms of breast tissue, angiograms of cardiac blood flow, and computed tomography images of entire portions of the body. This chapter focuses on the physics of X-ray image acquisition, provides a brief overview of several digital image processing techniques, and explores the clinical scenarios of stereotactic breast biopsy and blood flow visualization.

The use of X-rays as an imaging modality does have a major drawback. X-rays are ionizing radiation. These highly energetic photons can interact with atoms in tissues along the beam path, and these interactions can cause the removal of electrons from these atoms. Removal of an electron results in a charge imbalance in the molecule, transforming the molecule into an 'ion.' These ionized molecules can interfere with the normal working mechanisms of the cell, causing local cell death via apoptosis, and often cancer (Lyng et al., 2000). When the practitioners of X-ray imaging are cognizant of the risks involved in the use of X-rays and adjust their imaging needs accordingly, the benefits of avoiding exploratory surgery to determine the underlying causes of a patient's condition far outweigh the potential risks of exposure.

Medical Devices: Surgical and Image-Guided Technologies, First Edition.
Edited by Martin Culjat, Rahul Singh, and Hua Lee.
© 2013 John Wiley & Sons, Inc. Published 2013 by John Wiley & Sons, Inc.

This chapter first explains the physics underlying X-ray interactions at diagnostic imaging energies, the clinical production of X-rays, and their effects on patients. Once those basics have been covered, the formation and acquisition of 2D images using film, computed radiography (CR), and digital radiography (DR) are then covered, followed by an explanation of the clinical application of mammography and minimally invasive breast biopsy. Subsequently, a brief overview of image processing techniques is presented. Fluoroscopy and the clinical application of angiography are then discussed, and 3D techniques are briefly addressed as an extension to 2D techniques.

15.2   X-RAY PHYSICS

15.2.1   Photon Interactions with Matter

Photons are elementary physical particles. They have no mass, are entirely composed of energy and form the basic unit of light. Current quantum mechanical models models hold that the photon is both a particle and a wave, and as a wave, energy is represented by wavelength. X-ray wavelengths range from 10 to 0.01 nm. At these energies, X-rays penetrate through, are absorbed by, or scatter throughout the matter that they traverse. X-rays that do not interact with matter penetrate through that matter and can be absorbed on the opposite side, forming the basis of X-ray imaging. Those photons that do interact with matter do so through Rayleigh scattering, Compton scattering, or the photoelectric effect. High energy X-rays can also cause a phenomenon known as *pair production*, but those photons must be at higher energy ranges than that used in clinical X-ray imaging practice. This subsection is a brief introduction to a complex subject; further details can be found in textbooks such as the one by Bushberg et al. (2001).

The energy of electromagnetic radiation is given by the equation

$$E = h\upsilon = \frac{hc}{\lambda}$$

where h is Planck's constant (6.62×10^{-34} J s $= 4.13 \times 10^{-18}$ keV-s), υ is frequency, c is the speed of light (3×10^8 m/s), and λ is wavelength. X-ray wavelength is usually expressed in nanometers, and the energies of these photons are expressed in electron volts (eV). One electron volt is defined as the energy acquired by an electron as it traverses the electrical potential difference of one volt in a vacuum.

Rayleigh scattering occurs when a photon interacts with and excites an entire atom, instead of one of the atom's constituent electrons. This kind of scattering occurs mainly in low energy applications, such as those typically used in mammography (15–30 keV). The photon is essentially absorbed by the atom, which then oscillates in phase with the photon until the atom ejects another photon in a slightly different direction. Since the atom loses no electrons, this scattering is not an ionizing event.

Compton scattering occurs when a photon causes the atom to eject a photon and accounts for the majority of interactions of photons with matter in the diagnostic imaging range. This scattering results in the atom becoming ionized because of the loss of an electron. The angle of deflection is dependent on the energy of the incident photon, as more energetic photons tend to scatter electrons toward the same direction of the original photon's trajectory.

The photoelectric effect occurs when all of the photon energy is absorbed by an electron in the atom that subsequently ejects the electron as a photoelectron. The higher the energy of the incoming photon, the closer to the nucleus the ejected electron tends to be. The ejection of an inner shell electron causes a cascade effect as outer shell electrons move toward the inner shell, and this cascade releases characteristic photons. The probability of this event occurring decreases as atomic number decreases, so the effect is unlikely in soft tissue imaging.

The final result of absorption and scattering of X-rays as they pass through soft tissue is an attenuation of the original beam of photons. This attenuation is linear and follows the equation

$$n = \mu N \, \Delta thick$$

where N is the original number of photons in the beam, n is the number of photons removed from the beam, η is the linear attenuation coefficient of the traversed material (typically expressed in cm^{-1}), and $\Delta thick$ is the distance traveled through a material. However, as thickness increases, the attenuation is not linear. This equation can be expressed for large thicknesses of matter through calculus; as Δpos becomes infinitesimally small, the equation for a monoenergetic beam becomes

$$N = N_0 e^{-\mu thick}$$

where N is the number of photons leaving the material, N_0 is the number of photons entering the material, and *thick* is the thickness of the material.

Thus, X-rays passing through soft tissue are attenuated exponentially as they pass through, and the attenuation is dependent on the linear attenuation coefficient of the material. Bone will, therefore, allow fewer electrons to pass through than water, and water will allow fewer electrons through than air. This phenomenon forms the basis for differentiating tissues in an X-ray image.

15.2.2 Clinical Production of X-rays

X-ray are produced in the clinic via an X-ray tube, a simplified version of which appears in Figure 15.1, and a clinical installation appears in Figure 15.2. The tube itself consists of an anode, a cathode, and a window through which the produced photons can escape toward the imaging device. Electrons bridge the gap from the cathode to the anode to complete a circuit in the tube. Once these electrons arrive at the anode, the molecules in the anode temporarily absorb the electron and the added energy of the electron to transition from the ground state to an excited state. When

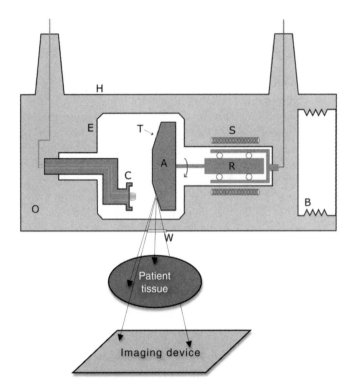

Figure 15.1   X-ray tube. This figure contains the basic schematic of an X-ray tube with a rotating anode (A). A circuit is completed when electrons are produced (or boiled off, colloquially) at the cathode (C) and then accelerate toward the target (T) on the anode (A). Induction stators (S) cause the rotor (R) to rotate the anode so that the anode does not overheat during this process. Both the cathode and anode are suspended in a vacuum envelope (E) to ensure that air particles do not interfere. The envelope itself is suspended in oil (O) to further facilitate cooling, with a Bellows system (B) to account for contractions and expansions of the oil. The entire apparatus is contained within a housing (H), and X-rays produced at the target escape the housing through the window (W). These X-rays then proceed toward the patient tissue, either to interact with that tissue or to pass through to interact with the detector.

these molecules relax from the excited state back to the ground state, the excess energy is released in the form of a photon. If this photon has a wavelength between 10 and 0.01 nm then the photon is termed an *X-ray*. X-rays are produced in every direction from the anode, but most are absorbed by the body of the tube. Those X-rays that strike the window of the tube can escape this apparatus to proceed toward the patient and the imaging detector. Some photons interact with the patient, and those photons that do not may also interact with the detector. The differences in the variations of the photons that are absorbed by the detector represent the differences in tissue density in the patient.

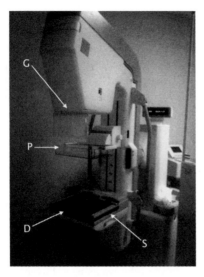

Figure 15.2 An X-ray generator and imaging device. This image shows a typical mammography X-ray acquisition station. The generator G produces X-rays that proceed through the patient and the imaging device D. When acquiring an image, the patient places the breast on D. This particular station has a compression paddle P to minimize the effects of X-ray scatter through the patient tissue by compressing the patient's tissue by 25–40 psi, thereby reducing $\Delta thick$. The apparatus can also rotate to acquire images from different directions, as discussed in Section 3.4. In this example, D can be a computed radiography (CR) or film cassette; direct radiography (DR) systems do not have the cassette slot S. As discussed in Section 3.1, D will also contain a *Bucky* placed between D and G that becomes active once the generator is acquiring an image. *Source*: Image courtesy of iCRco, Inc.

Large amounts of heat are generated through this process, so some form of cooling must be employed to prevent the anode from melting. The rotating tube in Figure 15.1 uses two forms of cooling. First, the anode itself rotates, so that no one region on the anode receives the full force of the electrons. Thus, the target region is actually a strip of material on the anode, rather than a single point. Second, vacuum envelope containing the anode and cathode is suspended in oil, so that the envelope can be continually cooled.

Many variations of this basic setup are employed depending on the needs of the imaging device. One alteration of the tube is the ability to change the cathode size. This alteration in turn affects the target spot size on the anode; a larger spot size can produce more photons but at the expense of losing spatial resolution at the detector. Another alteration is to change the cathode and anode materials. While these materials are generally molybdenum or tungsten, each has their own characteristics that will affect the nature of the produced X-ray beam. The specifics of these modifications are beyond the scope of this chapter; interested readers are again referred to the book by Bushberg et al. (2001).

The strength of the X-ray beam is determined through three factors: the peak voltage between the cathode and anode, the amperage of the circuit, and the total amount of time the tube is active. The first number is typically measured in kilovolts and is shortened to *kVp* or *peak kilo voltage*. A higher kVp produces photons with a higher potential energy volt. Very few of the produced photons will be at the peak; these photons will generally fall below the kVp of the tube. The X-ray technician can place pieces of metal, such as copper, in the beam path to harden the beam, causing lower energy photons to be absorbed in the metal. This effect is desirable for imaging studies that require higher energy photons, such as bone density studies and studies in which tissue contrast is not critical.

The amperage and time are usually combined as a single factor known as the *mAs*, a combination of milliampere and millisecond. Both the current strength and the time contribute to the number of photons produced; higher amperages and longer times will produce more photons. For example, when imaging a static patient or a nonmoving body part such as a hand, longer time and lower amperage can be used. That way, the number of photons that penetrate the tissue will be sufficient but will place less strain on the tube by using a lower current. Conversely, when imaging the chest or other moving body part, a higher amperage and lower time may be chosen so that respiration or other movements do not blur the final image. Typically, however, these numbers are combined into the single mAs factor.

The amount of radiation that the patient receives is called the *dose*. The SI unit of absorbed dose is either the Sievert (Sv) or the Gray (Gy). Both are equivalent to 1 J/kg of energy; however, Sieverts attempt to capture the level of danger of a particular dose by assigning a weighting factor to the different kinds of radiation. For instance, a photon or an electron will have a weighting factor of 1, so 1 Gy of absorbed energy will be the equivalent of 1 Sv. In the case of heavy ions or fission materials, 1 Gy of absorbed energy will be the equivalent of 20 Sv, as those materials are considered to be more dangerous.

15.2.3 Patient Dose Considerations

As discussed, X-rays are ionizing radiation, and this form of radiation is damaging to organic tissue. The damage falls into two basic categories: deterministic and nondeterministic. Deterministic in this context means that the amount of radiation received can be considered a direct cause of the resulting injury, even though the injury manifests in days to months. Skin receiving a large dose of radiation, for instance, can be permanently burned as the lower epithelial cells are damaged beyond repair and new skin cells cannot be formed. This effect was observed when surgeons used continuous X-ray exposures via fluoroscopy to remove a sewing needle from a patient's hand in 1978; the patient's hand was subsequently burned, and further bone necrosis prevented the patient from recovering full use of her hand (Lyng, 1978).

The deterministic damage effect is used in the treatment of cancer by directing large radiation pulses (on the order of 60–70 MeV) to cancerous tissue in order to permanently damage the cancer and prevent its further spread. Such applications of radiation must be carefully controlled, however, so that neighboring noncancerous tissue is not affected by the beam. Cancer treatment in this manner is termed 'radiation oncology;' interested readers are referred to Khan et al, 2012.

Long-term exposure to lower doses of radiation are considered to be dangerous as well, although those effects are more difficult to ascertain. These effects are called *nondeterministic*, in that there has been no direct causal link between the exposure and any deleterious effects later suffered by the patient. A patient who develops skin cancer may have done so after a series of X-ray images were acquired years in the past or may have done so after several sunburns. As previously discussed, the effects of each photon as it passes through the tissue are stochastic, and their ultimate effect is not known. The studies of the effects of radiation exposure are more statistically significant when a large number of people are exposed to the same amount of radiation at the same time. Thus, the atomic bombing of Hiroshima and Nagasaki at the end of World War II and the accident at Chernobyl offer distinct indications of the long-term effects of radiation exposure (Sauvaget et al., 2003; Moore et al., 1997; USNRC, 2011).

Radiation workers in the United States may not be exposed to more than 50 mSv of dose in any given 1 year time span (USNRC, 2011), which is half of the 100 mSv clearly linked to increased cancer risk (Brenner et al., 2003). Each radiation worker carries a badge that should absorb the same amount of radiation as the worker; thus, the dose for a worker can be tracked to ensure that overexposures are very limited.

Another effect that may have long-term effect on a patient's health is the use of dyes that absorb X-rays in the diagnostic range. Such materials and dyes are termed *radiopaque* for their absorptive properties. The linear absorption coefficient of most tissue is close to water, so distinguishing between different tissues is difficult, unless one tissue has been filled with or otherwise marked with such a dye. These contrast agents can be nephrotoxic (Andrew and Berg, 2004) and so must be used carefully. However, as shown in the X-ray image of the small intestine in Figure 15.3, these dyes can be very effective at providing contrast where the distinctions between tissues are otherwise indiscernible. Typical contrast materials are iodine or barium compounds.

The potentially adverse effects of using ionizing radiation as a means of acquiring an image combine to produce the various trade-offs that must be confronted to acquire a diagnostically useful image. Efforts to minimize such trade-offs have a great impact on patient well-being and constitute an area of active research.

In practice, each of the image acquisition devices described in this chapter must be calibrated to specifications described either by professional organizations, such as the American College of Radiology, or by law, depending on the location of the device. Such calibrations include ensuring that distances from the source to the detector are known and constant to ensuring that the X-ray tube output falls within expected values. These measurements are performed using a *phantom*, an object of known dimensions and characteristics. Phantoms range from ensuring that the system can produce images where small details are detectable to objects that describe image distortion. Each specific devices requires specific phantoms for calibration

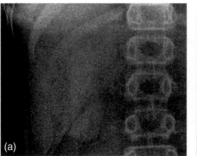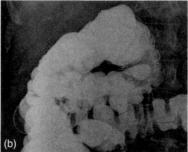

(a)   (b)

Figure 15.3 Contrast agents. The patient in (a) has received no contrast. The patient in (b) has received a barium contrast agent to more clearly delineate the intestines. *Source:* Images courtesy of iCRco, Inc.

purposes. In addition, image processing algorithms also employ phantom images to allow for ready comparison with other systems. These phantoms ensure that acquisition systems are performing within expected parameters. Deviations from these norms are generally considered cause to service or replace any portion of the imaging chain to preserve image quality and lessen patient dose.

15.3 TWO-DIMENSIONAL IMAGE ACQUISITION

Two-dimensional medical images can be taken using analog or digital devices. An *analog device* does neither use any form of discrete codification of signal nor have a numerical representation of the acquired data. Film images are acquired using analog techniques; these images use technologies that do not involve computers, but instead, film emulsion and chemicals produce the image. On the other end of the spectrum is a *digital device*, one that uses discrete spatial and temporal separation of signal and records that signal as a number into a computer. Some devices use a combination of the two technologies, keeping photon acquisition in the analog realm and then converting that stored analog signal to a digital result.

The fundamental unit of a two-dimensional digital image is the *picture element* or *pixel*. Typically, pixels are square, meaning that the horizontal and vertical dimensions of the pixel are identical. The size of the pixel, also referred to as the *pixel pitch*, is usually represented in units of square area, such as square micrometers (colloquially known as *microns*) or square centimeters. The specifics of the formation of each pixel will vary from device to device: analog film acquisition systems have no discrete pixels but instead have a variable grain size of the film crystals; CR systems, which use phosphor plates as the image acquisition device and a laser to retrieve stored information, have varying pixel pitches because of variations in laser spot size; DR systems, which use X-ray sensitive detectors to record the number of incident photons directly into a computer, have fixed pixel pitches determined at the time of manufacturing. The concepts behind each form of imaging are briefly explained in this chapter; interested readers can refer to the books by Bushberg et al. (2001), Knoll (2010), and Bick and Deikmann (2010) for details specific to each device.

Photon acquisition devices rely on one bedrock principle in order to produce an image: the *quantum efficiency* (QE) of the image acquisition device in Figure 15.1, colloquially known as the *detector*. The QE is the chance that the acquisition device will record an incident photon, meaning that the photon will collide with the device. For instance, a device with a QE of 1% will only convert 1 in 100 incoming photons into measurable signal, while a device with a QE of 10% will convert 10 times the number of photons and so require one-tenth the amount of signal per unit area to achieve a meaningful distinction between signal and noise. QE will vary with the configuration of the acquisition device, the material of the device itself, and the wavelength of the incoming photon; highly energetic photons will not interact with some materials.

Spatial resolution represents the ability of the recording device to capture detail in the recorded object. Spatial resolution is typically represented as *line pairs per millimeter* (lp/mm). A "line pair" is a dark linear region neighboring a white linear region; high resolution devices may have 7–10 lp/mm. The best resolution that a recording device can obtain is half of the number of pixels per unit distance. If the pixels in the device are 50 μm × 50 μm in area then the device has 20 pixel per millimeter and can distinguish at most 10 lp/mm. Even so, most devices rarely reach that level of spatial resolution, and instead, the difference in intensity at a variety of line pair measurements (for instance, at 0.5 lp/mm, 1 lp/mm, 2 lp/mm, etc.) is measured and combined into a device-specific metric known as the *modulation transfer function* (MTF). Typical MTFs for CR systems using various acquisition resolutions are shown in Figure 15.4.

Dynamic range represents the number of photons that can be stored in a unit area on the detector. Each system—film, CR, and DR—stores these photons in different ways and never as actual photons, but through some controllable secondary medium, such as electrons, and the conversion to that secondary medium is part of the QE. For the sake of explaining dynamic range, however, the received signal shall be called *stored photons*. Dynamic range of an image in a digital system is determined by the ratio of the maximum number of pixels that can be recorded in a unit area divided by the number of stored photons necessary to represent accurately a change in intensity over fluctuations because of noise in the detector. As the number of photons stored per unit area is generally considered to be uniform for a particular absorbing material, the larger the individual pixel pitch, the more dynamic range can be obtained from a single pixel. For example, consider a material that can store 1000 photons per square micrometer and an analog to digital converter that requires 10 photons to represent a measurable change in intensity. A detector made of this material with 6 μm × 6 μm pixels, comprising a surface area of 36 μm^2, would therefore have the potential to distinguish 6 × 6 × 1000/10 = 3600 different intensity levels.

One consequence of having a large dynamic range is that enough photons can be obtained to easily distinguish signal from noise. In general, the noise in a measurement will be proportional to the square root of the signal. If the incoming

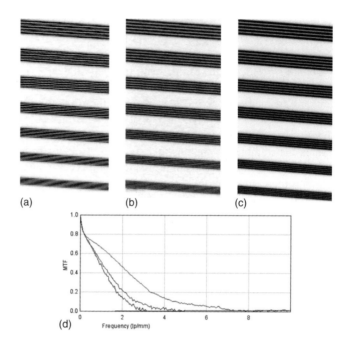

Figure 15.4 Modulation transfer functions. (a), (b), and (c) all depict the same line pair phantom imaged at 200, 100, and 50 μm pixel edge size, respectively. (a) and (b) have been enlarged to match (c). A line pair phantom is made of an alternating pattern of radiopaque material, and the seven sets of line pairs here depict resolutions of 2, 2.24, 2.5, 3.15, and 4 lp/mm, from the top of each image to the bottom. The chart in (d) shows the relative intensity difference between the pixel values in each set of line pairs for 200 μm (bottom line), 100 μm (middle line), and 50 μm (top line). The higher the MTF value, the more discernible are the differences between pixels at a given resolution. The chart states, and the images reflect, that at a pixel edge size of 200 μm, resolution begins to deteriorate at 2 lp/mm, but a 50-μm image has detail visible past 4 lp/mm. *Source*: Image courtesy of iCRco, Inc.

signal is composed of 100 photons then there is an expected range of measurements from 90 to 110 photons when the measurement is completed, or 10% of the signal intensity. If the incoming signal is composed of 10,000 photons, the square root of the signal is 100, or 1% of the signal range. Therefore, the greater the number of incoming photons, the better the signal-to-noise ratio (SNR), assuming that the measuring device can acquire all the photons from the incoming signal. In the previous example of a device with 36 μm^2 per pixel, if more than 36,000 photons are acquired in a single exposure, excess photons will not be recorded. The information that can be gleaned from such a pixel is that more photons than the device can measure have struck the detector at that location, but not how many more. Such pixels are referred to as *overexposed*, or, more colloquially, *pegged* or *blown*. Exposures of such an excessive amount of dose would be bad for the patient in two respects: first, because the patient will be exposed to that much radiation, and second, because the doctor will glean no useful information from the exposure and will likely require a second acquisition to obtain the information missing from the first.

The design of imaging devices strikes a balance between dynamic range and spatial resolution within the confines of QE. Devices with larger pixel pitches than counterparts made of the same material will have less spatial resolution but, in general, increased dynamic range. A larger pixel pitch will also increase the chance per unit area that a photon will be absorbed; a pixel with double the surface area of another pixel will also have double the absorbing material and so double the chances of an incident photon collision. The requirements of medical imaging devices vary according to clinical use. Increased spatial resolution is necessary in the case of mammograms as compared to chest X-rays; the clinical presentation of breast cancer mandates the use of a higher resolution imaging device to find early stages of the disease, while determining whether the patient has a collapsed lung will require far less spatial resolution. As such, patient dose for a mammogram is often orders of magnitude more than for a chest X-ray, simply because more radiation dose must be given to compensate for the decrease in relative sensitivity of the device because of very small pixel pitches.

15.4   IMAGE ACQUISITION TECHNOLOGIES AND TECHNIQUES

Three forms of 2D X-ray image acquisition are discussed in this chapter. The first, film, is an entirely analog approach, using chemicals and emulsions to produce an image. CR uses a phosphor plate to store an image and a laser to translate the image into a digital signal. DR uses a pixel grid of an X-ray sensitive material to absorb X-rays and produce a digital image.

15.4.1   Film

The state of the art of X-ray imaging is moving away from film as a means of acquiring images. Film is not easily copied to remote destinations, can be developed only once, and provides limited dynamic range in comparison to digital detectors. In terms of surgical applications, film can be used for positioning information and preprocedure planning, but compared to other two-dimensional digital X-ray imaging modalities, film is limited in both dynamic range and development time. Despite these limitations, film was the most advanced form of X-ray imaging until the advent of digital alternatives in the 1980s and 1990s, and many technologies were developed for film imaging that have been adopted by CR and DR. Essentially, each of these technologies is a separate solution to the "image acquisition device" shown in Figure 15.1, and later technologies reuse ideas from film.

The basics of film recording X-rays are shown in Figure 15.5. Two distinct problems have led to the configuration depicted in that figure, scatter and the

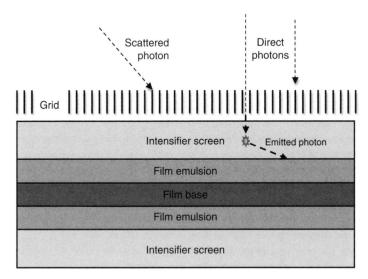

Figure 15.5 X-ray photons recorded in film. This figure describes the contents of Figure 15.2, parts D and S. The detector portion is composed of the grid, while the screens, emulsions, and base are part of the cassette and the film itself. Incoming photons that have been scattered by the molecules in the patient are stopped by the grid, while direct photons can proceed past the grid to interact with the intensifying screens. These screens then release photons that then subsequently interact with the film emulsion, producing an image. Some of these incoming direct photons are also stopped by the grid, requiring that the grid move during image acquisition to minimize this effect.

thickness of film as a recording medium. At medical diagnostic energies, a nontrivial percentage of incoming photons will have been scattered through previously described electromagnetic effects. These scattered photons would lower the contrast of the image, and since the effects of scatter are linearly proportional to the amount of dose received, increasing the patient dose would not improve the contrast of the image. Thus, an interstitial layer of a material that is relatively opaque to X-rays, such as lead, is introduced between the patient and the film. This layer, known as a *grid*, is composed of several thin strips oriented such that they are parallel to the incoming beam. The grid will attenuate any photons that are not oriented along the beam path. This arrangement ensures that only direct photons will be recorded.

The slats in the grid are not infinitesimally thin, however, and they will interfere with the final image, producing a moiré pattern shown in Figure 15.6. To solve this problem, the grid is continuously moved during the course of the exposure, so that all areas of the underlying film are exposed. This moving grid is called a *Bucky*. For the most part, all 2D X-ray image acquisition systems use a Bucky.

The second problem addressed in film imaging is the thickness of the film as a recording medium. Film consists of a nonreactive base layer and a reactive emulsion layer. When X-rays interact with this emulsion layer, they will induce a

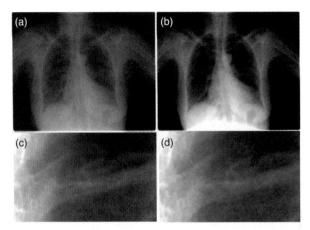

Figure 15.6   Grid pattern interfering with image quality. The image in (a) has a grid overlay causing a rippling moiré pattern. The direct pixel view in (c) of the patient's upper left clavicle shows the alternating pattern of the grid. (b,d) The same views, but with the grid effect removed using computational techniques briefly described in Section 15.4.3 *Source*: Image courtesy of iCRco, Inc.

chemical change via ionization, and these ionized locations are then developed to form the final image. The emulsion layer is quite thin, however, so quite a few X-rays will simply not be recorded. One solution to this problem is to increase the exposure time, so that more X-rays interact with the film, but that approach will cause more photons to be absorbed by the patient as well. Another is to produce film with emulsion layers on both sides of the base, thereby doubling the size of the volume with which X-rays can interact.

Intensifying screens can also increase the amount of radiation recorded in the film at the cost of spatial resolution in the final image. As depicted in Figure 15.5, X-rays that interact with the intensifying screens will emit photons via the photoelectric effect. Photons so produced will be emitted in a random direction; the photon could travel directly to the film, or the photon could travel at an angle towards the film, resulting in a local scatter effect that manifests as blur in the final image. The thicker the intensifying screen, the greater this blur effect, but the more chance that a photon has to interact with the layer. An X-ray that produces a photon that is far from the emulsion could travel quite a distance before it reaches the emulsion layer, while a photon that is produced directly adjacent to the emulsion layer will most likely be absorbed near where it was produced. In the final image, neither photon is distinct. Thus, intensifying screens can be chosen to adjust the trade-offs between spatial resolution in the final image and the amount of radiation necessary to produce a diagnostic image.

Film imaging served as the basis of radiology until the advent of technologies that were capable of producing similar images without the need to match intensifying screens to the emulsion layer. Depending on the clinical environment, each body part could be imaged using different film/intensifying screen combinations, resulting in a large number of distinct film types that had to be stocked and maintained by the clinic. Mammography film, extremity film, and chest films all have their own configurations formulated to preserve as much relevant spatial resolution while at the same time minimizing the X-ray dose necessary to produce a diagnostic image. In addition, film has a distinct dynamic range and can be overexposed much more easily than other digital acquisition devices. Chest imaging via film requires two exposures, one for the lungs (mostly air) and one for the upper abdomen (mostly water, in organs). More recent technologies address these shortcomings.

15.4.2 Computed Radiography

CR uses a photostimulable phosphor (PSP) plate as the storage medium for X-ray images, replacing film and the intensifying screen directly in the image acquisition chain.

Figure 15.7 briefly outlines the physics of the PSP. An incoming photon, called λ_1, strikes the phosphor, inducing an electron to enter an excited state. Many electrons immediately decay back to the ground state and release photons via the previously discussed photoelectric effect. When this decay back to the ground state occurs in the visible light spectrum, it is termed *fluorescence* and occurs on the order of nanoseconds. Some electrons, however, enter into a transitional configuration

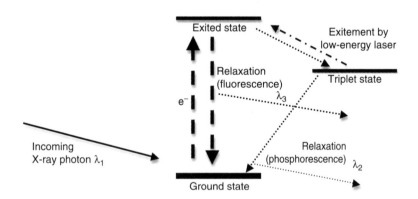

Figure 15.7 The physics of photostimulable phosphor (PSPs). An incoming X-ray, λ_1, induces an electron e^- to enter an excited state by striking a molecule in the phosphor. Some electrons immediately relax back to the ground state, releasing λ_3. Many electrons, however, transition to a triplet state, where they can decay back to the ground state over the course of several minutes to hours, releasing λ_2 in the process. If, however, the PSP is struck with a low energy laser, that photon brings the electron back to the excited state, where it can fluoresce back to the ground state, releasing λ_3. This photon will be of a predictable wavelength, and so can be filtered from other photons and recorded precisely.

called a *triplet state*. When the molecule relaxes from the triplet state back to the ground state, a second photon (λ_2), is emitted. This relaxation reaction occurs slowly, on the order of minutes, and is termed *phosphorescence*. PSPs will store their image information for about 30 min to an hour before completely returning to the ground state, allowing clinics to take multiple images on different plates before processing all of them.

If the PSP is exposed to photons from a laser or similar source after being exposed to an incident X-ray beam then the PSP will be induced to return to the initial excited state produced by being struck by release λ_1. As a result, electrons in the triplet state in the PSP can be returned to the original, excited state and then undergo a fluorescence reaction, releasing a more energetic photon, λ_3, as the PSP electron returns to the ground state. This reaction forms the basis of CR imaging. X-rays are stored in the phosphor via this excited/triplet state configuration and then released from the triplet state by a secondary excitation by a laser, as shown in Figure 15.8. An image is formed by moving the laser over the

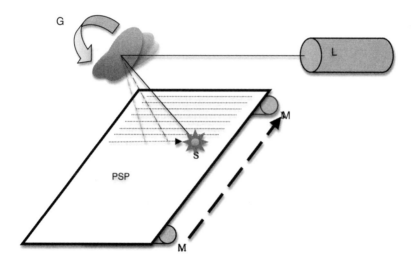

Figure 15.8 Scanning a CR PSP. As the PSP is moved through the machine via the motors (M), an excitation laser (L) shines on a mirror controlled by a galvanometer (G). The mirror moves constantly during the scanning process, so that the spot (S) where the laser strikes proceeds across the PSP linearly. The reaction described in Figure 15.7 then occurs at this spot, and photons more energetic than the excitation laser are emitted by the plate to be recorded by a photomultiplier tube, as depicted in Figure 15.9. The axis of motion controlled by the motors (M) is termed the *major scan direction*, while the axis of motion of the mirror is termed the *minor scan direction*. Each line from a single movement of the mirror is termed a *scanline*, and the resulting image is composed of these scanlines. Scanning in this manner is called *rastering*.

phosphor, releasing the stored information from the phosphor and capturing the released photons through a recording device such as a *photomultiplier tube* (PMT). This scanning technique is known as *rastering*. Increasing the size of the laser's excitation region on the plate will increase the number of photons released per pixel but at the cost of spatial resolution, as previously shown in the MTF charts in Figure 15.4. This size parameter is somewhat analogous to the use of intensifying screens of different sizes, as a large intensifying screen will also have the effect of reducing the spatial resolution to record more photons overall.

The QE of a CR system is limited both by the emission of the secondary photon and the detector used to acquire that photon. Practically, the secondary photon is released in an arbitrary direction, not necessarily in any direction related to the excitation laser. As such, only those photons that are emitted in the direction of the acquisition device can be recorded. Each manufacturer of CR devices uses different technologies to overcome this limitation. Some use light pipes, glass tubes that attempt to gather emitted light and guide it towards the detector. Regardless of the technology used, photons are emitted from the PSP in arbitrary directions, and 50% of photons will be emitted in a direction away from the detector. Some companies attempt to compensate for this signal halving by placing detectors on either side of the PSP.

The detector in CR systems is usually a PMT. Figure 15.9 shows the setup of a basic PMT. This device converts photons that strike the exterior surface of the detector into electrons. These electrons are then accelerated toward a final charge collector, and for every panel that the electrons strike along their journey, more electrons are released. The result of this cascade is that a single electron can be amplified to an exponentially large number of electrons by the time the electrons reach the final detector, thereby amplifying the incoming signal to readable amounts. Such devices also have limitations in QE, however, as the initial

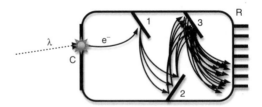

Figure 15.9 A photomultiplier tube (PMT). Photons (λ) emitted by the PSP as shown in Fig. 15.8 are corralled toward this detection device. Each manufacturer has their own particular technology to accomplish that delivery. Once the photon strikes the exterior surface of the PMT, the photon can be converted to an electron at location C. The electron (e^-) is then pulled toward the positively charged surface of stage 1 (1). Once the electron strikes that stage, multiple electrons are produced. These electrons are then pulled to the slightly more positively charged stage 2. The same multiplication effect occurs, and the new electrons proceed to stage 3. Typical modern PMTs will have 10–12 stages. After the final stage, electrons are accumulated toward the readout electronics (R), then leads to an analog to digital converter to produce a digital number.

conversion of a photon to an electron also suffers from the loss of 50% of the incoming photon signal. The produced electron can also be emitted in an arbitrary direction; 50% of electrons do not reach the interior of the PMT but instead emit in the other direction and are not measured.

These limitations are cumulative, such that the highest effective QE of a CR device with detectors along a single side of the PSP is 25%, not counting the QE of the PSP itself, which will vary with the amount of exposure used on the material of the PSP itself.

CR technology offers several distinct advantages over film. Film requires development via a chemical process, while CR requires development of the image via laser. The lack of development chemicals or space means a lower total cost of ownership of a CR system, especially when calculated over the course of years of operation. The use of a laser rastering device can also lessen the time to produce a clinical image; rather than developing individual films, PSPs can be scanned relatively quickly. CR images are purely digital, meaning that they can be processed to match the preferences of individual doctors or clinics, and as shall be discussed, processed in ways that film and analog technologies cannot easily match. In addition, the number of photons that can be stored per unit area in a PSP is much larger than that of film, resulting in an increased dynamic range. This increase in dynamic range means that only a single exposure is necessary to capture diagnostic information in both the lungs and upper abdomen, resulting in fewer exposures to the patient.

Practically, however, the increase in dynamic range did not result in a reduction reducing patient dose. Obtaining images from a CR device comparable to a film image should require a fraction of the dose of the film image, but technicians and radiologists quickly discovered that tripling the radiation dose would dramatically increase the SNR in the final image. This SNR increase, in turn, produces images from which diagnostic information is much more readily perceivable by the clinician. The increase in patient dose, however, is not considered to be a positive feature, especially in light of potential long-term nondeterministic adverse effects on patient health. As such, the concept of ALARA (as low as reasonably achievable) has been adopted as a standard of dose exposure in the United States and Europe (Willis and Solvis, 2005).

CR systems use the same configurations as film, meaning that clinics already invested in film systems do not require a complete change of X-ray exposure technologies to switch to a CR system. In clinics with less funding, such as those in developing nations, this similarity offers a distinct cost benefit. In terms of clinical impact, imaging configurations that were possible with film, such as placing two or three large films in a columnar configuration to image a patient's spine for scoliosis screening, are still possible with CR. These configurations have the added benefit of being done with digital images, so that processing techniques such as automatic image registration can be performed on these scoliosis images to produce a single large image for clinical analysis.

15.4.3 Digital Radiography

DR produces an image through a digital detector similar to a charge coupled device (CCD) or complementary metal-oxide semiconductor (CMOS)-based imaging device found in a digital camera. Two types of digital detectors are in clinical use today. The first, termed *direct digital detectors*, use materials such as amorphous selenium (a-Se) to directly convert incoming X-ray photons into electric charge, as shown in Figure 15.10. The second, called *indirect digital detectors*, uses an intermediate layer of X-ray sensitive crystals, such as cesium iodide (CsI) or gadolinium oxide (GadOx), to convert X-rays into photons in the visible spectrum for recording by the imaging device (Knoll, 2010), as shown in Figure 15.11. Either approach differs from most consumer electronic imaging devices in their pixel pitches; while most CCD- or CMOS-based imaging systems have pixel pitches on the order of 3 μm × 3 μm to 10 μm × 10 μm, these devices have much larger pixel pitches, generally near 75 μm × 75 μm to 100 μm × 100 μm. This increase in pixel pitch is necessary because the imaged area is much larger than that in a consumer-grade device; a 14 in × 17 in area is used to take a typical chest X-ray, while a consumer device can use a lens to focus incoming photons to a square centimeter of detector. Smaller pixels in such a large imaged area would drastically increase the size of the digital files produced, as well as requiring an increased patient dose to ensure

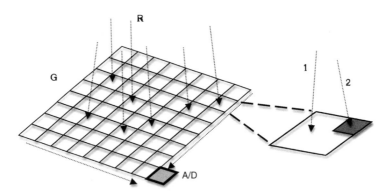

Figure 15.10 A generic digital radiography acquisition grid. Incoming X-rays (R) proceed to strike the detector grid (G). Each square in the grid represents a detector element. Two technologies, charge-coupled devices (CCDs) and complementary metal-oxide semiconductor (CMOS), can be chosen for the composition of the grid. In either case, a portion of the detector surface is given over to electronics. Ray 1 has struck the detector surface of the pixel, while ray 2 has struck the electronics portion of the pixel. The ratio of detector surface to total pixel surface is known as the *fill factor*, and the closer this value is to 1, the better the QE of the detector. In CMOS-based DR systems, each pixel can be individually addressed and accessed, while in CCD-based DR systems, accumulated charge is moved along the pixel grid until it can be read by the analog to digital (A/D) converter and translated into a digital image.

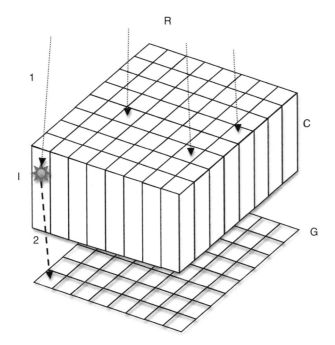

Figure 15.11   A generic indirect DR acquisition grid. For DR grids (G) that have poor QE in the X-ray energy range, an indirect strategy can be employed. In this instance, a grid of scintillator crystals (C) convert the incoming X-rays (R) into a lower energy photon, such as visible light, so that the grid can properly detect and measure the X-ray flux.

that each pixel has enough signal to have a diagnostically useful SNR. A larger digital detector can have practical problems with fabrication, requiring an increased cost to consistently produce larger image acquisition devices. Decreasing the pixel pitch can decrease the yields of usable detectors produced during manufacturing, further increasing costs.

In an indirect DR system, incoming X-ray photons are converted into photons in the visible spectrum via a network of CsI or GadOx crystals. These crystals also have the trade-off of being particularly sensitive to water absorption and so require replacement over the course of several years of use. The mechanism for this crystal conversion is similar to the technology in PSPs; incoming photons excite the crystal lattice, and the relaxation of the lattice back to the ground state produces a less energetic photon, typically in the visible spectrum. One practical consequence of this method is the lowering of the QE of the device because of the lack of directionality of this released photon; the secondary photon may be released in any direction, as discussed previously. As such, 50% of visible light photons that are produced in the crystal are emitted in a direction away from the plate, and the highest QE that an indirect digital measurement device can achieve is 50% of the QE of the absorbing crystal. Direct digital devices do not have this limitation; however, the crystal in an indirect digital device can be comparatively thick and, therefore, have a higher QE than a direct device.

DR systems do not require significant time to develop an image. Film uses a chemical process to produce the final image, CR uses a laser scanning process, but DR images are recorded by a digital device during the course of image formation, and so no further steps are necessary to obtain an image. This speed increases patient throughput in a clinical environment.

Acquiring a DR image also requires that the generator be coupled with the acquisition system. This detection system is active, requiring power to record the image, and that power becomes active when the generator is on. This approach has the benefit of allowing the acquisition system to record all of the imaging parameters, such as kVp and mAs, without operator action.

15.4.4 Clinical Applications of 2D X-ray Techniques

Each of the three 2D imaging techniques can be used to guide minimally invasive surgeries, as well as to determine whether the results of surgery were successful. Such applications include chest radiographs to determine the conditions of the lungs, spine, and other conditions in the thoracic cavity; dental radiographs to detect the presence of cavities; stitching together several radiographs or using a long PSP to form an image of the entire spine to prepare for corrective surgery for scoliosis; and radiographs of any body part to ascertain the presence and location of a broken bone.

Another common example is mammography, or acquisition of images of the breast. Close examination of properly acquired X-ray images can reveal the presence of cancerous or precancerous tissues. A typical screening mammography study involves four images of the patient, two for each breast. The images are taken in anatomically defined directions. One is taken in the *cranial-caudal* (CC) direction or from the head to the foot of the patient. The other is taken in the *mediolateral oblique* (MLO) direction, or positioned in the middle of the breast and taken from the left or right side of the patient. Mammographers examine these images for the presence of indicators of cancer or precancerous lesions. These indicators are clusters of *pleomorphic calcifications* or groupings of irregularly shaped calcium deposits; *spiculated masses*, or radiopaque regions with irregular borders; and *architectural distortions*, or regions of the breast that look warped or otherwise unevenly and uncharacteristically deformed. Two views are taken to corroborate one another: since X-ray imaging is a projectional medium, what appears as a potential mass in one view may be overlapping blood vessels or other normal breast densities in the other. The other reason to take these different views is to offer coordinates for the presence of the aforementioned indicators.

If a patient is deemed to have indications of potentially dangerous pathology, then the doctor will typically order a biopsy of the region in question. The biopsy can be performed using a stereotactic image, as shown in Figure 15.12. The biopsy

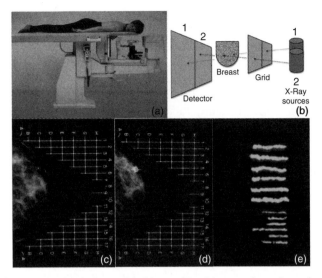

Figure 15.12   X-ray-guided stereotactic biopsy. For this procedure, the patient lies flat on a table (a), and the breast is placed between a grid and an X-ray detector (b). Two sources produce X-rays such that two images of the breast are taken at different angles, as shown in (b). Two images are produced, such as those in (c) and (d), from the different illumination angles of the two sources. Since the angle is known, the coordinates of the lesion can be deduced from the coordinates in the grid, resulting in proper localization of potentially cancerous tissue. The result is sampled tissue, shown in (e) (*Source*: Examination and X-ray images courtesy of Dr. Lawrence Bassett, MD, University of California, Los Angeles).

needle is guided to the coordinates determined during the screening mammograms and refined by the grid on the stereotactic images. The clinician then removes a series of tissue samples by means of a hollow needle. A pathologist then further evaluates this biopsy, and the mammographer and the pathologist will confer to determine the findings of the biopsy. In this comparison, the mammographer will refer to the previously mentioned features that are typically present in images of cancerous tissue, and the pathologist will refer to the results found in the biopsy. As such, the biopsy must capture enough tissue so that the doctors can be confident that the region was sampled properly, without taking enough to cause serious undue harm or cosmetic damage to the patient in the case of a false positive. Biopsies are then ideal candidates for minimally invasive surgical procedures, and stereotactic X-ray imaging provides guidance for the procedure.

15.5   BASIC 2D PROCESSING TECHNIQUES

The use of digital acquisition technologies allows for the use of digital processing and enhancement techniques. By recording each image as a series of numbers that correspond to the amount of photon flux at a particular point in space, mathematics can be applied to the image. Pixel processing tends to fall into one of the three following categories:

1. Pixels that are considered independently of their neighbors. One common operation employing this approach is the application of a lookup table (LUT) to map the actual, recorded values to a set of values that are displayed on the physician's screen.
2. Pixels that are grouped into clusters. Image sharpening commonly employs this approach.
3. Transforming the image into an entirely different mathematical domain, performing a procedure, and then inverting the original transformation. A common operation for this approach is frequency enhancement or reduction through the use of the Fourier transform or a wavelet transform.

The full extent of processing techniques that can be employed on digital imaging is beyond the scope of this chapter; a more complete general discussion can be found in the book by Gonzalez and Woods (2002) and techniques specific to mammography can be found in the book by Bick and Deikmann (2010).

15.5.1 Independent Pixel Operations

One of the major considerations of digital imaging is the transformation of the recorded pixel flux into an image that can be displayed to a doctor. Images recorded through digital means can have 16 or more bits of information at each pixel, while computer monitors have a dynamic range of 8 bits of gray scale display; that is, a typical monitor can show values from 0 to 255, while a medical image can range from 0 to 65535 for a 16 bit image. In modern monitors, color images are displayed through the use of independent red, green, blue, and alpha (opacity) channels, each of which contains 8 bits of information. Monitors specifically designed for radiology can have a much higher bit depth and can be capable of displaying 11 or 12 bits of information. Even so, these so-called high brightness monitors are not capable of displaying the entire dynamic range of the acquired digital image. Even if such a display device existed, the ability to emphasize one particular portion of the anatomy over another for closer inspection can be very useful, so limitations of the displayed dynamic range are often employed in clinical practice.

The limitation of this display range is accomplished through the use of an LUT. The LUT remaps pixel values for an image into the display range of the monitor. Figure 15.13 shows how LUTs can be applied to different regions in an image to bring out the full display range of a particular portion of anatomy. The patient shown in this figure has had corrective spinal surgery, and the implanted devices are much more radiopaque than the lungs. If the clinician were interested in viewing the details in the lung then the brightness and contrast can be adjusted for the lungs. In medical imaging, the relevant term is the *adjustment of the window and level*. The *window* is the difference between the maximum and minimum displayed

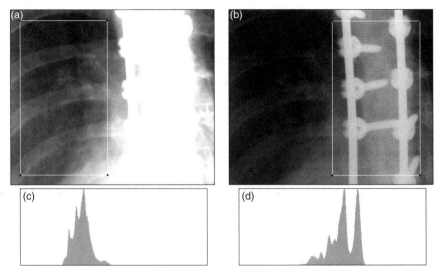

Figure 15.13   The use of regions of interest (ROIs) to determine anatomy-specific LUTs. (a) The spine has been selected, and the histogram of the values in the ROI is shown in (c). The horizontal axis in (c) and (d) ranges from 0 to 65,535, so in (c), values lower than 34,000 are shown as 0 (black) and values larger than 48,000 are displayed as 255 (white). The same image is shown in (b) and (c) but adjusted to display the lung field instead of the spine. Clinicians can use these operations to display regions of interest, and more complicated processing can make both fields visible at once without a loss of detail or depth in either. *Source*: Images courtesy of iCRco, Inc.

pixel values, and the *level* is the average between the maximum and minimum displayed pixel values. This basic operation can be quite powerful for isolating and emphasizing different regions in the same image. Traditional film imaging would require two exposures to acquire the same information: one exposure for the lungs and the other for the spine. In this case, the use of basic processing has spared the patient the dose of a second image.

15.5.2   Grouped Pixel Operations

Pixels can be grouped with one another to provide aggregate local statistics; these local statistics can then be used to enhance the image. The grouping operation is typically accomplished through an $N \times N$ predefined pixel operator, known as a *kernel* and is shown in Figure 15.14. The application of the kernel to the image is called a *convolution*. In the figure, a basic blurring kernel is used. Each pixel value is replaced with the weighted average of the pixel and its neighbors, thereby producing an image with less local variation in pixel intensity. In this case, the weights of the kernel are all equal to 1, but any weight can be used. Such local variation is the result either of fine variations in signal or from noise in the image formation process.

1	2	1
2	4	2
1	2	1

×

10	20	**20**	**45**	10
12	**22**	**40**	10	8
24	**43**	**11**	8	7
42	12	9	6	4
13	7	5	5	3

=

7.88	14.8	20.1	19.3	9.8
13.9	23.9	25.1	18.1	8.7
20.3	25.8	**18.4**	10.1	5.3
19.8	**18.9**	10.4	6.38	3.8
10.1	**8.69**	5	3.81	2.3

22 x 1 + 40 x 2 + 10 x 1 + 43 x 2 + 11 x 4 + 8 x 2 + 12 x 1 + 9 x 2 + 6 x 1 = 294

294 / (1 + 2 + 1 + 2 + 4 + 2 + 1 + 2 + 1) = **18**

Figure 15.14 Applying a kernel to an image. Each element in the kernel is multiplied to the current pixel and its neighbors, and then, the resulting number is divided by the summation of the kernel values to produce an image that is comparable in value to the original. Note: in this example, the high values of 40, 42, 43, and 45 are all averaged to 25, 19.8, 25.8, and 19.3, respectively—that sharp, bright edge has been averaged down. The largest difference between pixels is now 10.21 (bold italics), while in the initial image, the largest difference was 25 (bold italics). Therefore, this kernel has reduced both the local differences between pixels and the overall sharpness of the image.

These filtering operations can be performed sequentially. One such common operation is known as the *Laplacian of the Gaussian* (LoG) and is used as an edge detection filter. First, the image has a Gaussian filter applied. The Gaussian filter is a 1-sum kernel that, when convolved with an image, produces a blurred version of the original. The formula for a Gaussian curve in one dimension is

$$A = \frac{1}{2}\sigma\pi, \quad f(x) = Ae^{-A(x-x_0)^2}$$

The Gaussian kernel has several very interesting properties. One of the most useful from the perspective of image processing is that the Gaussian kernel is *linearly separable*, meaning that it can be applied along the columns of an image and then along the rows in two distinct steps. Any kernel applied directly to an image requires $O(N \times M)$ operations, where N represents the number of pixels in the image and M the number of pixels in the kernel. This type of formula is called *big O notation* and specifies the maximum number of individual operations that an algorithm will require to complete. Individual computers can be faster or slower than one another, but the performance of an algorithm on any given computer will scale with the formula designated by the big O notation. For large filters, where the diameter can approach or exceed 300 pixels, this separability drastically reduces the number of computational operations to be performed. If the algorithm is run in a single pass then the kernel size must be 300 × 300 to accommodate the larger kernel, so the time to perform the operation will scale at 90,000 times the size of the image. If the algorithm is linearly separable and can be run in two distinct passes then M will decrease to the size of the one-dimensional kernel. The final runtime will scale at (300 + 300 =) 600 times the size of the image, more than two orders of magnitude faster than a fully two-dimensional kernel.

The Laplacian kernel calculates the second derivative of the image and is used after applying the Gaussian kernel to avoid the interference of noise in the final image. In its simplest form, the Laplacian kernel is as follows:

$$
\begin{array}{ccc}
0 & 1 & 0 \\
1 & -4 & 1 \\
0 & 1 & 0
\end{array}
$$

This kernel follows from applying the first derivative in multiple directions. Recall that the definition of the first derivative of a continuous function is

$$ m = \frac{(f(x+h) - f(x))}{h} $$

where m is the slope, h is an infinitesimally small difference, and $f(x)$ indicates the value of the function f at x and $f(x+h)$ indicates the value of f at $x+h$. In a digital image, the smallest increment is the width of a pixel, so the kernel of a first derivative looks like

$$
\begin{array}{ccc}
0 & 0 & 0 \\
0 & -1 & 1 \\
0 & 0 & 0
\end{array}
$$

This kernel is directional, meaning that it will find differences in one direction of the image (in this case, the horizontal direction), but will not find differences in the vertical direction. The first derivative can be extended to cover both directions in the following summation kernel:

$$
\begin{array}{ccc}
0 & 1 & 0 \\
0 & -2 & 1 \\
0 & 0 & 0
\end{array}
$$

This kernel will produce an image that is highly directional. Running the first derivative in all directions, then, will yield a symmetric kernel:

$$
\begin{array}{ccc}
0 & 1 & 0 \\
1 & -4 & 1 \\
0 & 1 & 0
\end{array}
$$

This kernel is the equivalent of calculating the second derivative in both directions. Recall the formula for the second derivative:

$$ f''(x) = \frac{f(x+h) - 2f(x) + f(x-h)}{h^2} $$

If this formula is then extended to cover the y-direction independently of the x-direction, the previous symmetric kernel has been derived. This kernel is zero

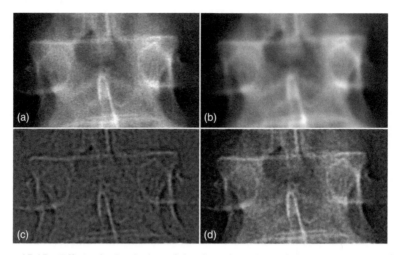

Figure 15.15   Effect of a Laplacian of the Gaussian edge enhancement. (a) An original image of a vertebra. (b) The result of a Gaussian blur with a radius of 3 to (a), (c) The result of applying the Laplacian kernel to (b). (d) The final result of adding (a) and (c), with edges heavily emphasized. *Source*: Images courtesy of iCRco, Inc.

summed and so will produce a value of zero wherever the image is homogeneous in intensity and will have peaks wherever the image has the largest variations in intensity. Adding this edge image back to the original image will produce a sharpened image, as shown in Figure 15.15.

15.5.3   Image Transformation Operations

Several operations exist that can transform an image from its standard presentation to an alternative mathematical domain. The most common of these transformations is the Fourier transform. This operation transforms the image from what is termed the *spatial domain* into the *temporal domain*. The full mathematics of the operation is beyond the scope of this chapter; more information can be found in the book by Gonzalez and Woods (2002). The basic formula of the Fourier transform is

$$F\left(\xi\right) = \int f\left(x\right)^{-2\pi i x \xi} \, dx$$

and the inverse is

$$f\left(x\right) = \int F\left(\xi\right)^{2\pi i x \xi} \, d\xi$$

One benefit of this transform is the ability to remove periodic patterns from the image, as shown in Figure 15.16.

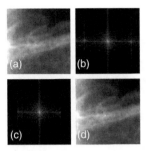

Figure 15.16 Fourier transformation to remove repeating grid lines. (a) A grid pattern overlaid on lung tissue. (b) The result of a Fourier transform on the image; note that the repeating vertical pattern appears as two vertical lines near the left and right edges of (b), with periodic dashes. (c) depicts (b) with those bars replaced with zeros, and (d) shows the result of the inverse transform performed on (c). Note that the effect of the grid has been somewhat ameliorated. A similar procedure was used to remove grid lines in Figure 15.6. *Source*: Images courtesy of iCRco, Inc.

15.6 REAL-TIME X-RAY IMAGING

Fluoroscopy is an X-ray imaging modality used to obtain real-time images rather than a single exposure. Historically, the first radiologists used fluoroscopy exclusively to visualize the inner physiology of their patients, using such positioning as shown in Figure 15.17. In this figure, the X-ray tube is a sphere in the foreground, and the patient's foot lies between the tube and the radiologist. The radiologist is holding a mask with a phosphor plate that immediately reflects any changes to the patient's interior. These early radiologists tended to get radiation burns, as well as cancers of the head and neck due to extensive exposure from standing in the X-ray beam (Weber, 2001; Brown, 1936).

As these dangers were revealed, X-ray imaging began to adopt static imaging with film, but uses of live X-ray imaging both as a surgical guidance tool and as a means to provide immediate feedback in patient status persisted. Minimally invasive surgery techniques involving vasculature are greatly enhanced by the ability to see such vasculature. Fluoroscopy affords the surgeon the ability to see such internal detail during the course of surgery by acquiring video rate X-ray images. As shown in Figure 15.18, fluoroscopy images can be used in such techniques as pacemaker lead placement and blood flow analysis, also known as an *angiogram*.

15.6.1 Fluoroscopy Technology

The technology to accomplish this task is based on similar techniques used in 2D X-ray imaging, with one variant similar to CR technology and the other using DR technology. The DR approach uses a C-arm gantry to hold the digital flat panel detector and the generator on opposite sides of the patient, and the resulting digital signal can be both recorded and played back in the operating room. The

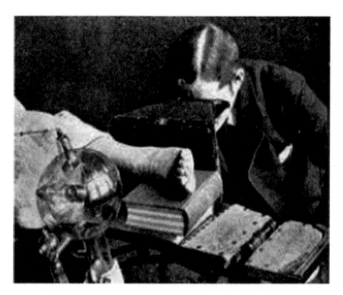

Figure 15.17 An early radiologist. This radiologist is using a mask with a phosphor screen held up to his face. The patient's foot is between the mask and the X-ray source. Note that the source is unshielded, meaning that X-rays are produced in all directions. Modern X-ray sources, as shown in Figure 15.1, have shielding to prevent unnecessary X-ray exposure to both the doctor and the patient. In addition, many of these early radiologists stood directly in the path of the X-ray beam and died of radiation-induced tumors.

analog variant, shown in Figure 15.19, uses a phosphor plate that produces light when excited by X-ray photons, an intensifier to enhance the faint light from the phosphor in the place of the digital detector. Thus, doctors can guide devices through the patient's vasculature using the X-ray image as real-time feedback to device location.

Fluoroscopy centers on the idea of a rapid conversion from photons to electrons that can then be visualized at video rate. An analog system shown in Figure 15.20 uses an image intensifier coupled to a set of optics that then feed input signal into a video camera, and the output is then displayed to the surgeon in the operating theater, as well as recorded for future review. The intensifier itself has a vacuum tube that houses the input layer that converts X-ray photons into electrons, focusing electronics that guide emitted electrons toward the output layer, and the output layer that converts the focused electrons into visible light photons.

The technology behind the input layer of an analog system is similar to the technology behind CR imaging. Crystals sensitive to X-rays, such as cesium iodide, are used to produce electrons. These electrons are, in turn, accelerated toward the output layer. The output layer translates those electrons to photons in the visible spectrum using the same physics as a cathode ray tube monitor. This process has the side benefit of *minification gain* — as the input layer is larger than the output

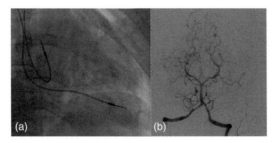

Figure 15.18   Some uses of fluoroscopy. The image in (a) depicts two pacemaker leads placed during surgery. The leads are opaque to X-rays and appear as dark regions in the image. The image in (b) depicts cerebral blood flow via injection of contrast. Obstructions to blood flow would be apparent as blockages in the passage of the dye. *Source*: Part (b) courtesy of Dr. Gregory Marcus, MD, MAS, FACC.

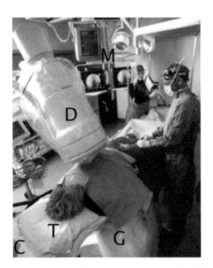

Figure 15.19   A fluoroscopy-equipped operating theater. With the patient lying on table T, the operating room staff will position the C-arm C (so named because of it shape, somewhat obscured from this angle) such that the generator G and the detector D bracket the patient. The fluoroscopy image appears on the monitors M (*Source*: Courtesy of UCLA).

layer, the density of visible photons is increased in proportion to the decrease in area from one layer to the next.

These analog solutions have been mostly obviated through the introduction of digital imaging technology. In general, digital solutions tend to be faster, more compact, and lacking artifacts produced by the image intensifier. Digital solutions also allow for the use of mathematics during image capture to enhance image contrast and appearance, providing further utility for surgical routines.

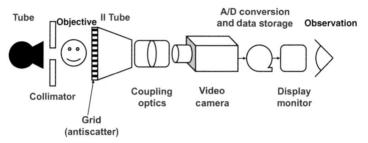

Figure 15.20   The analog fluoroscopy imaging chain. The main difference between this approach and the film imaging chain shown in Figure 15.1 is the use of technology to acquire multiple images through time. The image intensifier tube serves to increase the signal from the phosphor, and the camera output can then be stored and visualized during the course of surgery. These devices are also mounted on C-arms, as show in Figure 15.19, but their increased weight because of several more steps in the image acquisition steps mean that these C-arms have to be more heavily reinforced, and wobble can become a concern. Courtesy of Dr. Daniel J Valentino, PhD.

15.6.2   Angiography

Angiography refers to obtaining images of blood vessels; the word itself is derived from the Greek *angio* (blood vessel) and *graphy* (writing). Obtaining images of the function of a vascular system allows physicians to pinpoint problems in the patient's blood flow that are not evident through a purely static image. Such visualization can be used to depict coronary artery disease, stroke, brain aneurysms and vascular malformations, abdominal aneurysms, and peripheral artery disease or to find angiogenesis associated with malignant tumor growth. Furthermore, interventional therapy can place stents, retrieval devices, coils, or deliver drugs to a particular region in the vascular network.

Angiography through fluoroscopy is accomplished through digital, rather than analog, imaging. The nature of digital imaging allows for the signal from the camera to be intercepted and manipulated before its display to the doctor; this manipulation can enhance the presence of blood vessels through the use of contrast agents. Contrast agents, as previously discussed, can be nephrotoxic; so, they must be employed to maximal effect, and avoided in patients who would otherwise experience adverse or allergic effects. An image taken before the administration of the agent is subtracted from the image acquired after the agent is applied. The result is an enhanced depiction of the patient's vascular system.

Vasculature that has been blocked or is leaking is readily apparent in an angiogram. Minimally invasive surgical techniques can correct these problems. Blockages, referred to medically as an *embolism* and as *thrombosis* when specifically caused by a blood clot, can be removed by devices such as the MERCI (*Mechanical Embolus Removal in Cerebral Ischemia*) retriever (Fig. 15.22)

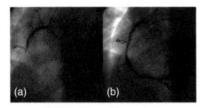

Figure 15.21   Restoration of cardiac blood flow. The angiogram on (a) depicts occluded blood flow and (b) shows that the blood flow has been restored. *Source*: Image courtesy of Dr. J Heuser, MD.

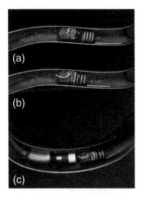

Figure 15.22   The MERCI retrieval device. This is one of many devices that is used in minimally invasive surgeries enabled by fluoroscopy. (a) The surgeon guides the device past the location of a cerebral embolus that has been found using the dye techniques shown in Figure 15.14b. The coil is then deployed. (b) The coil is pulled back over the embolus. (c) A balloon in the catheter is then inflated to pull the ebolus out of the area and restore blood flow. *Source*: © 2008 Neil C. Barman, used under a Creative Commons Attribution-ShareAlike license: http://creativecommons.org/licenses/by-sa/3.0.

that restore blood flow to an affected area. Leakages, referred to medically as *hemorrhages*, can be addressed through the use of a *stent*, an artificial tube that is placed over the leak to prevent further blood loss into surrounding tissue. Figure 15.21 shows an angiogram with a heart occlusion that has been restored through interventional techniques. If this occlusion is caused by a thrombus then a retrieval device such as the MERCI retriever can be employed to remove the clot, as shown in Figure 15.18, although, as the name suggests, this device is typically used in cerebral rather than cardiac events. If a narrowing of the arteries from an atherosclerotic lesion causes the occlusion then a stent can be placed at the point of occlusion to widen the artery, as discussed in Chapter 10. In either case, blood flow to the affected area will be restored.

15.7 THREE-DIMENSIONAL X-RAY IMAGING

True three-dimensional (3D) imaging has become an important tool in medical diagnosis of a variety of conditions, and in providing a blueprint to plan surgeries and other therapies that require a 3D depiction of the patient. A variety of these techniques exist—computed tomography (CT), 3D fluoroscopy, and breast tomosynthesis, for instance—and they are derived from extensions to the technique employed in stereotactic breast biopsy discussed in Section 3.4. This section briefly touches on these methods, but interested readers can refer to the books by Bushberg et al. (2001), Green et al. (2004), and Smith (2005). In stereotactic biopsy, two images of the same tissue are taken with a slight offset used in the second image. That offset results in a different angle of projection through the breast tissue. Since an X-ray image is the projection of the X-rays onto a final recording medium, this change in angle affords a different view of the same tissue. Rotating the source and recording device around the tissue (or in the case of CT and 3D fluoroscopy imaging, around the entire patient) while periodically recording images of the tissue will result in different images, each taken at different angles. The result of this process is a *sinogram*, where the result of each angle of projection corresponds to

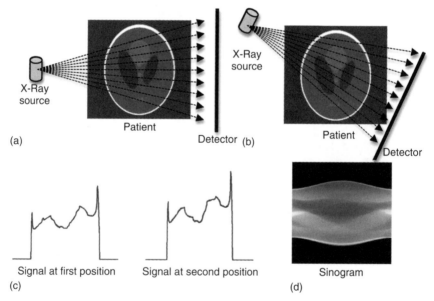

Figure 15.23 Construction of the sinogram in CT. (a) As with all X-ray imaging, the patient is placed between the source and the detector; in this case, the Shepp-Logan phantom is shown to simulate the presence of a patient (Shepp and Logan, 1974). The signal recorded at the detector in the upper left is shown in (b). The source and detector are then rotated about the patient, producing the signal in (c). If 180 rotations are made in increments of 1°, the result is a sinogram, shown in (d).

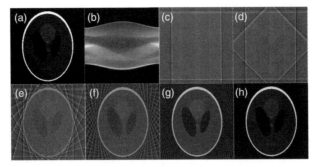

Figure 15.24 Inverting the radon transform. (a,b) The original Shepp-Logan phantom, followed by the sinogram of producing 180 projections at $1°$ increments. Reconstruction is then performed using 2 projections in (c), 4 in (d), 15 in (e), 30 in (f), 60 in (g), and 180 in (h). At least 15 projections are necessary to visualize any detail within the body of the phantom, and using more projections increases the contrast and detail of the final image. Increasing the number of projections linearly increases the dose that a patient receives.

a different column. This image is a physical manifestation of a mathematical construct known as the *radon transform*; the inverse of the radon transform produces a 3D image. Figure 15.23 shows this process using a simplified image of a phantom.

The inversion of the radon transform produces the final 3D image that a doctor can use for clinical practice, as shown in Figure 15.24. The trade-off between patient dose and image quality is readily apparent in the creation of CT images. Each projection correlates to a single X-ray image; the more projections that are acquired, the higher the resolving power and contrast of the resulting image. Unfortunately, the patient will also receive a concomitantly larger dose. Image processing techniques, such as filtering the sinogram before reconstruction, a process known as *filtered back projection*, can aid in reducing the number of images required to obtain a reasonable result.

As mentioned previously, CT images are one of several 3D reconstruction techniques. Fluoroscopes mounted on C-arms can also produce 3D images by rotating the C-arm around the patient during surgery. Breast tomosynthesis can produce a 3D image of the breast by taking several full 2D images around the breast tissue. Each of these approaches use different sampling and exposure techniques according to the clinical requirements of the imaging involved. Nevertheless, they all rely on this basic approach to produce images.

15.8 CONCLUSION

X-ray imaging provides the clinician with extremely powerful tools to explore, diagnose, and treat the conditions of patients. While X-rays themselves are not without their risks, the images that can be produced with X-rays are an important tool to the modern physician. Two-dimensional X-ray images are the most acquired images in the clinical setting because of the relative simplicity of acquisition, but 3D and real-time imaging techniques are powerful tools now available to physicians as well.

An important aspect of X-ray imaging research is the improvement to the devices themselves; any improvement that can reduce patient X-ray dose will likely improve patient outcomes. Materials, optics, mechanical, and other fields of hardware engineering will be utilized to improve the devices described in this chapter to further advance the clinical and diagnostic utility of the X-ray.

Potentially more exciting, however, are the advances afforded by cheap computational power. Initial production of CT images via inverse radon transforms would require banks of computers, and now single workstations can accomplish the same task. Powerful noise reduction, reconstruction, and real-time visualization techniques will provide clinicians with reliable and useful tools that can minimize patient risk while maximizing treatment options and bettering patient outcome. In addition, computational storage of images allows doctors both to compare a patient's condition with any previously acquired images of the same patient and to compare different patients to one another across different hospitals and continents. Machine learning has been employed to automatically detect the presence of lesions and other patient pathologies, termed *computer-aided diagnosis* (CAD); such approaches are only viable with data readily available in a form that a computer can read and manipulate. As computing power becomes even more readily available, other software-based image manipulation techniques will probably be developed to further increase the clinical utility of the X-ray.

CHAPTER 16

Navigation in Neurosurgery

JEAN-JACQUES LEMAIRE

Department of Neurosurgery, Auvergne University, Clermont-Ferrand

ERIC J. BEHNKE, ANDREW J. FREW, and ANTONIO A. F. DESALLES

Department of Neurosurgery and Radiology Oncology, University of California, Los Angeles, CA

16.1 BASICS OF NEUROSURGERY

16.1.1 General Technical Issues in Neurosurgery

Neurosurgery focuses on the surgical treatment of diseases affecting the central nervous system (CNS) and the peripheral nervous system (PNS). Examples of pathologies treated by neurosurgery of the head and spine include tumors, degenerative diseases such as Parkinson's disease, traumatic lesions, epilepsy, neuropathic pain, skull lesions, vertebral lesions, disc lesions, congenital malformations, vascular diseases, and cerebrospinal fluid circulatory problems.

Open head surgery is the most common approach to neurosurgery and can be broadly categorized into five surgical phases: (i) craniotomy, in which the skull is opened by creating a bone flap; (ii) opening and fixation of the superficial envelope containing the brain, known as the *dura mater*; (iii) intracerebral surgery; (iv) control steps, such as final validation of hemostasis and the decision to stop the operation; and (v) repositioning and fixation of the bone flap to close the braincase. During open head surgery, the patient's head is typically immobilized by affixing a head holder or surgical clamp, which is locked to the surgical table.

Closed head surgery is another common approach, in which a burr hole is created for minimally invasive tasks, such as introduction of a needle, probe, or catheter. Stereotactic surgery (described below) and placement of cerebrospinal fluid catheters are typically closed head surgeries.

Medical Devices: Surgical and Image-Guided Technologies, First Edition.
Edited by Martin Culjat, Rahul Singh, and Hua Lee.
© 2013 John Wiley & Sons, Inc. Published 2013 by John Wiley & Sons, Inc.

Neurosurgeries are categorized according to the associated anatomic topography, with surgeries within the braincase being classified as *intracranial*, and superficial procedures, such as those of the skull, skin, or orbit, classified as *extracranial*. Intracranial procedures can be further categorized, with *brain surgery* referring to the tasks performed within the brain, such as procedures involving tumors, angiomas, hydrocephalus, and hematomas; and *extracerebral surgery* referring to the tasks performed within the braincase but outside the brain, such as procedures involving aneurysms and meningiomas. *Cranial base surgery* is defined as extracerebral surgery within the cranium but through the base of the skull toward neighbor regions such as the pharynx, larynx, sinuses, and neck. Cranial base surgery includes procedures involving neurinomas, meningiomas, malformations, and fractures.

16.1.2   Instrumentation in Neurosurgery

Various tools are used to aid in visualization during neurosurgery. Optical magnification, such as surgical microscopes, binocular magnifying glasses, or monocular endoscopes, are often used to precisely explore the CNS and view anatomical details down to the submillimeter scale during open head surgery. Endoscopes can be introduced directly into the braincase and the brain during open head surgery or through a burr hole during closed head surgery. Majority of neuroendoscopes that are used are rigid, although flexible neuroendoscopes are also available. Neuroendoscopy is primarily performed within the cerebral ventricles because the cerebrospinal fluid improves both visibility and maneuverability of surgical tools manipulated through the endoscope (Ahmad and Sandberg, 2010).

Numerous tools are used in neurosurgery, with many of them such as forceps, manipulators, and coagulators similar to those used in general surgery. Microsurgical instruments (Fig. 16.1) are often required because of the small size of anatomical components, including vessels, nerves, and thin membranes, and are better suited for use with optical microscopes and neuroendoscopes. Microsurgical instruments have various shapes and functions and include microscissors, micromanipulators, bipolar microcoagulators, and micro clip-holders. These instruments are used during dissection and manipulation and are frequently used in conjunction with cotton pads or sponge pieces to protect fragile anatomic structures. Microsurgical instruments are also used to manipulate vascular clips.

Powered surgical tools are also available for neurosurgery, including ultrasonic aspirators and CO_2 laser scalpels. The ultrasonic aspirator, among the most common powered surgical instruments in neurosurgery, is used for brain tumor removal. The energy delivered at the tip of the ultrasonic probe selectively dissociates components within the nervous tissue using different energy settings. Thus, it is often possible to remove tumor tissue while protecting vessels, nerves, or bone. This instrument is also used to remove lesions located elsewhere within the head and spine. Recent advances in CO_2 laser scalpel technology will likely spread its utilization as an additional tool for tissue removal or dissection (Ryan et al., 2010).

(a)

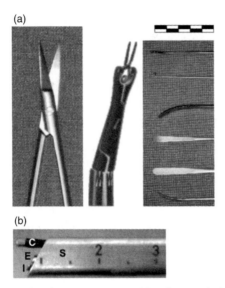

(b)

Figure 16.1 Instrumentation in neurosurgery: (a) microsurgical instruments: scissors (left), vascular clip holder (center), and several manipulators and dissectors (right); horizontal bar is 10 mm. (b) Distal extremity of a neuroendoscope (E) emerging from its metallic introducer sheath (S, values are in cm); the tip of a monopolar coagulator (C) and the distal aperture of a second channel for a second instrument (I) are visible.

16.1.3 Complications

Beyond classical infections and hemorrhage, which are not specific to neurosurgery, the complications most often associated with neurosurgery are neurological deficits. Some of the most common neurological deficits include motor deficits such as hemiplegia, language disorder or aphasia, visual field loss, and behavioral problems. These postoperative neurological deficits typically result from difficulties in differentiating a tumor from its environment or result from surgical trauma during tumor dissection. Therefore, the challenge during neurosurgery is to protect healthy tissue while removing pathologic tissue such as tumors or dead tissue (Talacchi et al., 2010).

Several methods have been proposed to assist the operator to differentiate a tumor from healthy tissue. The first group of methods features intraoperative contrast enhancement, such as protoporphyrin IX fluorescence; the second group uses neuroimaging coupled with neuronavigation. Of these methods, neuronavigation has become one of the most common tools used to locate and identify the tumor among adjacent tissues. Neuronavigation is a technique that allows the surgeon to guide a neurosurgery by matching pre- or intraoperative imagery with views of the surgical field. Surgical trauma during dissection is always possible, but the risk can be reduced by advanced planning before surgery, precise navigation during surgery, a priori knowledge of the surgical anatomy, and experience. However, intraoperative brain shifts, or local shifting of the brain during surgery, must also be taken into account during any type of operation. Brain shifts are usually related to cerebrospinal fluid leakage or tissue removal and can cause misinterpretation of brain topography during neuronavigation. This misinterpretation can result in trauma during surgery, and solutions have been proposed to address this limitation (Kajiwara et al., 2010; Valdés et al., 2010).

16.1.4 Functional Neurosurgery

Functional neurosurgery is a type of surgery used mainly to alleviate severe neurological symptoms, such as tremors, muscle rigidity, epilepsy, and pain, generated by abnormal neurocircuitries within the CNS. Therefore, the symptoms, and not the disease itself, are treated. Functional neurosurgery uses neuromodulation techniques that interfere directly or indirectly with such pathologic circuitries within the gray matter territories of the brain (cortex or deep brain nuclei) or within the white matter fascicles of the spinal cord. The primary neuromodulation techniques in place today include lesioning (sectioning, coagulation, and irradiation), electrical stimulation (implantation of electrodes and chronic stimulation), and noninvasive, repetitive transcranial magnetic stimulation of the cortex. Several electrophysiological techniques are also used intraoperatively to localize functional areas, such as neuronal activity recordings. The field of functional neurosurgery also sometimes involves surgical techniques concerning the PNS, such as neurotomy, in which peripheral nerves are sectioned to relieve pain.

16.1.5 Stereotactic Neurosurgery

Stereotactic surgery is an image-guided surgical technique in which an external frame is attached to the head to orient the surgeon during neurosurgery and to support surgical tools. A stereotactic frame is securely fixed to the skull bone and is used to establish accurate geometric locations of targets, such as the thalamus or a tumor. The frame is typically a rigid metallic structure (Fig. 16.2). Location boxes, also called *coordinate indicators*, are placed on the frame before imaging. Location boxes are geometrically calibrated, with fiducial markers embedded within the walls, and are used to calculate the geometric position (X-, Y-, and Z-coordinates in millimeters) of targets; specific boxes are used with X-ray, computed tomography (CT), or magnetic resonance imaging (MRI) systems (Fig. 16.2). After definition of the target and calculation of coordinates, an instrument holder is fixed to the frame. The instrument holder is also incorporated onto the frame and consists of a stereotactic arc and a guiding instrument that support tools such as biopsy probes or electrodes (Fig 16.2). This technique is used to track specific probes as they are navigated to reach the target with submillimeter accuracy. Similar techniques that use fiducial markers that are not fixed to the skull, called *nonrigid fixation*, have also been developed; these techniques are often referred to as *frameless stereotactic surgery* (Owen and Linskey 2009; Ringel et al., 2009). More and more stereotactic surgeries are making use of very sophisticated planning software, very similar to those used for neuronavigation planning (described in Section 16.4).

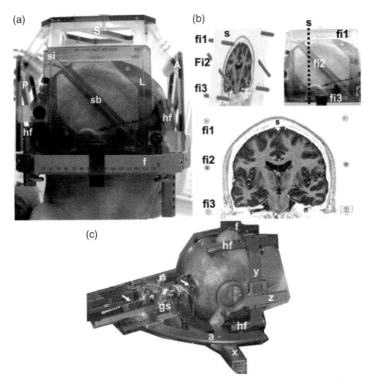

Figure 16.2 Stereotactic neurosurgery: (a) Stereotactic frame (Leksell G Frame, Elekta, Sweden): frame (f) with rigid head fixations (hf); stereotactic box or coordinate indicator (sb; here MRI compatible) with anterior (A), posterior (P), superior (S), and lateral (L) plates. Stereotactic coordinate indicators (fiducials) are visible (si; fine linear tubes embedded within plate and filled with MRI contrast agent). (b) Calculation of stereotactic coordinates: X-, Y-, and Z-coordinates are calculated from stereotactic fiducials (e.g., fi1, fi2, and fi3) visible on the MRI slice (s). (c) Stereotactic devices are mounted on an instrument holder: a stereotactic arc (a) is mounted on the frame (f) fixed on the head (hf); X, Y, and Z target coordinates are adjusted (x, y, and z); a guiding system (gs) allows placing a probe (here an electrode, white arrows). *Source*: Courtesy of the Department of Neurosurgery, Auvergne University—University Hospital, Clermont-Ferrand, France.

16.1.6   Neuroimaging for Neurosurgery

Neuroimaging is performed preoperatively, intraoperatively, and postoperatively in order to direct, guide, and assess a procedure, respectively. The most common neuroimaging techniques include MRI, CT, positron emission tomography (PET), angiography, and ultrasound. These techniques are used to view cranial and brain anatomy, blood vessels, and cellular activity.

MRI is a nonionizing technique that takes advantage of the magnetic properties of water protons in soft tissues. A paramagnetic agent, such as gadolinium, is injected into the blood stream, to visualize the rupture of the blood–brain barrier in tumors or following head injury. Functional magnetic resonance imaging (fMRI) is a specialized MRI acquisition technique that reveals brain regions activated during specific tasks performed by the patient, such as speaking or moving fingers. Diffusion tensor imaging (DTI) is another specific technique using MRI that quantifies the movements of water molecules within the white matter and allows displaying interconnectivity pathways using fiber tracking techniques. Both fMRI and DTI techniques necessitate post-acquisition processing that provides images usable for medical applications.

PET is a functional, rather than anatomical, imaging technique, relying on radiation properties of isotopes. A radiotracer containing an isotope is injected intravenously just before the examination. It binds specific molecules within targeted regions. The most common radiotracer, fluorodeoxyglucose, is an indicator of metabolism, or cellular activity, of a tissue and allows differentiation of tissues such as tumors from its environment. Post-acquisition processing of image data sets is mandatory before medical utilization.

CT relies on X-rays that travel through living tissues and ionize molecules and allows for viewing image slices of the brain and cranial anatomy. In CT angiography, radiopaque contrast agents, such as iodine, can be injected into the blood stream and used to visualize blood circulating in vessels of the brain. Standard angiography is also used and relies on X-rays like CT but is not a slice imaging technique. Angiography provides pictures acquired sequentially, like movies, after direct injection of radiopaque agents into arteries. This is the gold standard used to visualize the blood vessels and allows vascular malformations such as aneurysms to be viewed. In modern angiography rooms, two orthogonal X-ray tubes work simultaneously, rotating around the patient. This double acquisition allows reconstructing 3D rendering of vessels, almost in real time, using very fast online computer analysis of images.

Ultrasonography allows for the visualization of biological boundaries between anatomic structures or lesions. An ultrasonic probe is placed at the brain surface following craniotomy, and lesions can be located within the brain. This technique has historically been the most common method to guide navigation within the brain and continues to be used because of its real-time capabilities, low cost, and ease of use. However, ultrasound provides limited anatomical information compared to MRI and has low contrast between tumor and healthy tissues. Recent neuronavigation advances have primarily been made with other imaging modalities, with the focus of ultrasound research on neuroimaging now toward the evaluation of the degree of tumor removal rather than navigation.

Neuroimaging, as most other medical imaging fields, utilizes a worldwide archive format called *Digital Imaging and Communications in Medicine* (DICOM). This standard allows the transfer and archive of images independent of the manufacturer of the imaging equipment and to different users, such as radiological departments, operating rooms (OR), and archive systems through medical servers. Different levels of authorization and security are used to safeguard from data loss and to guarantee that data is only sent to authorized users. Most medical servers use a standardized communication process called *Picture Archiving and Communication Systems* (PACS).

16.2 INTRODUCTION TO NEURONAVIGATION

In the past 20 years, the most important technological advancement in neurosurgery has been the utilization of advanced imaging techniques to guide the surgeon within the cranium or spinal column. These advances have greatly improved the classical surgical field of view by providing 3D views of anatomy in real time. As a consequence, the surgeon's interpretation of the surgical anatomy and positions of instruments have been made significantly easier. Neuronavigation became possible in the early 1980s when the coregistration of CT images with the patient's head became feasible, ushering in a new era of computer-aided surgery (Kelly et al., 1985). Immediately, neuronavigation provided significant pre- and intraoperative improvements in the accuracy of location of lesions within the head. This allowed the bone flap to be optimally positioned during craniotomy, facilitating the access and approach route to lesions. The first tangible improvement was the reduction of the size of the skull flap and therefore reduced trauma during surgery. Advances in neuroimaging, mechatronics, and in particular, computer science, quickly improved navigation systems and the technology has not had since spread worldwide. There are currently a number of fields that have also adopted image guidance, including surgery, radiology, radiotherapy, and endoscopy. Computer-aided surgery is an allied concept that includes preoperative, intraoperative, or postoperative methods of computer assistance in these and other fields.

Neuronavigation has become the *de facto* gold standard when performing both noninvasive and invasive neurosurgical procedures. Almost all modern operative rooms are equipped with at least one navigation system. However, navigation is not always used during neurosurgery, and its usage depends on the clinical and local context. For example, computer-aided navigation may not be required for location of large superficial lesions. The choice in using alternative or historical techniques, such as open head surgery guided by the experience of the surgeon or by basic fiducial markers, depends on the difficulty of the procedure, the surgeon's skill level, the availability of a neuronavigation system, and in the case of emergency surgeries, the time required to launch the navigation system. Paradoxically, efforts to evaluate the clinical benefit of neuronavigation have not shown convincing results. This is likely because the specific impact of this technique has been difficult to assess among the numerous clinical parameters such as tumor invasiveness, effects of other therapeutic modalities such as radiotherapy and chemotherapy, and reorganization of brain functions.

16.3 NEURONAVIGATION SYSTEMS

The most common intracranial neuronavigation systems are based on optical tracking and consist of three principal components (Fig. 16.3): a tracking system, a

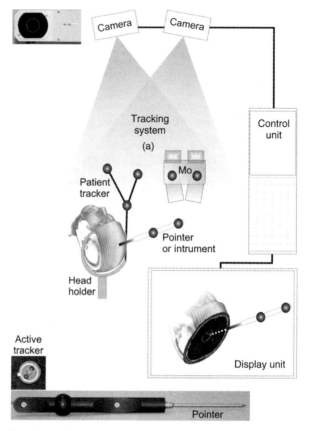

Figure 16.3 Overview of a common navigation system. a. The tracking system is composed of (i) two or three cameras mounted on a console; (ii) a patient tracker usually fixed on the head holder (fixed to the surgical table); (iii) a pointer (or other surgical tools, or instruments, tracked, that is, with trackers affixed to them); and (iv) a tracked surgical microscope (Mo). b. The control unit manages the tracking system, the display unit, and several interfaces with medical networks. c. The display unit displays in real time the pointer location according to the patient's medical images; an interactive system allows for real-time modification of navigation parameters.

display unit, and a control unit. The tracking system is used to track surgical instruments, surgical devices, and the patient's head in 3D space. Meanwhile, the display unit shows medical images oriented and reconstructed according to the tracking configuration, which is computed and updated by the control unit.

16.3.1 The Tracking System

The tracking system comprises either two or three infrared cameras that are mounted onto a device and used to track several objects concurrently in real time.

The device holding these cameras is an articulated arm fixed to the ceiling of the OR or carried by a mobile system. Between two to five fiducial markers, or trackers, are attached to or embedded within each object. Either passive or active trackers are used, where passive trackers reflect the light emitted by the infrared cameras and active trackers emit their own light that is received by the cameras.

Tracking systems can typically achieve submillimeter accuracy throughout the surgical field. By manually manipulating the location and orientation of the cameras, the surgical team centers the working space of the tracking system onto the surgical field. Since it is a visual system, something placed between the cameras and the trackers can interrupt operation of the system. Indicators on the control unit indicate whether there is a failure resulting from the camera or tracker, but whether the cause is due to failure of the specific component or blockage is not specified.

Several objects are tracked: (i) a patient tracker, (ii) surgical instruments, (iii) a surgical microscope, and (iv) a pointer. The patient tracker is fixed securely to the patient's head, usually by attaching it to the surgical clamp, leading to rigid relationship between head, tracker, and the surgical table. Various surgical instruments may be used, including ultrasonic aspirators, biopsy probes, and cannulae, as well as imaging devices, such as endoscopes or ultrasound probes. Trackers are also placed on the pointer, and on the surgical microscope. The pointer is used for registration, calibration, and guidance.

The trackers are usually locked onto surgical instruments just prior surgery, so the tracker's position in space must be calibrated to the navigation system before utilization. Owing to size and shape constraints, microsurgical instruments cannot currently bear trackers. Sterilization also precludes the use of some trackers that are not sterilizable.

16.3.2   The Display Unit

The display unit consists of two parts, the display itself and an input system. A computer keyboard is a common input system; however, touch screen monitors are now becoming available. The display unit is placed near the surgical field, well visible by the operator, the assistant, and OR nurses. Sophisticated systems with multiple screens and advanced touch screen capabilities are available. The most sophisticated systems display images of virtual objects, such as tumor contours, simultaneously within the optics of the surgical microscope and on the screens in OR; this is referred to as *image injection* and enhances the information available within the surgical field (described below).

Several types of images can be displayed on the display unit, including neuroimagery (MRI, CT, angiography, PET, fMRI, ultrasound) and the optical view of the surgical field. The surgical field view can be provided from the surgical microscope or neuroendoscope, or a larger OR view is provided from an overhead camera. Modern neuronavigation systems are capable of fusing different neuroimaging modalities using advanced 3D renderings, with the modality selected for fusing depending on the surgeon's aim. For instance, during aneurysm surgery it can be useful to overlay angiography with CT or MRI. Only preoperatively coregistered

(Section 16.5) images can be overlaid. Optical images are not usually coregistered or overlaid (Fig. 16.5). Finally, different virtual objects, such as tumor contours or planned trajectories, can be displayed and overlaid on images as well (Section 16.4).

Most surgical teams use only preoperative neuroimaging, usually captured a few hours or days before surgery; however, intraoperative neuroimaging is also occasionally performed for research purposes. Intraoperative neuroimaging, typically using CT or MRI, has historically been performed directly inside the CT or MRI system, requiring expensive imaging systems to be located within the OR. More recently, the tendency has been to place the imaging systems in hybrid rooms, shared with a neuroradiology department, where the patient is transferred as needed.

Neuroimaging is most often presented using one of two slice orientation conventions (Fig 16.4). The most common is the classical medical imaging convention, in which the axial, coronal, and sagittal planes are used. This convention is generally used when raw neuroimages are acquired. A second convention is used to provide a perspective relative to a surgical instrument, pointer, or microscope. In this case, images are reconstructed from raw neuroimages along the longitudinal and perpendicular axes of these objects. Thus, pseudoaxial, pseudocoronal, and pseudosagittal views are available and displayed in real time when the tracked tool moves within the surgical field. This reconstructed view allows the surgeon to navigate within the virtual radiological anatomy while linked to the real surgical field, according to the orientation of the tool placed in his or her hand.

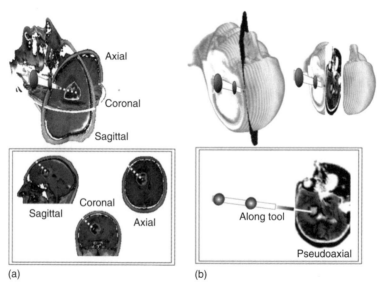

Figure 16.4 Display modalities of medical imaging during navigation. (a) Common display of MRI slices according to classic axial, coronal, and sagittal planes, used in medicine. (b) Reconstructed slice (pseudoaxial) along the surgical tool (here a pointer). The tumor is visible within the left hemisphere of the brain, as a black volume (tumor necrosis) surrounded by a white contour (contrast enhancement of tumor vessels).

16.3.3 The Control Unit

The control unit links the tracking system and display unit and interfaces with medical information networks. It exchanges the import–export of neuroimages, usually through PACS servers; exchanges images to the display unit; manages the real-time analysis of tracking data; and manages robotic control signals.

Most navigation systems feature passive navigation, with the surgical instruments, microscope, or pointer moved manually by the user. Images are then updated in real time according to the location of the tracked object within the surgical field. Motorized microscopes, which move using manual control independent of the control unit, are also used during passive navigation.

Robotic control of instrument holders and operative microscopes is available in advanced navigation systems and is usually applied during minimally invasive procedures. Robotic control is realized when the orientation and location of tracked objects are mastered by the control unit, using the tracking system, allowing completely automated targeting.

There are two different robotic systems used during neuronavigation: a robotic arm, which serves as an instrument holder, and a robotic microscope, which guides the surgical microscope and can also be converted into an instrument holder (Fig 16.5). These systems position the instrument or microscope along a planned trajectory determined preoperatively or intraoperatively in real time. This

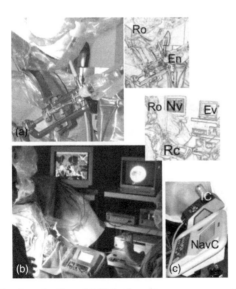

Figure 16.5 Robotized navigation. (a) Robotic microscope converted into a robotic arm (Ro) carrying a neuroendoscope (En). (b) Navigation of the robotic arm (Ro), display unit (Nv), display of the endoscope view (Ev), and remote control (Rc). (c) Remote control comprises a lever for adjusting neuroendoscope progression (Invasive control, Ic) and the neuronavigation control unit (NavC). *Source*: Courtesy of the Department of Neurosurgery, Auvergne University—University Hospital, Clermont-Ferrand, France.

technology is useful during the navigation phase where a determined trajectory of a tool into the brain is to be followed accurately. The main limitations of robotic control are geometric in nature since the robotic system cannot always achieve all of the desired positions along the planned trajectory, because of the overall bulk of the system within the surgical environment. During robotic procedures, the navigation system is used to verify the motions and trajectories and to adjust them accordingly. The functional organization of navigation systems is described in the following section.

16.4   IMPLEMENTATION OF NEURONAVIGATION

Neuronavigation consist of three steps: *surgical planning*, in which the imagery is prepared according to the surgical aim; *patient registration*, where the patient's head is matched with image data before surgery; and *surgical navigation*, which enhances the surgical content available to the neurosurgeon and allows the procedures to be performed minimally invasively. The overall geometrical accuracy during surgical navigation is strongly dependent on the quality of the preparation of the operation during surgical planning and patient registration. Another important parameter affecting geometric accuracy is the brain shift. Image characteristics, such as pixel size and image quality, must be taken into account as well: pixel size influences the geometric accuracy of navigation and registration, and image quality influences the ability to identify anatomic structures and lesions. The images most appropriate to the surgical aim must be selected before the procedure and may include the pertinent MRI sequences and images from one of several additional imaging modalities such as CT or PET.

16.4.1   Surgical Planning

Historically, neurosurgical planning has almost exclusively been used to provide the surgeon with a mental picture of the intraoperative environment and to define the appropriate approach route into the brain. Manual methods were used to fit preoperative imagery to clinical data to aid in surgical planning, and this technique was highly dependent on the expertise of the surgeon. The advent of computer assistance has enabled neurosurgeons to better visualize and understand image data, to more accurately register patient data, and to track surgical instruments and devices.

Modern surgical planning is the first step performed during neuronavigation, in which a surgeon and/or an assistant organizes morphological image information according to the surgical aim. DICOM image data sets are imported into a workstation, which is usually the display unit, and reformatted according to the manufacturer's software. Post-acquisition processed image data sets (fMRI, DTI, and PET) are also supported by most navigation systems. These advanced morphological and functional data allow extensive analysis of the target and its environment.

For advanced planning, it is often necessary to contour a target, such as an anatomic structure or a tumor and its environment, on images. Contouring is performed manually or semiautomatically after manual selection of a region-of-interest on the target. The selection of pixels belonging to the target is achieved, slice by slice, by manually drawing on the image using visual identification of the target and/or by automated selection of pixels according to the value of signal (MRI) or density (CT) after manual adjustment of a threshold. After acceptation by the user, these contours are transformed into 3D virtual objects and overlaid onto images (Fig. 16.6). Fiducial registration is also performed during the planning phase when the navigation system incorporates spatial information from fiducial markers that are positioned on the patient's head before the acquisition of at least one series of images. This overall planning phase is also referred to as *virtual reality surgery planning*.

16.4.2   Patient Registration

During patient registration, the patient's head is matched with image data before operation of the navigation system. Patient registration can be performed by using fiducial markers affixed to the skin, a fiducial tracker secured on the patient's head (usually with an elastic headband), or by using anatomical landmarks on the

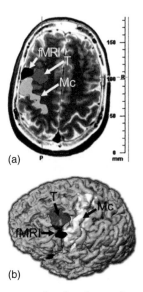

(a)

(b)

Figure 16.6   Virtual reality surgery planning (example, tumor removal). (a) The tumor contours (T), the motor cortex (Mc), and the functional area involved in the motor activity of the face (revealed using fMRI) are overlaid onto a morphologic MRI slice. (b) Three-dimensional representation of the patient's brain with tumor (T), motor cortex (Mc), and face-related fMRI (fMRI) objects. *Source*: Courtesy of the Department of Neurosurgery, University of California—Los Angeles, US.

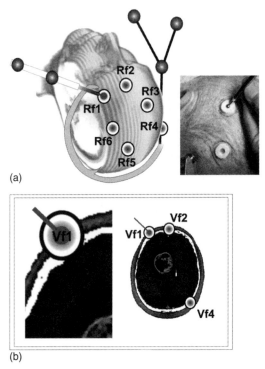

(a)

(b)

Figure 16.7 Patient registration. In this example, the registration is realized using six fiducials (f1–f6), with the fiducial registration, or specification of the location of each fiducial by the user using software functionalities, already complete. (a) The pointer is successively precisely positioned on each real fiducial (from Rf1 to Rf6) on the patient's head (on the right, e.g., of real patient registration). (b) According to the location of the equivalent virtual fiducial, only fiducials visible on the MRI slice are visible, Vf1, Vf2, and Vf4; each real fiducial is matched with its equivalent virtual fiducial (e.g., Rf1 with Vf1). When all virtual fiducials are correctly registered with all real fiducials, the user is asked to validate the geometric precision of matching before beginning the navigation.

surface of the head. When patient registration relies on fiducial markers or trackers, fiducial registration on the images must be completed before patient registration. This matching procedure coregisters the patient's head with images within the work space of the navigation system. This mechanical step is operator dependent because the surgeon or assistant must point manually to each fiducial or anatomic landmark, using the pointer (Fig. 16.7).

Any region within the surgical work space that is selected with a tracked instrument (such as the pointer or any other tracked tool) is identifiable on the display unit and is usually indicated as a cross or a circle overlaid on the slice images. If the surgical microscope is tracked, its focal point is displayed and is updated when the focal distance is changed. The direction of a tracked tool is also displayed, with

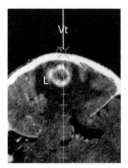

Figure 16.8   Virtual trajectory. In this real example, the virtual trajectory (Vt) is represented as a line displayed on a reconstructed image (CT) where a lesion (L, here a metastasis) is visible (gray mass circled with a white ring). The virtual trajectory is projected within the head and scaled (cm) to facilitate the navigation.

its longitudinal axis shown as a line representing its virtual trajectory. Figure 16.8 shows a real example of a virtual trajectory.

16.4.3 Navigation

Surgical navigation begins after the user (a surgeon or an assistant) validates the anatomical match obtained during patient registration. This final validation is relatively subjective. The user compares the pointer location on the patient's head and its projection onto the neuroimages. Overall geometric accuracy is verified by comparing obvious superficial anatomic landmarks, such as the nose, ears, and eyes. During this anatomic calibration, a global geometric accuracy value is provided by the neuronavigation system. When this value exceeds about 5 mm, it is generally recommended to redo the patient registration.

Navigation enhances the surgical content available to the neurosurgeon and allows the procedures to be performed minimally invasively. Navigation allows the placement of a craniotomy to be optimized in order to facilitate the approach route and avoid critical functional regions of the brain. As a consequence, both the size of the bone flap and the approach route can be reduced. Further, the navigation enables precise target location and allows the surgeon to anticipate its visualization. During the removal of a tumor, the control of the degree of debulking, or partial removal, is also facilitated by navigation.

Surgical planning can be modified in real-time intraoperatively in the event that unanticipated data is encountered during navigation. This is performed by selecting new anatomical landmarks with the pointer to reregister data in the surgical field. For example, pertinent anatomic landmarks, such as a vascular loop or the bottom of a sulcus, can be registered if they are observed during the approach into the brain. This can facilitate the interpretation of navigation imagery and can also assist the user in evaluating the brain shift.

Brain shifts may also require patient registration and planning to be updated during surgery. This is required because the brain shifts during surgery, typically as a result of tumor removal, cerebrospinal fluid leakage, or intracranial hemorrhage. The brain can shift from a few millimeters to several centimeters, depending on the progression of the surgical procedure, with brain shift increasing as a larger volume of tissue is removed. Patient registration can be updated intraoperatively using ultrasound, CT, or MRI and can allow for quantification of the volume of lesion removed by comparing preoperative with intraoperative imagery. However, this updating procedure is rarely used since very few teams have intraoperative imaging facilities (Uhl et al., 2009).

Geometric accuracy remains the primary limitation of navigation systems, and inaccuracies can result from misinterpretation of anatomy or lesion boundaries. For this reason, it is strongly recommended to consider navigation information to be complementary to direct visual information, on which surgical decisions are made. In spite of this limitation, navigation serves a critical purpose in neurosurgery and greatly facilitates the interpretation of information within the surgical field for all neurosurgeons, regardless of experience or skill level.

16.5 AUGMENTED REALITY AND VIRTUAL REALITY

The information available to a neurosurgeon during an operation comes from pre-operative data, in particular image data integrated during surgical planning, as well as the surgical content, which represents all of the information contained within the operating field. All techniques that can enhance the surgical content are valuable to a surgeon, whose interest is to improve the intraoperative decision workflow. The first significant enhancement to the surgical content was optical magnification, which provided improved visualization of anatomic details and more accurate dissection. Navigation, as well as the development of more modern optical instruments such as surgical microscopes and endoscopes, has subsequently provided dramatic improvement. The enhancements provided by these tools are often described by the terms *augmented reality*, *virtual reality (virtuality)*, or *augmented virtuality*, which are often confused in the context of computer-aided surgery.

The term *reality* represents the actual surgical field, including optical images of the real patient's anatomy. Future devices giving nonoptical information such as temperature or texture profiles of brain tissue could theoretically be included in the concept of reality when validated for surgical applications. Reality implies real-time data, at least for actions such as hemostasis; thus, remote surgery, if performed in the future, must keep high speed interaction between the surgeon and the OR (Marescaux et al., 2001).

Augmented reality most often refers to improvements in image quality and can depend on technical issues such as camera resolution, or data fusion of multiple image modalities, which are superimposed within the surgical field. Navigation systems with image-injection capabilities can also provide a form of augmented reality, where virtual objects are injected into a surgical microscope viewing eye piece. The display of virtual objects within the optics of the microscope is controlled by a focus unit that allows the image-injection system to display in real time the exact outlines of virtual objects according to the depth of the field of view.

Virtual reality, or *virtuality*, is the real-time display of the representation of reality, built from medical images; it refers to the creation of data sets of medical images and virtual objects such as tumor contours and instrument representations. The medical images are displayed according to the surgeon's preference and surgical goals. Virtual objects are created by outlining a surgical target (see above) or, in recent navigation systems, by automatic processes that can perform tasks such as extracting blood vessels or bones from other tissues. However, the lack of standardized techniques to transfer virtual object data precludes data transfer between medical terminals, such as between computer stations and navigation systems.

Most navigation systems tend to propose *augmented virtuality*, or reality merged with virtuality on the display unit. Augmented virtuality refers specifically to the overlay of reality on virtuality. However, commonly, it is *pseudoaugmented virtuality* with two separate windows on the same display unit: the first to display the virtuality and the other to display the reality. The real-time fusion of two imaging modalities, such MRI and endoscopic imagery, is available on advanced systems but currently only in research conditions. At a minimum, one can consider that the real-time display of images according to tracked instruments enhances virtuality; several options to display images according to the operating context also dramatically improve the understanding of surgical content. Figure 16.9 summarizes these concepts in the context of neuronavigation.

16.6 SUMMARY/FUTURE

Neuronavigation has become widely used and is the new gold standard in neurosurgery. Although neuronavigation is mainly used for cranial surgery, it is rapidly being extended to spinal surgery. The principle of image guidance is under development in almost all medical areas, and in particular in surgery, radiology, and radiotherapy. Endovascular surgery and radiosurgery, both closely related to neurosurgery, also largely benefit from image guidance.

However, there are still challenges related to patient (body or organ) registration, which should be partially mitigated by using updated registration procedures. MRI, X-ray, CT, and ultrasound devices are the techniques most frequently used for this purpose and are already available in advanced interventional suites where intraoperative facilities are shared by surgical and radiological teams. Each imaging technique has advantages and disadvantages related to the conditions of acquisition. Some of these include image acquisition time, size of the machine, portability, presence of a magnetic field or ionizing radiation, image quality, repeatability, reproducibility, and cost.

The trend in the production of future neuronavigation systems is toward devices that are lighter and easier to use, such as those that depend on anatomical markers rather than fiducial markers. Magnetic navigation, allowing navigation of the tool

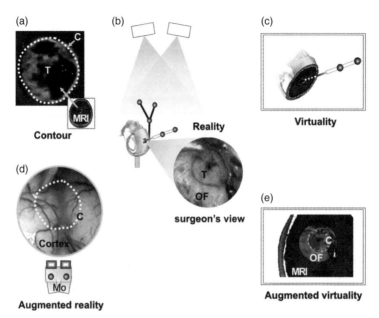

Figure 16.9 Concepts of surgical content enhancement. (a) The tumor (T) is outlined (contour, C) on an MRI image. (b) Reality is typically the surgeon's view (operating field, OF). (c) Virtuality, or virtual reality, is typically the real-time display of the representation of reality, here an instrument representation on a medical image. (d) Example of augmented reality, in which the tumor contour (C, or virtual object) is injected (overlaid) in the surgical microscope (MO) field of view showing the cortex. (e) Augmented virtuality corresponds to the enhancement of virtuality; here, the operating field is overlaid on MRI.

tip within the brain, is expected to be available shortly (Sangra et al., 2009). With simpler neuronavigation systems, utilization of navigation will increase in neurosurgery and other disciplines, such as otolaryngology. Multimodality imaging, which combines modalities such as MRI, fMRI, and PET, has also become common in neurosurgery but requires considerable effort and time to prepare data for surgical planning and navigation. Many other improvements await further technological advances, such as those for tracking, fiducials, patient registration, virtual reality, augmented reality, and augmented virtuality, as well as robotic capabilities.

Beyond serving as a surgical aid, neuronavigation systems allow for the collection of data that is beneficial to learning and teaching. Virtual reality surgery planning is already possible for this purpose (Kockro et al., 2007; Lemole et al., 2007). Neurosurgical simulators, developed in parallel with navigation systems, are still rudimentary, and in the future, it is likely that these tools will also benefit from advances in navigation (Yang et al., 2009).

Finally, neuronavigation technology, when connected with high speed communication networks, is expected to enable remote surgery or telesurgery (Mendez et al., 2005). Applications of telesurgery will be in long-distance medical assistance such as for military or civilian applications in remote areas, ships, or even spacecraft. In addition, telesurgery could be used to provide mean distance assistance for inadequately trained surgeons in vital emergency situations, such as for extradural hematoma.

The reader is referred to visit web sites of medical companies providing clinical equipment (see references, e.g., Brainlab; Elekta; General Electric; Integra-Radionics; Isis; Medtronic; Philips; and Storz).

▰▰▰ REFERENCES

Adamis AP, Shima DT, Tolentino MJ, Gragoudas ES, Ferrara N, Folkman J, D'Amore PA, Miller JW. Inhibition of vascular endothelial growth factor prevents retinal ischemia—associated iris neovascularization in a nonhuman primate. Arch Ophthalmol 1996;114:66–71.

Advincula AP. A preliminary comparison of mechanical compression amongst three electrosurgical devices. J Minim Invasive Gynecol 2005;12:S43–S44.

Agarwala VS, Kim KY, et al. *An Electrochemical Methods for Investigating Corrosion in Rubbing Surfaces. Materials Evaluation under Fretting Conditions*. West Conshohocken (PA): American Society for Testing and Materials; 1982. p 106–124.

Ahčan U, Zorman P, Recek D, Ralca S, Majaron B. Port wine stain treatment with a dual-wavelength Nd:Yag laser and cryogen spray cooling: a pilot study. Lasers Surg Med 2004;34:164–167.

Ahmad F, Sandberg DI. Endoscopic management of intraventricular brain tumors in pediatric patients: a review of indications, techniques, and outcomes. J Child Neurol 2010;25(3):359–367.

Ahmed A, Raab S, et al. Metal/cement interface strength in cemented stem fixation. J Orthop Res 1984;2(2):105–118.

Albrektsson T, Albrektsson B. Osseointegration of bone implants. A review of an alternative mode of fixation. Acta Orthop Scand 1987;58(5):567–577.

Albrektsson T, Jacobsson M. Bone-metal interface in osseointegration. J Prosthet Dent 1987;57(5):597–607.

Al-Sakere B, Andre F, Bernat C, et al. Tumor ablation with irreversible electroporation. PLoS One 2007;2(11):e1135.

American Academy of Ophthalmology. *Basic and Clinical Science Course: Lens and Cataract. Basic Principles of Ophthalmic Surgery*. San Francisco (CA): American Academy of Ophthalmology; 2006.

American Cancer Society. 2009. The history of cancer. Available at http://www.cancer.org /docroot/CRI/content/CRI_2_6x_the_history_of_cancer_72.asp. Accessed 2009 Sep 3.

Amstutz HC. *Hip Arthroplasty*. New York: Churchill Livingstone; 1991.

Amstutz HC, Jacobs JJ, Ebramzadeh E. Current status of metal-on-metal hip resurfacing. Orthop Clin North Am 2011;42(2).

Anderson NJ, Beran RF, Schneider TL. Epi-LASEK for the correction of myopia and myopic astigmatism. J Cataract Refract Surg 2002;28:1343–1347.

Anderson R, Parrish J. Selective photothermolysis: precise microsurgery by selective absorption of pulsed radiation. Science 1983;220:524–527.

Andrew E, Berg KJ. Nephrotoxic effects of X-ray contrast media. J Toxicol Clin Toxicol 2004:325–332.

Medical Devices: Surgical and Image-Guided Technologies, First Edition.
Edited by Martin Culjat, Rahul Singh, and Hua Lee.
© 2013 John Wiley & Sons, Inc. Published 2013 by John Wiley & Sons, Inc.

Ang W, Khosla P, Riviere C. (2001). An intelligent hand-held microsurgical instrument for improved accuracy. In: Proceedings of the 23rd Annual International Conference IEEE Engineering in Medicine and Biology Society; 2001 Oct. p 3450–3453.

Antonelli PJ, Lundy LB, Kartush JM, Burgio DL, Graham MD. Mechanical versus CO2 laser occlusion of the posterior semicircular canal in humans. Otol Neurotol 1996;17:416–420.

Anvari B, Tanenbaum BS, Milner TE, Kimel S, Svaasand LO, Nelson JS. A theoretical study of the thermal response of skin to cryogen spray cooling and pulsed laser irradiation: implications for treatment of port wine stain birthmarks. Phys Med Biol 1995;40:1451.

Applegate RA, Howland HC, Sharp RP, Cottingham AJ, Yee RW. Corneal aberrations and visual performance after radial keratotomy. J Refract Surg (Thorofare, N.J.: 1995) 1998;14:397–407.

d'Arsonval M. Action physiologique des courants alternatifs. Comptes Rendus des Séances et Mémoires de la Société de Biologie 1891;43:283–286.

Bagley DH. Expanding role of ureteroscopy and laser lithotripsy for treatment of proximal ureteral and intrarenal calculi. Curr Opin Urol 2002;12:277–280.

Bagley DS, Grasso M. *Flexible Ureteroscopy with the Flex-X2 Ureteroscope*. Tuttlingen: Endo Press; 2009.

Bailey MR, Khokhlova VA, Sapozhnikov OA, Kargl SG, Crum LA. Physical mechanisms of the therapeutic effect of ultrasound (a review). Acoust Phys 2003;49(4):369–388.

Baim DS, Grossman W. *Grossman's Cardiac Catheterization, Angiography, and Intervention*. Lippincott Williams & Wilkins; 2006.

Ballantyne G. *Minimally invasive surgery for diseases of the colon & rectum: The legacy of an ancient tradition*. In: Jager RM, Wexner S, editors. *Laparoscopic Colectomy*. New York: Churchill & Livingstone; 1995. p 13–23.

Bargar WL, et al. Primary and revision total hip replacement using the Robodoc(R) system. Clin Orthop Relat Res 1998;354:82–91.

Barrack R. Modularity of prosthetic implants. J Am Acad Orthop Surg 1994;2:16–25.

Barron HV, Every NR, Parsons LS, Angeja B, Goldberg RJ, Gore JM, Chou TM. The use of intra-aortic balloon counterpulsation in patients with cardiogenic shock complicating acute myocardial infarction: Data from the National Registry of Myocardial Infarction 2. Am Heart J 2001;141(6):933–9.

Basov NG, Danilychev VA, Popov YM. Stimulated emission in the vacuum ultraviolet region. Sov J Quantum Electron 1971;1:18–22.

Bates M, Campbell J, Campbell J. Late complication of stent fragmentation related to the "lever-arm effect". J Endovasc Ther 2008;15:224–230.

Bayr H. Reactive oxygen species. Crit Care Med 2005;33:S498–S501. DOI: 10.1097/01 .CCM.0000186787.64500.12.

Becquemin JP, Favre JP, Marzelle J, Nemoz C, Corsin C, Leizorovicz A. Systematic versus selective stent placement after superficial femoral artery balloon angioplasty: a multicenter prospective randomized study. J Vasc Surg 2003;37:487–494.

Berci G. History of endoscopy. In: Berci G, editor. *Endoscopy*. NY: Appleton-Century-Crofts; 1976. p 19–23.

Bick U, Deikmann F, editors. *Digital Mammography*. Berlin: Springer; 2010.

Bird AC, Grey RH. Photocoagulation of disciform macular lesions with krypton laser. Br J Ophthalmol 1979;63:669–673.

Bishop WJ. *The Early History of Surgery*. London: Hale; 1960.

Bobyn J, Tanzer M, et al. Concerns with modularity in total hip arthroplasty. Clin Orthop 1994;298:27–36.

Boggan R, Lemons J, et al. Clinical and laboratory investigations of fretting and corrosion of a three-component modular femoral stem design. J Long Term Eff Med Implants 1994;4:177–191.

Boggs B, Greiner C, Wang T, Lin H, Mossberg TW. Simple high-coherence rapidly tunable external-cavity diode laser. Opt Lett 1998;23:1906–1908.

Bonhoeffer P, et al. Transcatheter implantation of a bovine valve in pulmonary position: a lamb study. Circulation 2000;102:813–816.

Bonjer HJ, Hazebroek EJ, Kazemier G, et al. Open versus closed establishment of pneumoperitoneum in laparoscopic surgery. Br J Surg 1997;84:599–602.

Brace CL. Microwave ablation technology: what every user should know. Curr Probl Diagn Radiol 2009;38(2):61–67.

Brainlab. n.d. Brainlab cancer treatment and minimally invasive surgery. Available at http://www.brainlab.com. Accessed 2011, Sep.

Branemark PI, Adell R, et al. Intra-osseous anchorage of dental prostheses. I. Experimental studies. Scand J Plast Reconstr Surg 1969;3(2):81–100.

Breasted JH. *The Edwin Smith Surgical Papyrus: Hieroglyphic Transliteration, Translation and Commentary*. Chicago: Chicago University Press; 1930.

Brenner DJ, Doll R, Goodhead DT, Hall EJ, Land CE, Little JB, Lubin JH, Preston DL, Preston RJ, Puskin JS, Ron E, Sachs RK, Samet JM, Setlow RB, Zaider M. Cancer risks attributable to low doses of ionizing radiation: Assessing what we really know. Proc Natl Acad Sci U S A 2003:13761–13766.

Bressler NM. Treatment of Age-related macular degeneration with photodynamic therapy study, photodynamic therapy of subfoveal choroidal neovascularization in age-related macular degeneration with verteporfin: two-year results of 2 randomized clinical trials-tap report 2. Arch Ophthalmol 2001;119:198–207.

Brett H, Doarn C, Rosen J, Hannaford B, Broderick TJ. Evaluation of unmanned airborne vehicles and mobile robotic telesurgery in an extreme. Environment 2008;14(6):534–544.

Brinkman WT, Mack MJ. Transcatheter cardiac valve interventions. Surg Clin N Am 2009;89:951–966.

Brinkmann R, Hansen C, Mohrenstecher D, Scheu M, Birngruber R. Analysis of cavitation dynamics during pulsed laser tissue ablation by optical on-line monitoring. IEEE J Select Top Quantum Electron 1996;2:826–835.

Brown P. *American Martyrs to Science through Roentgen Rays*. Springfield (IL): Charles C Thomas Publisher Ltd.; 1936.

Brown SB, Brown EA, Walker I. The present and future role of photodynamic therapy in cancer treatment. Lancet Oncol 2004;5:497–508.

Brown SA, Flemming C, et al. Fretting corrosion accelerates crevice corrosion of modular hip tapers. J Appl Biomater 1995;6:19–26.

Brunner E. *Ultrasound System Considerations and Their IMPACT on Front-End Components*. Norwood (MA): Analog Devices; 2002.

Bulsara KR, Sukhla S, et al. History of bipolar coagulation. Neurosurgical Review 2006;29(2):93–96.

Burbridge BE, Walker DR, Millward DF. Incorporation of the Gunter temporary inferior vena cava filter into the caval wall. J Vasc Interv Radiol 1996;7:289–290.

Burdio F, Guemes A, Burdio JM, et al. Large hepatic ablation with bipolar saline-enhanced radiofrequency: an experimental study in in vivo porcine liver with a novel approach. J Surg Res 2003;110(1):193–201.

Busch J, Johnson A, Lee C, Stevenson D. Shape-memory properties in Ni-Ti sputter-deposited film. J Applied Phys 1990;68(12):6224–6228.

Bushberg JT, Seibert JA, Leidholdt EM Jr, Boone JM. *The Essential Physics of Medical Imaging*. 2nd ed. Philadelphia (PA): Lippincott Williams & Wilkins; 2001.

Bussiere RL. *Principles of Electrosurgery*. Edmonds (WA): Tektran Incorporated; 1997.

Butler KC, Dow JJ, Litwak P, Kormos RL, Borovetz HS. Development of the nimbus/university of Pittsburgh innovative ventricular assist system. Ann Thorac Surg 1999;68:790–4.

Bystritsky A, Korb AS, Douglas PK, Cohen MS, Melega WP, Mulgaonkar AP, DeSalles A, Min BK, Yoo SS. A review of low-intensity focused ultrasound pulsation. Brain Stimul 2011;4(3):125–36.

Callstrom MR, Kurup AN. Percutaneous ablation for bone and soft tissue metastases–why cryoablation? Skeletal Radiol 2009;38(9):835–839.

Carbonare MD, Pathak MA. Skin photosensitizing agents and the role of reactive oxygen species in photoaging. J Photochem Photobiol B: Biol 1992;14:105–124.

Cardella JF, Fox PS, Lawler JB. Interventional radiologic placement of peripherally inserted central catheters. Vasc Interv Radiol 1993;4(5):653–660.

Catheter design and construction. 2010. (Online). Available at http://www.nc3rs.org.uk /bloodsamplingmicrosite/page.asp?id=1121. Accessed 2010 July 15.

Challoner J. *1001 Inventions That Changed the World*. Barron's Educational Series. Barron's; 2009.

Champault G, Cazacu F, Taffinder N. Serious trocar accidents in laparoscopic surgery: a French survey of 103,852 operations. Surg Laparosc Endosc 1996;6:367–370.

Chan KC, Godman MJ, Walsh K, Wilson N, Redington A, Gibbs JL. Transcatheter closure of atrial septal defect and interatrial communications with a new self expanding nitinol double disc device (Amplatzer septal occluder): multicentre UK experience. Heart 1999;82(3):300–306.

Chandler JG, Corson SL, Way LW. Three spectra of laparoscopic entry access injuries. J Am Coll Surg 2001;192:478–491.

Chang C-J, Nelson JS. Cryogen spray cooling and higher fluence pulsed Dye laser treatment improve port-wine stain clearance while minimizing epidermal damage. Dermatol Surg 1999;25:767–772.

Char DH. Metastatic choroidal melanoma. Am J Ophthalmol 1978;86:76–80.

Charles S. Debating the pros and cons of 23-g vs 25-g vitrectomy: the pros of 25-g vitrectomy. Retin Physician 2006;3:24–25.

Charnley J. *The Closed Treatment of Common Fractures*. Cambridge: Cambridge University Press; 1950.

Cheatham JP. Stenting of coarctation of the aorta. Catheter Cardiovasc Interv 2001;54(1):112–125.

Chessa M, Carminati M, Cao QL, Butera G, Giusti S, Bini RM, Hijazi ZM. Transcatheter closure of congenital and acquired muscular ventricular septal defects using the amplatzer device. J Invasive Cardiol 2002;14(6):322–327.

Clowes AW, Zacharias RK, Kirkman TR. Early endothelial coverage of synthetic arterial grafts: porosity revisited. Am J Surg 1987;153:501–504.

Collier J, Mayor M, et al. Mechanisms of failure of modular prostheses. Clin Orthop 1992;285:129–139.

Commission, United States Regulatory. 2011. Part 20—Standards for protection against radiation. Available at www.nrc.gov; http://www.nrc.gov/reading-rm/doc-collections/cfr/part020. Accessed 2011 Sep.

Cope JT, Kaza AK, Reade CC, Shockey KS, Kern JA, Tribble CG, Kron IL. A cost comparison of heart transplantation versus alternative operations for cardiomyopathy. Ann Thorac Surg 2001;72:1298–305.

Cotton RT, Tewfik TL. Laryngeal stenosis following carbon dioxide laser in subglottic hemangioma. Report of three cases. Ann Otol Rhinol Laryngol 1985;94:494–497.

Cowan NJ, Goldberg K, Chirikjian GS, Fichtinger G, Alterovitz R, Reed KB, Kallem V, Park W, Misra S, Okamura AM. Robotic needle steering: design, modeling, planning, and image guidance. In: Rosen J, Hannaford B, Satava RM, editors. *Surgical Robotics-Systems, Applications, and Visions*. 1st ed. New York: Springer; 2011. Chapter 23.

Cowin SC. Wolff's law of trabecular architecture at remodeling equilibrium. J Biomech Eng 1986;108(1):83–88.

Crowninshield RD, Brand RA, et al. An Analysis of Femoral Component Stem Design in Total Hip Arthroplasty. J Bone Joint Surg 1980a;62-A(1):68–78.

Crowninshield RD, Brand RA, et al. The effect of femoral stem cross-sectional geometry on cement stresses in total hip reconstruction. Clin Orthop Relat Res 1980b;(146):71–77.

Culjat MO, Bennett DB, Lee M, Brown ER, Lee H, Grundfest WS, Singh RS. Polyimide-based conformal ultrasound transducer array for needle guidance. IEEE Sensor J 2009;9(10):1244–1245.

Culjat MO, Goldenberg D, Tewari P, Singh RS. A review of tissue substitutes for ultrasound imaging. Ultrasound Med Biol 2010;36(6):861–873.

Curley SA. Radiofrequency ablation of malignant liver tumors. Oncologist 2001;6(1):14–23.

Cushing HI. The control of bleeding in operations for brain tumors: with the description of silver "clips" for the occlusion of vessels inaccessible to the ligature. Annals of Surgery 1911;54(1):1–19.

Cushing H, Bovie WT. Electro-surgery as an aid to the removal of intracranial tumors. Surg Gynecol Obstet 1928;47:751–784.

Dastjerdi MH, Soong HK. LASEK (laser subepithelial keratomileusis). Curr Opin Ophthalmol 2002;13:261–263.

De S, Rosen J, Dagan A, Swanson P, Sinanan MN, Hannaford B. Assessment of tissue damage due to mechanical stresses. Int J Robot Res 2007;26(11–12):1159–1171.

DeBakey M. The odyssey of the artificial heart. Artif Organs 2000;24:405.

Dee K, Puleo DA, Bizios R. *Tissue-Biomaterial Interactions*. Hoboken (NJ): John Wiley & Sons Inc; 2002.

Deitch EA. *Tools of the Trade and Rules of the Road: A Surgical Guide*. Lippincott Williams & Wilkins; 1997. p 190–201.

Deodhar A, Hamilton W, Single G, Rubinsky B, Jaybody M, Solomon S, et al. Irreversible electroporation near the heart: ventricular arrhythmias can be prevented with EKG synchronization; RSNA Scientific Paper. Annual RSNA meeting; 2009. December 2009.

Desilets CS, Fraser JD, Kino GS. The design of efficient broad-band piezoelectric transducers. IEEE Trans Son Ultrason 1978;25:115–125.

Devgan U. *Phaco Fundamentals for the Beginning Phaco Surgeon. Bausch and Lomb Ophthalmology World Report Series*. New Delhi: ILX MEDIA GROUP; 2009. p 1–54.

Devries W. The permanent artificial heart: four case reports. JAMA 1988;259:849–59.

Dhalla MS, Shah GK, Blinder KJ, Ryan EHJ, Mittra RA, Tewari A. Combined photodynamic therapy with verteporfin and intravitreal bevacizumab for choroidal neovascularization in Age-related macular degeneration. Retina 2006;26:988–993. DOI: 10.1097/01.iae.0000247164.70376.91.

Dichek DA, Neville RF, Zweibel JA, Freeman SA, Leon MB, Anderson WF. Seeding of intravascular stents with genetically engineered endothelial cells. Circulation 1989;80(5):1347–1353.

Diener-West M, Hawkins BS, Markowitz JA, Schachat AP. A review of mortality from choroidal melanoma: II. A meta-analysis of 5-year mortality rates following enucleation, 1966 through 1988. Arch Ophthalmol 1992;110:245–250.

Dierickx CC, Casparian JM, Venugopalan V, Farinelli WA, Anderson RR. Thermal relaxation of port-wine stain vessels probed in vivo: the need for 1-10-millisecond laser pulse treatment. J Investig Dermatol 1995;105:709–714.

Dillon B, Dmochowski R. Radiofrequency for the treatment of stress urinary incontinence in women. Curr Urol Rep 2009;10(5):369–374.

DiMaio S, Hanuschik M, Kreaden U. The da vinci surgical system. In: Rosen J, Hannaford B, Satava RM, editors. *Surgical Robotics-Systems, Applications, and Visions*. 1st ed. New York: Springer; 2011. Chapter 7.

Dodd GD 3rd, Soulen MC, Kane RA, Livraghi T, Lees WR, Yamashita Y, Gillams AR, Karahan OI, Rhim H. Minimally invasive treatment of malignant hepatic tumors: at the threshold of a major breakthrough. Radiographics 2000;20(1):9–27.

Dolmans DEJGJ, Fukumura D, Jain RK. Photodynamic therapy for cancer. Nat Rev Cancer 2003;3:380–387.

Dougherty PJ, Wellish KL, Maloney RK. Excimer laser ablation rate and corneal hydration. Am J Ophthalmol 1994;118:169–176.

Du ZD, Hijazi ZM, Kleinman CS, Silverman NH, Larntz K, Investigators A. Comparison between transcatheter and surgical closure of secundum atrial septal defect in children and adults: results of a multicenter nonrandomized trial. J Am Coll Cardiol 2002;39(11):1836–1844.

Dubose T. Poly-autobiography of diagnositic medical sonography. Proceedings of American Association of History and Computing; 2000 Apr 13–15; Waco, TX. 2000.

Ebramzadeh E. *On Factors Affecting the Long-Term Outcome of Total Hip Replacement*. Gothenburg: Biomaterials Group, University of Gothenburg Sweden; 1995.

Ebramzadeh E, Normand PL, et al. Long-term radiographic changes in cemented total hip arthroplasty with six designs of femoral components. Biomaterials 2003;24:3351–3363.

Ebramzadeh E, Sangiorgio SN, et al. Initial stability of cemented femoral stems as a function of surface finish, collar, and stem size. J Bone Joint Surg 2004;86-A(1):106–115.

Ebramzadeh E, Sangiorgio SN, et al. Greater expectations and greater joint loads in modern total joint arthroplasty patients. In: Garino J, Beredjiklian P, editors. *Core Knowledge in Orthopedics: Adult Reconstruction and Arthroplasty*. Philadelphia: Elsevier; 2007.

Ebramzadeh E, Sarmiento A, et al. The cement mantle in total hip arthroplasty. Analysis of long-term radiographic results. J Bone Joint Surg Am 1994;76(1):77–87.

Edd JF, Horowitz L, Davalos RV, Mir LM, Rubinsky B. In vivo results of a new focal tissue ablation technique: irreversible electroporation. IEEE Trans Biomed Eng 2006;53(7):1409–1415.

Elekta. n.d. Intra-operative imaging for neurosurgery. Available at http://www.elekta.com /healthcare_international_intra_operative_imaging.php. Accessed 2010.

Elhawary H, Tse ZTH, Hamed A, Rea M, Davies BL, Lamperth MU. The case for MR-compatible robotics: a review of the state of the art. Int J Med Robot Comput Assist Surg 2008;4(2):105–113.

Ellis LM, Curley SA, Tanabe KK, editors. *Radiofrequency ablation for cancer: current indications, techniques, and outcomes*. New York: Springer; 2004.

Engh JA, Kondziolka D, Riviere CN. Percutaneous intracerebral navigation by duty-cycled spinning of flexible bevel-tipped needles. Neurosurgery 2010;67(4):1117–1123.

Englehardt J, Englehardt S, et al. *The Orthopaedic Industry Annual Report: 2008-2009*. OrthoWorld: Chagrin Falls; 2009. p 1–154.

Esmore D, Kaye D, Spratt P, Larbalestier R, Ruygrok P, Tsui S, Meyers D, Fiane AE, Woodard J. A prospective, multicenter trial of the VentrAssist left ventricular assist device for bridge to transplant: safety and efficacy. J Heart Lung Transplant 2008;27(6):579–88.

Espicom Healthcare Intelligence. The Medical Device Market: USA. 2011. Pub ID: ESPI6418303.

Fantes FE, Hanna KD, Waring GO III, Pouliquen Y, Thompson KP, Savoldelli M. Wound healing after excimer laser keratomileusis (photorefractive keratectomy) in monkeys. Arch Ophthalmol 1990;108:665–675.

Faraz A, Payandeh S. *Engineering Approaches to Mechanical and Robotic Design for Minimally Invasive Surgeries*. Boston: Kluwer Academic Publishers; 2000.

FDA. *Approval Order—PO20008*. U.S. Food & Drug Administration; 2002. Available at http://www.accessdata.fda.gov/cdrh_docs/pdf2/P020008a.pdf. Accessed 2011 Dec.

FDA. *Karl Storz Photodynamic Diagnostic D-Light C (PDD) System—P050027I, Summary of Safety and Effectiveness*. U.S. Food & Drug Administration; 2010. Available at http://www.accessdata.fda.gov/cdrh_docs/pdf5/P050027b.pdf. Accessed 2010 Jun 14.

FDA Device Advice website. 2011a. Quality System (QS) Regulation/Medical Device Good Manufacturing Practices. Available at http://www.fda.gov/MedicalDevices /DeviceRegulationandGuidance/PostmarketRequirements/QualitySystemsRegulations /default.htm. Accessed 2011.

FDA Medical Device Databases website. 2011b. CFR—Code of Federal Regulations Title 21. Available at http://www.accessdata.fda.gov/scripts/cdrh/cfdocs/cfcfr/CFRSearch .cfm?CFRPart=58.

FDA, US Food and Drug Administration. Diathermy. Inspection Guidelines. US Food and Drug Administration; 2009.

FDA, U.S. Food and Drug Administration. 2010. Available at www.fda.gov/medical devices. Accessed 2012.

Feller ED, Sorensen EN, Haddad M, Pierson RN, Johnson FL, Brown JM, Griffith BP. Clinical outcomes are similar in pulsatile and nonpulsatile left ventricular assist device recipients. Ann Thorac Surg 2007;83(3):1082–88.

Fiandesio D, Valobra G. The therapeutic power of heat and its use in the course of ages. Minerva Medica 1968;59(21):1206–1212.

Fichtinger G, Kazanzides P, Okamura AM, Hager GD, Whitcomb LL, Taylor RH. Surgical and interventional robotics: Part II. IEEE Robot Autom Mag 2008;15(3):94–102.

Fitzpatrick RE, Tope WD, Goldman MP, Satur NM. Pulsed carbon dioxide laser, trichloroacetic acid, baker-Gordon phenol, and dermabrasion: a comparative clinical and histologic study of cutaneous resurfacing in a porcine model. Arch Dermatol 1996;132: 469–471.

Food and Drug Administration (FDA). *Information for Manufacturers Seeking Marketing Clearance of Diagnostic Ultrasound Systems and Transducers. Guidance for Industry and FDA Staff*. 2008. Available at www.fda.gov. Accessed 2011 Oct.

Fornasier VL, Cameron HU. The femoral stem/cement interface in total hip replacement. Clin Orthop 1976;(116):248–252.

Framme C, Walter A, Prahs P, Regler R, Theisen-Kunde D, Alt C, Brinkmann R. Structural changes of the retina after conventional laser photocoagulation and selective retina treatment (SRT) in spectral domain OCT. Curr Eye Res 2009;34:568–579.

Franco ML, King CH, Culjat MO, Lewis CE, Bisley JW, Holmes EC, Grundfest WS, Dutson EP. An integrated pneumatic tactile feedback actuator array for robotic surgery. Int J Med Robot Comput Assist Surg 2009;5(1):13–19.

Fraser AG, Daubert J-C, Van de Werf F, Estes NAM III, Smith SC Jr, Krucoff MW, Vardas PE, Komajda M. Clinical evaluation of cardiovascular devices: principles, problems, and proposals for European regulatory reform. Eur Heart J 2011;32(13):1673–1686.

Frazier OH, Kirklin JK. *Mechanical Circulatory Support*. Elsevier; 2006.

Frazier OH, Rose EA, McCarthy P, Burton NA, Tector A, Levin H, Kayne HL, Poirier VL, Dasse KA. Improved mortality and rehabilitation of transplant candidates treated with a long-term implantable left ventricular assist system. Ann Surg 1995;222(3):327–36.

Fried MP. Complications of CO2 laser surgery of the larynx. Laryngoscope 1983;93: 275–278.

Gallagher AG, O'Sullivan GC. *Fundamentals of Surgical Simulation: Principles and Practice*. Springer; 2011.

Gallo A, de Vincentiis M, Manciocco V, Simonelli M, Fiorella ML, Shah JP. CO2 Laser cordectomy for early-stage glottic carcinoma: a long-term follow-up of 156 cases. Laryngoscope 2002;112:370–374.

Garcia P, Rosen J, Kapoor C, Noakes M, Elbert G, Treat M, Ganous T, Hanson M, Manak J, Hasser C, Rohler D, Satava R. Trauma Pod: a semi-automated telerobotic surgical system. Int J Med Robot Comput Assist Surg 2009;5(2):136–146.

Garrison BJ, Srinivasan R. Laser ablation of organic polymers: microscopic models for photochemical and thermal processes. J Appl Phys 1985;57:2909–2914.

General Electric. n.d. Imagination at work. Available at http://www.ge.com. Accessed 2010.

Gewillig M, Boshoff DE, Dens J, Mertens L, Benson LN. Stenting the neonatal arterial duct in duct-dependent pulmonary circulation: new techniques, better results. J Am Coll Cardiol 2004;43:107–112.

Gillinov AM, Liddicoat JR. Percutaneous mitral valve repair. Semin Thorac Cardiovasc Surg 2006;18:115–121.

Giuntini RE. Developing safe, reliable medical devices. MD&DI 2000;22(10).

Glover JL, Bendick PJ, et al. The use of thermal knives in surgery: electrosurgery, lasers, plasma scalpel. Current Problems in Surgery 1978;15(1):1–78.

Goldberg SN, Gazelle GS, Mueller PR. Thermal ablation therapy for focal malignancy: a unified approach to underlying principles, techniques, and diagnostic imaging guidance. Am J Roentgenol 2000;174(2):323–31.

Goldberg SN, Grassi CJ, Cardella JF, Charboneau JW, Dodd GD 3rd, Dupuy DE, Gervais D, Gillams AR, Kane RA, Lee FT Jr, Livraghi T, McGahan J, Phillips DA, Rhim H, Silverman SG, Society of Interventional Radiology Technology Assessment Committee; International Working Group on Image-Guided Tumor Ablation. Image-guided tumor ablation: standardization of terminology and reporting criteria. Radiology 2005;235(3):728–39.

Goldberg L, Mehuys D. High power superluminescent diode source. Electron Lett 1994;30:1682–1684.

Golden MA, Hanson SR, Kirkamn TR, Schneider PA, Clowes AW. Healing of polytetrafluoroethylene arterial grafts is influenced by graft porosity. J Vasc Surg 1990;11:838–845.

Goldman L, Nath G, et al. High-power neodymium-YAG laser surgery. Acta Dermato-Venereologica 1973;53(1):45–49.

Gomes P. Surgical robotics: reviewing the past, analyzing the present, imagining the future. Robot Comput Integr Manuf 2011;27(2):261–266.

Gonzalez R, Woods R. *Digital Image Processing*. Upper Saddle River (NJ): Prentice-Hall; 2002.

Goodrich JA, Volcan IJ, editors. *eXtreme Lateral Interbody Fusion (XLIF)*. St. Louis: Quality Medical Publishing, Inc; 2008.

Gordon JP, Zeiger HJ, Townes CH. The maser—New type of microwave amplifier, frequency standard, and spectrometer. Phys Rev 1955;99:1264–1274.

Gough-Palmer AL, Gedroyc WMW. Laser ablation of hepatocellular carcinoma–A review. World J Gastroenterol 2008;14(47):7170–7174.

Granot Y, Ribinsky B. Mass transfer model for drug delivery in tissue cells with reversible electroporation. Int J Heat Mass Tran 2008;51(23–24):5610–5616.

Green NE, Chen S-Y, Messenger JC, Groves BM, Carroll JD. Three-dimensional vascular angiography. Curr Probl Cardiol 2004:104–142.

Green PS, Hill JW, Jensen JF, Shah A. Telepresence surgery. IEEE Eng Med Biol Mag 1995;14(3):324–329.

Greenfield LJ, Michna BA. Twelve-year experience with the Greenfield vena cava filter. Surgery 1988;104:706–712.

Greenhill WA, FDA, U.S. Food and Drug Administration. Chirurgia. In: William S, editor. *A Dictionary of Greek and Roman Antiquities*. London: John Murray; 1875. p 272–275. Available at www.fda.gov/medicaldevices. Accessed 2012.

Greenwood J Jr. Two point coagulation: a new principle and instrument for applying coagulation current in neurosurgery. The American Journal of Surgery 1940;50(2):267–270.

Guell JL, Muller A. Laser in situ keratomileusis (LASIK) for myopia from -7 to -18 diopters. J Refract Surg (Thorofare, N.J.: 1995) 1996;12:222–228.

Gutensohn K, Beythien C, Bau J, Fenner T, Grewe P, Koester R, Padmanaban K, Kuehnl P. In vitro analyses of diamondlike carbon coated stents: Reduction of metal ion release, platelet activation and thrombogenicity. Thromb Res 2000;99:577–585.

Guyer RD, McAfee PC, et al. Prospective, randomized, multicenter Food and Drug Administration investigational device exemption study of lumbar total disc replacement with the CHARITE artificial disc versus lumbar fusion: five-year follow-up. Spine J 2009;9(5):374–386.

terHaar G. Therapeutic applications of ultrasound. Progr Biophys Mol Biol 2007;93(1–3): 111–129.

Haberkamp TJ, Harvey SA, Khafagy Y. Revision stapedectomy with and without the CO2 laser: an analysis of results. Otol Neurotol 1996;17:225–229.

Hagag B, Abovitz R, Kang H, Schmitz B, Conditt M. RIO: robotic-arm interactive orthopedic system MAKOplasty: user interactive haptic orthopedic robotics. In: Rosen J, Hannaford B, Satava RM, editors. *Surgical Robotics-Systems, Applications, and Visions*. 1st ed. New York: Springer; 2011. Chapter 10.

Hager GD, Okamura AM, Kazanzides P, Whitcomb LL, Fichtinger G, Taylor RH. Surgical and interventional robotics: Part III: surgical assistance systems. IEEE Robot Autom Mag 2008;15(4):84–93.

Hagiike M, Phillips EH, Berci G. Performance differences in laparoscopic surgical skills between true high-definition and three-chip CCD video systems. Surg Endos 2007;21(10):1849–1854.

Haid RW, Foley KT, et al. The Cervical Spine Study Group anterior cervical plate nomenclature. Neurosurg Focus 2002;12(1):E15.

Hall JC. Electrosurgery: a short history. Australian and New Zealand Journal of Surgery 1976;46(4):400–401.

Hall RN, Fenner GE, Kingsley JD, Soltys TJ, Carlson RO. Coherent light emission from GaAs junctions. Phys Rev Lett 1962;9:366–368.

Halliwell B. Reactive oxygen species and the central nervous system. J Neurochem 1992;59:1609–1623.

Hanajiri K, Maruyama T, Kaneko Y, et al. Microbubble-induced increase in ablation of liver tumors by high-intensity focused ultrasound. Hepatol Res 2006;36(4):308–314.

Harrop JS, Youssef JA, et al. Lumbar adjacent segment degeneration and disease after arthrodesis and total disc arthroplasty. Spine (Phila Pa 1976) 2008;33(15):1701–1707.

Hashizume M, Sugimachi K. Needle and trocar injury during laparoscopic surgery in Japan. Surg Endosc 1997;11:1198–1201.

Hasson HM. Open laparoscopy vs. closed laparoscopy. A comparison of complication rates. Adv Planned Parenthood 1978;13:41–50.

Hazell W, Heaven D, Kazemi A, Fourie D. Atrio-oesophageal fistula: an emergent complication of radiofrequency ablation. Emerg Med Australas 2009;21(4):329–32.

HeartMate II LVAS Operating Manual. Thoratec Corporation; Pleasanton, California. 2008.

HeartMate XVE LVAS Operating Manual. Thoratec Corporation; Pleasanton, California. 2008.

Hedrick W, Hykes D, Starchman D. *Ultrasound Physics and Instrumention*. 3rd ed. St. Louis (MO): Mosby; 1995. Chapter 2.

Heitler W. *The Quantum Theory of Radiation, 1*. New York: Dover Publications Inc; 1984.

Henderson BW, Dougherty TJ. How does photodynamic therapy work? Photochem Photobiol 1992;55:145–157.

Henderson E, Nash DH, Dempster WM. On the experimental testing of fine nitinol wires for medical devices. J Mech Behav Biomed Mater 2011;4:261–268.

Henein M, Birks EJ, Tansley P, Bowles CT, Yacoub MH. Temporal and spatial changes in left ventricular pattern of flow during continuous assist device "HeartMate II". Circulation 2002;105(19):2324–5.

Heniford BT, Mathews B. Basic instrumentation for laparoscopic surgery. In: Greene FL, Heniford BT, editors. *Minimally Invasive Cancer Management*. New York: Springer-Verlag New York Inc; 2001.

Herberts P, Malchau H. Long-term registration has improved the quality of hip replacement: a review of the Swedish THR Register comparing 160,000 cases. Acta Orthop Scand 2000;71(2):111–121.

Herrmann H, Feldman T. Percutaneous mitral valve edge-to-edge repair with the Evalve MitraClip System: rationale and phase I results. EuroIntervention Supplements 2006;1(Suppl A):A36–A39.

Hersh PS, Scher KS, Irani R. Corneal topography of photorefractive keratectomy versus laser in situ keratomileusis. Ophthalmology 1998;105:612–619.

Hilibrand AS, Yoo JU, et al. The success of anterior cervical arthrodesis adjacent to a previous fusion. Spine (Phila Pa 1976) 1997;22(14):1574–1579.

Hills DA, Nowell D, editors. *Mechanics of Fretting Fatigue. Solid Mechanics and Its Applications*. Dordrecht, The Netherlands: Clure Academic Publishers; 1994.

Hlatky L, Hahnfeldt P, Folkman J. Clinical application of antiangiogenic therapy: microvessel density, what It does and Doesn't tell Us. J Natl Cancer Inst 2002;94: 883–893.

Ho K, Carman G. Sputter deposition of NiTi thin film shape memory alloy using a heated target. Thin Solid Films 2000;370:18–29.

Holmes DR, Davis BJ, Bruce CJ, Robb RA. 3D visualization, analysis, and treatment of the prostate using trans-urethral ultrasound. Comput Med Imaging Graph 2002;27(5): 339–349.

Holsinger FC, Prichard CN, Shapira G, Weisberg O, Torres DS, Anastassiou C, Harel E, Fink Y, Weber RS. Use of the photonic band Gap fiber assembly CO2 laser system in head and neck surgical oncology. Laryngoscope 2006;116:1288–1290.

Hong TE, Thaler D, Brorson J, Heitschmidt M, Hijazi ZM, Amplatzer PFO Investigators. Transcatheter closure of patent foramen ovale associated with paradoxical embolism using the amplatzer PFO occluder: initial and intermediate-term results of the U.S. multicenter clinical trial. Catheter Cardiovasc Interv 2003;60(4):524–528.

Huang N, Yang P, Leng YX, Chen JY, Sun H, Wang J, Wang GJ, Ding PD, Xi TF, Leng Y. Hemocompatibility of titanium oxide films. Biomaterials 2003;24:2177–2187.

Hunter DG, Repka MX. Diode laser photocoagulation for threshold retinopathy of prematurity. A randomized study. Ophthalmology 1993;100:238–244.

Hutchisson B, Baird MG, et al. Electrosurgical safety. AORN Journal 1998;68(5):829–837.

Hynynen K, Clement G. Clinical applications of focused ultrasound—the brain. Int J Hyperthermia 2007;23(2):193–202.

Iannitti DA, Martin RC, Simon CJ, Hope WW, Newcomb WL, McMasters KM, Dupuy D. Hepatic tumor ablation with clustered microwave antennae: the US Phase II Trial. HPB (Oxford) 2007;9(2):120–124.

Ikuta K, Hayashi M, Matsuura T, Fujishiro H. Shape memory alloy thin film fabricated by laser ablation. An Investigation of micro structures, sensors, actuators, machines and robotic systems. In: Proceedings IEEE Micro Electro Mechanical Systems; 1994 Jan 25–28; Oiso. IEEE; 1994. p 355–360.

Ilizarov GA. Clinical application of the tension-stress effect for limb lengthening. Clin Orthop Relat Res 1990;(250):8–26.

Integra-Radionics. n.d. For the neurosurgeon at Integra—medical device company—medical technologies. Available at http://www.integralife.com/Neurosurgeon/. Accessed 2010.

International Organization for Standardization (ISO). (2010a). ISO 10555-1-6 Sterile, single-use intravascular catheters. Retrieved from http://www.iso.org/iso/catalogue _detail.htm?csnumber=18640. Accessed 2010 July 15.

International Organization for Standardization (ISO). (2010b). ISO 11070 Sterile, single-use intravascular catheter introducers. Retrieved from http://www.iso.org/iso/catalogue _detail.htm?csnumber=19052. Accessed 2010 July 15.

Intuitive Surgical, Inc. 2011. Available at http://www.intuitive.com.

Isis. n.d. Accueil—ISIS—Intelligent Surgical Instruments & Systems. Available at http://www.isis-robotics.com/english.html

ITA, International Trade Administration. 2009. U.S. Department of Commerce, Medical Devices Industry Assessment. Available at www.ita.doc.gov.. Accessed 2012.

Jacques SL, Nelson JS, Wright WH, Milner TE. Pulsed photothermal radiometry of port-wine-stain lesions. Appl Opt 1993;32:2439–2446.

Jasim ZF, Handley JM. Treatment of pulsed dye laser–resistant port wine stain birthmarks. J Am Acad Dermatol 2007;57:677–682.

Javan A, Bennett WR Jr, Herriott DR. Population inversion and continuous optical maser oscillation in a Gas discharge containing a He-Ne mixture. Phys Rev Lett 1961;6:6–110.

Ji B, Ündar A. An evaluation of the benefits of pulsatile versus nonpulsatile perfusion during cardiopulmonary bypass procedures in pediatric and adult cardiac patients. Am Soc Artif Intern Organs J 2006;52(4):357–61.

Jingkuang C. Capacitive micromachined ultrasonic transducer arrays for minimally invasive medical ultrasound. J Micromech Microeng 2010;20(2):023001. DOI: 10.1088/0960-1317/20/2/023001.

Johnson DE, Cromeens DM, Price RE. Use of the holmium:YAG laser in urology. Lasers Surg Med 1992;12:353–363.

Jones MI, McColl IR, Grant DM, Parker KG, Parker TL. Protein adsorption and platelet attachment and activation, on TiN, TiC, and DLC coatings on titanium for cardiovascular applications. J Biomed Mater Res A 2000;52(2):413–21.

Kabir AM, Selvarajah A, Seifalian AM. How safe and how good are drug-eluting stents? Future Cardiol 2011;7:251–270.

Kahlenberg MS, Volpe C, Klippenstein DL, Pentrante RB, Petrelli NJ, Rodriguez-Bigas MA. Clinicopathologic effects of cryotherapy on hepatic vessels and bile ducts in a porcine model. Ann Surg Oncol 1998;5:713–718.

Kahn AM, Ostad A, Moy RL. Grafting following short-pulse carbon dioxide laser de-epithelialization. Dermatol Surg 1996;22:965–967. Discussion 967–968.

Khan F, Gerbi B. *Treatment Planning in Radiation Oncology*. Lippincott Williams & Wilkins; 2012.

Kajiwara K, Yoshikawa K, Ideguchi M, Nomura S, Fujisawa H, Akimura T, Kato S, Fujii M, Suzuki M. Navigation-guided fence-post tube technique for resection of a brain tumor: technical note. Minim Invasive Neurosurg 2010;53(2):86–90.

Kallmes DF, McGraw JK, Evans AJ, Mathis JM, Hergenrother RW, Jensen ME, et al. Thrombogenicity of hydrophilic and nonhydrophilic microcatheters and guiding wires. Am J Neuroradiol 1997;18:1243–1251.

Kalso E. A short history of central venous catheterization. Acta Anaesthesiol Scand 1985;29(81): 7–10.

Kaneko Y, Maruyama T, Takegami K, et al. Use of a microbubble agent to increase the effects of high intensity focused ultrasound on liver tissue. Eur Radiol 2005;15(7): 1415–1420.

Kannan RY, Salacinski HJ, Butler PE, et al. Current status of prosthetic bypass grafts: a review. J Biomed Mater Res B Appl Biomater 2005;74:570–581.

Kantelhardt SR, Martinez R, Baerwinkel S, Burger R, Giese A, Rohde V. Perioperative course and accuracy of screw positioning in conventional, open robotic-guided and per-cutaneous robotic-guided, pedicle screw placement. Eur Spine J 2011;20(6):860–868. DOI: 10.1007/s00586-011-1729-2.

Kaplan AV, et al. Medical devices: a regulatory overview. Circulation 2004Available at http://circ.ahajournals.org/content/109/25/3068.full. Accessed 2011.

Karas CS, Chiocca EA. Neurosurgical robotics: a review of brain and spine applications. J Robotic Surg 2007;1:39–43.

Karino T, Goldsmith HL. Role of blood cell-wall interactions in thrombogenesis and athero-genesis: a microrheological study. Biorheology 1984;21:587–601.

Karrer L, Duwe J, Zisch AH, Khabiri E, Cikircioglu M, Napoli A, et al. PPS-PEG surface coating to reduce thrombogenicity of small diameter ePTFE vascular grafts. Int J Artif Organs 2005;28:993–1002.

Kawalec JS, Brown SA, et al. Mixed-metal fretting corrosion of Ti6Al4V and wrought cobalt alloy. J Biomed Mater Res 1995;29(7):867–73.

Kazanzides P. Robot for orthopedic joint reconstruction. In: Faust RA, editor. *Robotics in Surgery: History, Current and Future Applications*. New York: Springer; 2007. p 61–94. Chapter 5.

Kazanzides P, Fichtinger G, Hager GD, Okamura AM, Whitcomb LL, Taylor R. Surgical and interventional robotics: core concepts, technology, and design. IEEE Robot Autom Mag 2008;15(2):122–130.

Kelly PJ, Earnest F, Kall BA, Goerss SJ, Scheithauer B. Surgical options for patients with deep-seated brain tumors: computer-assisted stereotactic biopsy. Mayo Clin Proc 1985;60(4):223–229.

Kelman C. Symposium: phacoemulsification. History of emulsification and aspiration of senile cataracts. Transactions—American Academy of Ophthalmology and Otolaryngol-ogy 1974;78(1):OP5.

Kelman CD. The history of phacoemulsification. In: Agarwal S, et al., editors. *Phacoemulsi-fication Laser Cataract Surgery and Foldable IOL's*. 1 st ed. New Delhi: Jaypee Brothers; 1998. p 123–125.

Kelman GJ, Krakauer JD, et al. Two- to five-year follow-up of 100 total hip arthroplasties using DF-80 implants. Orthop Rev 1994;23(5):420–425.

Kennedy MB, Ruth DJ, Martin EJ. Module 6: Intrapartum fetal monitoring. In: *Intrapartum Management Modules: A Perinatal Education Program*. 4th ed. Baltimore: Lippincott, Williams & Wilkins; 2008. p 144–151.

Kennedy JS, Stranahan PL, Taylor KD, Chandler JG. High- burst-strength feedback controlled bipolar vessel sealing. Surgical Endoscopy 1998;12(6):876–878.

Kertesz I, Fenyö M, Mester E, Bathory G. Hypothetical physical model for laser biostimulation. Opt Laser Technol 1982;14:31–32.

Khambadkone S, Bonhoeffer P. Nonsurgical pulmonary valve replacement: why, when, and how? Catheter Cardiovasc Interv 2004;62:401–408.

Kilby W, et al. The CyberKnife® robotic radiosurgery system in 2010. Technol Cancer Res Treat 2010;9(5):433–452.

Kim YS, Rhim H, Choi MJ, Lim HK, Choi D. High-intensity focused ultrasound therapy: an overview for radiologists. Korean J Radiol 2008;9(4):291–302.

Kirklin JK. Mechanical circulatory support as a bridge to pediatric cardiac transplantation. Semin Thorac Cardiovasc Surg Pediatr Card Surg Annu 2008;2008:80–85.

Knoll GF. *Radiation Detection and Measurement*. 4th ed. Danvers (MA): John Wiley & Sons, Inc; 2010.

Ko T, Adler D, Fujimoto J, Mamedov D, Prokhorov V, Shidlovski V, Yakubovich S. Ultrahigh resolution optical coherence tomography imaging with a broadband superluminescent diode light source. Opt Express 2004;12:2112–2119.

Kobak D. Diathermy in medicine and surgery. IMJ. Illinois Medical Journal 1925;47: 276–284.

Koch DD, Liu JF, Hyde LL, Rock RL, Emery JM. Refractive complications of cataract surgery after radial keratotomy. Am J Ophthalmol 1989;108:676–682.

Kockro RA, Stadie A, Schwandt E, Reisch R, Charalampaki C, Ng I, Yeo TT, Hwang P, Serra L, Perneczky A. A collaborative virtual reality environment for neurosurgical planning and training. Neurosurgery 2007;61(5 Suppl 2):379–391.

Konishi H, Antaki JF, Amin DV, Boston JR, Kerrigan JP, Mandarino WA, Litwak P, Yamazaki K, Macha M, Butler KC, Borovetz HS, Kormos RL. Controller for an axial flow blood pump. Artif Organs 1996;20(6):618–20.

Kornilovsky IM. Clinical results after subepithelial photorefractive keratectomy (LASEK). J Refract Surg (Thorofare, N.J.: 1995) 2001;17:S222–S223.

Korver JG, Oosterhuis JA, Kakebeeke-Kemme HM, Wolff-Rouendaal D. Transpupillary thermotherapy (TTT) by infrared irradiation of choroidal melanoma. Doc Ophthalmol 1992;82:185–191.

Krauss JM, Puliafito CA. Lasers in ophthalmology. Lasers Surg Med 1995;17:102–159.

Krötz F, Sohn H-Y, Pohl U. Reactive oxygen species. Arterioscler Thromb Vasc Biol 2004;24:1988–1996.

Kubasova T, Horvath M, Kocsis K, Fenyo M. Effect of visible light on some cellular and immune parameters. Immunol Cell Biol 1995;73:239–244.

Kumpati GS, Cook DJ, Blackstone EH, Rajeswaran J, Abdo AS, Young JB, Starling RC, Smedira NG, McCarthy PM. HLA sensitization in ventricular assist device recipients: does type of device make a difference? J Thorac Cardiovasc Surg 2004;127(6):1800–7.

Kurtz S, Edidin A, editors. *Spine Technology Handbook*. Academic Press: New York; 2006.

Land RI. On the physical behavior of flashing incandescent light sources. Leonardo 1975;8:232–234.

Laredo J, Xue L, Husak VA, Ellinger J, Singh G, Zamora PO, et al. Silyl-heparin bonding improves the patency and in vivo thromboresistance of carbon coated polytetrafluoroethylene vascular grafts. J Vasc Surg 2004;39:1059–1065.

Lee JL, Billi F, et al. Wear of an experimental metal-on-metal artificial disc for the lumbar spine. Spine (Phila Pa 1976) 2008;33(6):597–606.

Lee FT, Haemmerich D, Wright AS, Mahvi DM, Sampson LA, Webster JG. Multiprobe radiofrequency ablation: pilot study in an animal model. J Vasc Intervent Radiol 2003;14(11):1437–1442.

Lee EW, Loh CT, Kee ST. Imaging guided percutaneous irreversible electroporation: ultrasound and immunohistological correlation. Tech Canc Res Treat 2007;6(4):287–294.

Lee EW, Prieto V, Chen C, et al. A novel hepatic ablation technique creating complete cell death: irreversible electroporation. Radiology 2010 Forthcoming.

van Leeuwen TG, van der Veen MJ, Verdaasdonk RM, Borst C. Noncontact tissue ablation by Holmium: YSGG laser pulses in blood. Lasers Surg Med 1991;11:26–34.

Lehnert T, Tixier S, Boni P, Gotthardt R. A new fabrication process for Ni-Ti shape memory thin films. Mater Sci Eng 1999;A273–275:713–716.

Lemole GM, Banerjee PP, Luciano C, Neckrysh S, Charbel FT. Virtual reality in neurosurgical education: part-task ventriculostomy simulation with dynamic visual and haptic feedback. Neurosurgery 2007;61(1):142–148.

Leroy J, Dutson E, Henri M. Access and trocar complications. WeBSurg.com 2005;5(3) Available at http://www.websurg.com/ref/doi-ot02en274.htm.

Leroy J, Marescaux J. Double plastic wound protection for specimen extraction in colorectal surgery. WeBSurg.com 2009;9(4) Available at http://www.websurg.com/ref/doi-vd01en2589.htm.

Lethaby A, Hickey M, Garry R, Penninx J. *Endometrial resection/ablation techniques for heavy menstrual bleeding*. Cochrane Database Syst Rev 2009;Oct 7(4).

Levy JH, Dutton RP, et al. Multidisciplinary approach to the challenge of hemostasis. Anesthesia and Analgesia 2010;110(2):354–364.

Liang P, Wang Y. Microwave ablation of hepatocellular carcinoma. Oncology 2007;72(Suppl. 1):124–131.

Lichtenberg BHM. (2010). Wiedereröffnung eines akut thrombotisch verschlossenen Aortenstentgrafts unter Einsatz der Rotations thrombektomie, Gefäßmedizin. Fallbericht. p 16–21.

Life Science Intelligence. *U.S. Markets for General and Pelvic Endoscopic Surgery Products*. Life Science Intelligence, Inc; 2010. Available at www.lifescienceintelligence.com/market-reports-page.pho?id=A568. Accessed 2010.

Linford GJ. Time-resolved xenon flash-lamp opacity measurements. Appl Opt 1994;33: 8333–8345.

Ling RS. The use of a collar and precoating in cemented femoral stems is unnecessary and detrimental. Clin Orthop 1992;(285):73–83.

Livraghi T, Goldberg SN, Monti F, et al. Saline-enhanced radio-frequency tissue ablation in the treatment of liver metastases. Radiology 1997;202(1):205–210.

Lloyd-Jones D, Adams R, Carnethon M, De Simone G, Ferguson TB, Flegal K, Ford E, Furie K, Go A, Greenlund K, Haase N, Hailpern S, Ho M, Howard V, Kissela B, Kittner S, Lackland D, Lisabeth L, Marelli A, McDermott M, Meigs J, Mozaffarian D, Nichol G, O'Donnell C, Roger V, Rosamond W, Sacco R, Sorlie P, Stafford R, Steinberger J, Thom T, Wasserthiel-Smoller S, Wong N, Wylie-Rosett J, Hong Y. Heart disease and stroke statistics—2009 update: a report from the American Heart Association Statistics Committee and Stroke Statistics Subcommittee. Circulation 2009;119(3):e182.

Lockwood GR, Turnbull DH, Foster FS. Fabrication of high frequency spherically shaped ceramic transducers. IEEE Trans Ultrason, Ferroelectr Freq Control 1994;41(2):231–235.

Loisance D. Mechanical circulatory support: a clinical reality. Asian Cardiovasc Thorac Ann 2008;16:419–431.

Long et al. Long-Term Destination Therapy with the HeartMate XVE Left Ventricular Assist Device: Improved Outcomes since the REMATCH Study. Congestive Heart Failure 2005;11(3):133–138.

Lorber A, Gazit AZ, Khoury A, Schwartz Y, Freudental F. Percutaneous transaortic occlusion of patent ductus arteriosus using a new versatile angiographic and delivery catheter. Pediatr Cardiol 2003;24(5):482–483.

Lovisolo CF, Fleming JF. Intracorneal ring segments for iatrogenic keratectasia after laser in situ keratomileusis or photorefractive keratectomy. J Refract Surg (Thorofare, N.J.: 1995) 2002;18:535–541.

Lum M, Friedman D, King H, Donlin R, Sankaranarayanan G, Broderick T, Sinanan M, Rosen J, Hannaford B. (2007). Teleoperation of a surgical robot via airborne wireless radio and transatlantic internet links. In: The 6th International Conference on Field and Service Robotics; Chamonix, France; 2007 July 9–12.

Lum MJH, Rosen J, King H, Friedman DCW, Lendvay T, Wright AS, Sinanan MN, Hannaford B. (2009). Teleopeartion in surgical robotics—network latency effects on surgical performance. In: 31th Annual International Conference of the IEEE Engineering in Medicine and Biology Society EMBS; Minneapolis MN.

Lurz P, et al. Percutaneous pulmonary valve implantation: impact of evolving technology and learning curve on clinical outcome. Circulation 2008;117:1964–1972.

Lyng FM, Seymour CB, Mothersill C. Production of a signal by irradiated cells which leads to a response in unirradiated cells characteristic of initiation of apoptosis. Br J Cancer 2000:1223–1230.

Machemer R, Parel JM, Norton EW. Vitrectomy: a pars plana approach: technical improvements and further results. Trans Am Acad Ophthalmol Otolaryngol 1972;76: 462–466.

Madhani AJ, Niemeyer G, Salisbury JK Jr. (1998). The Black Falcon: a teleoperated surgical instrument for minimally invasive surgery. In: IEEE/RSJ International Conference on Intelligent Robots and Systems, Proceedings Volume 2:936–944; Victoria, BC, Canada; 1998 Oct 13–17.

Maitino AJ, Levin DC, Parker L, Rao VM, Sunshine JN. Nationwide trends in rates of utilization of noninvasive diagnostic imaging among the Medicare population between 1993 and 1999. Radiology 2003:113–117.

Majaron B, Kimel S, Verkruysse W, Aguilar G, Pope K, Svaasand LO, Lavernia EJ, Nelson JS. Cryogen spray cooling in laser dermatology: Effects of ambient humidity and frost formation. Lasers Surg Med 2001;28:469–476.

Makaroun MS, Dillavou ED, Wheatley GH, Cambria RP, Gore TAG Investigators. Five-year results of endovascular treatment with the Gore TAG device compared with open repair of thoracic aortic aneurysms. J Vasc Surg 2008;47(5):912–918.

Makino E, Uenoyama M, Shibata T. Flash evaporation of NiTi shape memory thin film for microactuators. Sensors Actuator Phys 1998;A71(3):187–192.

Mala T, Samset E, Aurdal L, et al. Magnetic resonance imaging-estimated three-dimensional temperature distribution in liver cryolesions: a study of cryolesion characteristics assumed necessary for tumor ablation. Cryobiology 2001;43(3):268–275.

Malchau H, Herberts P, et al. The Swedish total hip replacement register. J Bone Joint Surg 2002;84-A(Suppl 2):2–20.

Malek AM, Alper SL, Izumo S. Hemodynamic shear stress and its role in stherosclerosis. JAMA 1999;282:2035–2042.

Malis LI. Electrosurgery. Journal of Neurosurgery 1996;85(5):970–975.

Mandel L. Photon degeneracy in light from optical maser and other sources. J Opt Soc Am 1961;51:797–798.

Mani G, Feldman MD, Patel D, Agrawal CM. Coronary stents: a materials perspective. Biomaterials 2007;28:1689–1710.

Marcos S, Barbero S, Llorente L, Merayo-Lloves J. Optical response to LASIK surgery for myopia from total and corneal aberration measurements. Investig Ophthalmol Vis Sci 2001;42:3349–3356.

Marescaux J, Leroy J, Gagner M, Rubino F, Mutter D, Vix M, Butner SE, Smith MK. Transatlantic robot-assisted telesurgery. Nature 2001;413(6854):379–380.

Mario CD, Dangas G, Barlis P. *Interventional Cardiology: Principles and Practice*. 1. Aufl ed. John Wiley & Sons; 2010.

Martin J, Friesewinkel O, Benk C, Sorg S, Schultz S, Beyersdorf F. Improved durability of the HeartMate XVE left ventricular assist device provides safe mechanical support up to 1 year but is associated with a high risk of device failure in the second year. J Heart Lung Transplant 2006;25(4):384–90.

Martinez DA. *Therapeutic Ultrasound: A Review of the Literature*. Chiro Access; 2010. Available at http://www.chiroaccess.com/Articles/Therapeutic-Ultrasound-A-Review-of-the-Literature.aspx?id=0000210. Accessed 2011 Oct.

Mason WP. Piezoelectricity, its history and applications. Journal of the Acoustical Society of America 1981;70(6):1561–1566.

Massarweh NN, Cosgriff N, et al. Electrosurgery: history, principles, and current and future uses. Journal of the American College of Surgeons 2006;202(3):520–530.

Mathews HH, Lehuec JC, et al. Design rationale and biomechanics of Maverick Total Disc arthroplasty with early clinical results. Spine J 2004;4(6 Suppl):268S–275S.

Matthews IP, Gibson C, et al. Sterilisation of implantable devices. Clin Mater 1994;15(3):191–215.

McAfee PC, Phillips FM, et al. Minimally invasive spine surgery. Spine 2010;35(26S):S271–73.

McAllister IL, Douglas JP, Constable IJ, Yu D-Y. Laser-induced chorioretinal venous anastomosis for nonischemic central retinal vein occlusion: evaluation of the complications and their risk factors. Am J Ophthalmol 1998;126:219–229.

McDonald MB, Davidorf J, Maloney RK, Manche EE, Hersh P. Conductive keratoplasty for the correction of low to moderate hyperopia: 1-year results on the first 54 eyes. Ophthalmology 2002a;109:637–649.

McDonald MB, Durrie D, Asbell P, Maloney R, Nichamin L. Treatment of presbyopia with conductive keratoplasty(R): Six-month results of the 1-year united states FDA clinical trial. Cornea 2004;23:661–668.

McDonald MB, Hersh PS, Manche EE, Maloney RK, Davidorf J, Sabry M. Conductive keratoplasty for the correction of low to moderate hyperopia: U.S. Clinical trial 1-year results on 355 eyes. Ophthalmology 2002b;109:1978–1989.

McGahan JP, Dodd GD. Radiofrequency ablation of the liver. Current status. AJR 2001;176:3–16.

McGahan JP, Browning PD, Brock JM, Tesluk H. Hepatic ablation using radiofrequency electrocautery. Investig Radiol 1990;25:267–270.

McKeighen RE. Volume 2, Proceedings of the Annual International Conference of the IEEE Engineering in Engineering in Medicine and Biology Society; 1989 Nov 9–12; Seattle, WA. 1989. p 402–404.

McKeighen RE. Design guidelines for medical ultrasound arrays. Volume 3341, Proceedings of SPIE Medical Imaging. 1998. p 2.

McKellop H, Ebramzadeh E, et al. Stability of subtrochanteric femoral fractures fixed with interlocking intramedullary rods. In: Harvey JP, et al., editors. *Intramedullary Rods: Clinical Performance and Related Laboratory Testing*. West Conshohocken: ASTM; 1989.

McKellop, H, Ebramzadeh E, et al. Stem-bone micromotion in noncemented hip prostheses. In: Proceedings of the European Congress on Biomaterials; 1990; Bologna, Italy.

McKellop H, Shen F, et al. Development of an extremely wear resistant ultra high molecular weight polyethylene for total hip replacements. J Orthop Res 1999;17(2):157–167.

Medical Device Databases website. 2011b. CFR—Code of Federal Regulations Title 21. Available at http://www.accessdata.fda.gov/scripts/cdrh/cfdocs/cfcfr/CFRSearch.cfm ?CFRPart=58.

Medtronic. n.d. Medtronic, the world leader in medical technology and pioneering therapies. Available at http://www.medtronic.com.

Megadyne. Principles of electrosurgery. Megadyne 2005. p 2–17.

Melodelima D, N'Djin WA, Parmentier H, et al. Thermal ablation by high-intensity-focused ultrasound using a toroid transducer increases the coagulated volume. Results of animal experiments. Ultrasound Med Biol 2009;35(3):425–435.

Menciassi A, Valdastri P, Harada K, Dario P. Single and multiple robotic capsules for endoluminal diagnosis and surgery. In: Rosen J, Hannaford B, Satava RM, editors. *Surgical Robotics-Systems, Applications, and Visions*. 1st ed. Springer; 2011.

Mendez I, Hill R, Clarke D, Kolyvas G, Walling S. Robotic long-distance telementoring in neurosurgery. Neurosurgery 2005;56(3):434–440.

Mesana TG. Rotary blood pumps for cardiac assistance: a "must"? Artif Organs 2004;28(2): 218–225.

Meyer G. Die Architektur der Spongiosa. Archiv fur Anatomie Physiologie und wissenschaftlich Medizin 1867;34:615–628.

Michael G. Experience with the holmium laser as an endoscopic lithotrite. Urology 1996;48:199–206.

Michiardi A, Aparicio C, Ratner BD, Planell JA, Gil FJ. The influence of surface energy on competitive protein adsorption on oxidized NiTi surfaces. Biomaterials 2007;28: 586–594.

Miller JP, Cohen AR. Intraventricular endoscopy. In: Goodrich J, editor. *Pediatric Neurosurgery*. Noida, India: Thieme Medical & Scientific Publishers (Pvt) Ltd; 2008.

Miller LW, Pagani FD, Russell SD, John R, Boyle AJ, Aaronson KD, Conte JV, Naka Y, Mancini D, Delgado RM, MacGillivray TE, Farrar DJ, Frazier OH. Use of a continuous-flow device in patients awaiting heart transplantation. N Engl J Med 2007;357(9):885–96.

Miller JW, Schmidt-Erfurth U, Sickenberg M, Pournaras CJ, Laqua H, Barbazetto I, Zografos L, Piguet B, Donati G, Lane A-M, Birngruber R, van den Berg H, Strong HA, Manjuris U, Gray T, Fsadni M, Bressler NM, Gragoudas ES. Photodynamic therapy with verteporfin for choroidal neovascularization caused by Age-related macular degeneration: results of a single treatment in a phase 1 and 2 study. Arch Ophthalmol 1999;117:1161–1173.

Mina IG, Kim H, Kim I, Park SK, Choi K, Jackson TN, Tutwiler RL, Trolier-McKinstry S. High frequency piezoelectric MEMS ultrasound transducers. IEEE Trans Ultrason Ferroelectr Freq Control 2007;54(12):2422–2430.

Misra S, Ramesh KT, Okamura AM. Modeling of tool-tissue interactions for computer-based surgical simulation: a literature review. Presence (Camb) 2008;17(5):463–491.

Mitchell JP, Lumb GN. The principles of surgical diathermy and its limitations. British Journal of Surgery 1962;50(221):314–320.

Mjoberg B. Theories of wear and loosening in hip prostheses. Wear-induced loosening vs loosening-induced wear—a review. Acta Orthop Scand 1994;65(3):361–371.

Mooney MG, Wilson MG. Linear array transducers with improved image quality for vascular ultrasonic imaging. Hewlett Packard J 1994;45(4):43–51.

Moore DH II, Tucker JD, Jones IM, Langlois RG, Pleshanov P, Vorobtsova I, Jensen R. A study of the effects of exposure on cleanup workers at the Chernobyl nuclear reactor accident using multiple end points. Radiat Res 1997:463–475.

Moreno-Barriuso E, Lloves JM, Marcos S, Navarro R, Llorente L, Barbero S. Ocular aberrations before and after myopic corneal refractive surgery: LASIK-induced changes measured with laser Ray tracing. Investig Ophthalmol Vis Sci 2001;42:1396–1403.

Morrone G, Guzzardella GA, Tigani D, Torricelli P, Fini M, Giardino R. Biostimulation of human chondrocytes with Ga-Al-as diode laser: 'In Vitro' research. Artif Cells Blood Sub Biotechnol 2000;28:193–201.

Mow VC, Hayes WC. *Basic Orthopaedic Biomechanics*. Philadelphia: Lippincott-Raven; 1997.

Mulvaney WP, Beck CW. The laser beam in urology. J Urol 1968;99:112–115.

Munnerlyn CR, Koons SJ, Marshall J. Photorefractive keratectomy: a technique for laser refractive surgery. J Cataract Refract Surg 1988;14:46–52.

Murrey D, Janssen M, et al. Results of the prospective, randomized, controlled multicenter Food and Drug Administration investigational device exemption study of the ProDisc-C total disc replacement versus anterior discectomy and fusion for the treatment of 1-level symptomatic cervical disc disease. Spine J 2009;9(4):275–286.

Mutter D, Garcia A, Jourdan I. Laparoscopic instruments. WeBSurg.com 2005;5(7) Available at http://www.websurg.com/ref/doi-ot02en320.htm.

Mutter D, Marescaux J. SILS single port cholecystectomy. WeBSurg.com 2009;9(9) Available at http://www.websurg.com/ref/doi-vd01en2722.htm.

Nagy Z, Takacs A, Filkorn T, Sarayba M. Initial clinical evaluation of an intraocular femtosecond laser in cataract surgery. J Refract Surg 2009;25(12):1053–1060.

Natarajan S, Marks LS, Margolis DJA, Huang J, Macairan ML, Lieub P, Fenster A. Clinical application of a 3D ultrasound-guided prostate biopsy system. Urol Oncol 2011;29(3):334–342.

National Centre for the Replacement, Refinement and Reduction of Animals in Research (NC3Rs). (n.d.). *Catheter design and construction*. Retrieved from http://www.nc3rs.org.uk/bloodsamplingmicrosite/page.asp?id=1121. Accessed 2010 July 15.

National Electrical Manufacturers Association (NEMA). *Acoustic Output Measurement Standard for Diagnostic Ultrasound Equipment, Revision 3*. 2009. Standards Publication UD 2-2004 (R2009). Available at www.nema.org. Accessed 2011 Oct.

National Electrical Manufacturers Association (NEMA). *Standard for Real-Time Display of Thermal and Mechanical Acoustic Output Indices on Diagnostic Ultrasound Equipment, Revision 2*. 2010. Standards Publication UD 3-2004 (R2009). Available at www.nema.org. Accessed 2011 Oct.

Nedeltchev K. Pre- and in-hospital delays from stroke onset to intra-arterial thrombolysis. Stroke 2003;34(5):1230–1234.

Neel HBKA. Requisites for successful cryogenic surgery of cancer. Arch Surg 1971;102: 45–48.

Nelson JS, Majaron B, Kelly KM. Active skin cooling in conjunction with laser dermatologic surgery. Semin Cutan Med Surg 2000;19:253–266.

Nelson JS, Milner TE, Anvari B, Tanenbaum BS, Svaasand LO, Kimel S. Dynamic epidermal cooling in conjunction with laser-induced photothermolysis of port wine stain blood vessels. Lasers Surg Med 1996;19(2):224–229.

Newcomb WL, Hope WW, Schmelzer TM, et al. Comparison of blood vessel sealing among new electrosurgical and ultrasonic devices. Surgical Endoscopy 2009;23:90–96.

Nezhat C. *Let There Be Light: A Historical Analysis of Endoscopy's Ascension Since Antiquity*. Society of Laparoendoscopic Surgeons; 2008. Available at http://laparoscopy .blogs.com/endoscopyhistory/chapter_02/. Accessed 2011 Dec.

Nezhat C, Page B. *Nezhat's History of Endoscopy*. Tuttlingen: EndoPress; 2011. ISBN: 978-3-89756-916-4.

Nguyen CM, Yohn JJ, Huff C, Weston WL, Morelli JG. Facial port wine stains in childhood: prediction of the rate of improvement as a function of the age of the patient, size and location of the port wine stain and the number of treatments with the pulsed dye (585 nm) laser. Br J Dermatol 1998;138:821–825.

Nietlispach F, Wijesinghe N, Wood D, Carere RG, Webb JG. Current balloon-expandable transcatheter heart valve and delivery systems. Catheter Cardiovasc Interv 2010;75(2):295–300 [Epub ahead of print].

Nishihara S, et al. Clinical accuracy evaluation of femoral canal preparation using the ROBODOC system. J Orthop Sci 2004;9(5):452–461.

Nkomo VT, Gardin JM, Skelton TN, Gottdiener JS, Scott CG, Enriquez-Sarano M. Burden of valvular heart diseases: a population based study. Lancet 2006;368:1005–1011.

Norgren L, Hiatt WR, Dormandy JA, Nehler MR, Harris KA, Fowkes RG. Inter-society consensus for the management of peripheral arterial disease (TASC II). J Vasc Surg 2007;45(1):S5A–S67A.

O'Connor JL, Bloom DA, William T. Bovie and electrosurgery. Surgery 1996;119(4): 390–396.

Odén A, Fahlén M. Oral anticoagulation and risk of death: a medical record linkage study. BMJ 2002;325(7372):1073–75.

O'Doherty M, Kirwan C, O'Keeffe M, O'Doherty J. Postoperative pain following epi-LASIK, LASEK, and PRK for myopia. J Refract Surg (Thorofare, N.J.: 1995) 2007;23:133–138.

Oi S. *EndoWorld: The OI HandyPro™ Neuroendoscope*. Tuttlingen: KARL STORZ GmbH & Co. KG; 2009.

Oliva VL, Soulez G. Sirolimus-eluting stents versus the superficial femoral artery: second round. J Vasc Interv Radiol 2005;16:313–315.

Onda Corporation. *Tables of Acoustic Properties of Materials*. 2011. Available at www.ondacorp.com. Accessed 2011 Oct.

Oosterhuis JA, Journee-de Korver HG, Kakebeeke-Kemme HM, Bleeker JC. Transpupillary thermotherapy in choroidal melanomas. Arch Ophthalmol 1995;113:315–321.

Oshika T, Klyce SD, Applegate RA, Howland HC, El Danasoury MA. Comparison of corneal wavefront aberrations after photorefractive keratectomy and laser in situ keratomileusis. Am J Ophthalmol 1999;127:1–7.

Ottensmeyer MP, Kerdok AE, Howe RD, Dawson S. The effects of testing environment on the viscoelastic properties of soft tissues. ISMS 2004;2004:9–18.

Ottensmeyer MP, Salisbury JK Jr. In vivo data acquisition instrument for solid organ mechanical property measurement. MICCAI 2001;2001:975–982.

Overholt BF, Panjehpour M, Haydek JM. Photodynamic therapy for Barrett's esophagus: follow-up in 100 patients. Gastrointest Endosc 1999;49:1–7.

Owen CM, Linskey ME. Frame-based stereotaxy in a frameless era: current capabilities, relative role, and the positive- and negative predictive values of blood through the needle. J Neurooncol 2009;93(1):139–149.

Ozgur BM, Aryan HE, et al. Extreme Lateral Interbody Fusion (XLIF): a novel surgical technique for anterior lumbar interbody fusion. Spine J 2006;6(4):435–443.

Pallikaris IG, Siganos DS. Excimer laser in situ keratomileusis and photorefractive keratectomy for correction of high myopia. J Refract Corneal Surg 1994;10:498–510.

Palmaz JC. Bring that pioneering spirit back! A 25year perspective on the vascular stent. Cardiovasc Intervent Radiol 2007;30:1095–1098.

Palmaz JC, Bailey S, Marton D, Sprague E. Influence of stent design and material composition on procedure outcome. J Vasc Surg 2002;36(5):1031–1039.

Papaioannou TG, Stefanadis C. Basic principles of the intraaortic balloon pump and mechanisms affecting its performance. ASAIO J 2005;51(3):296–300.

Parer JT, King T. Fetal heart rate monitoring: is it salvageable? Am J Obstet Gynecol 2000;182(4):982–987.

Parodi JC, Palmaz JC, Barone HD. Transfemoral intraluminal graft implantation for abdominal aortic aneurysms. Ann Vasc Surg 1991;5(6):491–499.

Pasquale K, Wiatrak B, Woolley A, Lewis L. Microdebrider versus CO2 laser removal of recurrent respiratory papillomas: a prospective analysis. Laryngoscope 2003;113:139–143.

Pass HI. Photodynamic therapy in oncology: mechanisms and clinical Use. J Natl Cancer Inst 1993;85:443–456.

Pass RH, Hijazi Z, Hsu DT, Lewis V, Hellenbrand WE. Multicenter USA Amplatzer patent ductus arteriosus occlusion device trial: initial and one-year results. J Am Coll Cardiol 2004;44(3):513–519.

Patel AA, Brodke DS, et al. Revision strategies in lumbar total disc arthroplasty. Spine (Phila Pa 1976) 2008;33(11):1276–1283.

Pavlica P, Menchi I, Barozzi L. New imaging of the anterior male urethra. Abdom Imag 2003;26:180–186.

Pearson M, McClurken M, Thompson R. Saline enhanced thermal sealing of tissue: potential for bloodless surgery. Minimally Invasive Therapy & Allied Technologies 2002;11(5/6):265–270.

Pellicci P, Salvati E, et al. Mechanical failures in total hip relplacement requiring reoperation. J Bone Joint Surg Am 1979;61(1):28–36.

Peng Q, Juzeniene A, Chen J, Svaasand LO, Warloe T, Giercksky K-E, Moan J. Lasers in medicine. Rep Prog Phys 2008;71:056701.

Pennes HH. Analysis of tissue and arterial blood temperature in the resting human forearm. J Appl Physiol 1948;1:93–122.

Peters TM, Cleary KR. *Image-Guided Interventions: Technology and Applications*. Springer; 2008.

Petrou I. (2009). I, Robot. Dermatology Times: 66. Available at http://www.restoration robotics.com/pdf/Dermatology_Times_Aug_2009.pdf.

Philips. n.d. Sense and simplicity. Available at http://www.usa.philips.com. Accessed 2010.

Phillips FM, Garfin SR. Cervical disc replacement. Spine (Phila Pa 1976) 2005;30(17 Suppl):S27–S33.

Piazza N, Asgar A, Ibrahim R, Bonan R. Transcatheter mitral and pulmonary valve therapy. J Am Coll Cardiol 2009;53:1837–1851.

Pimenta L, McAfee PC, et al. Superiority of multilevel cervical arthroplasty outcomes versus single-level outcomes: 229 consecutive PCM prostheses. Spine (Phila Pa 1976) 2007;32(12):1337–1344.

Portner P, Oyer P, McGregor C. First human use of an electrically powered implantable ventricular assist system. Artif Organs 1985;9:36.

Prutchi D, Norris M. *Design and Development of Medical Electronic Instrumentation*. Wiley Online Library; 2005.

Radovancevic B, Vrtovec B, de Kort E, Radovancevic R, Gregoric I, Frazier OH. End-organ function in patients on long-term circulatory support with continuous- or pulsatile-flow assist devices. J Heart Lung Transplant 2007;26(8):815–18.

Razvi HA, Denstedt JD, Chun SS, Sales JL. Intracorporeal Lithotripsy With the Holmium:YAG Laser. J Urol 1996;156:912–914.

Reddy VY, et al. View-synchronized robotic image-guided therapy for atrial fibrillation ablation experimental validation and clinical feasibility. Circulation 2007;115: 2705–2714.

Reichel E, Berrocal AM, Ip M, Kroll AJ, Desai V, Duker JS, Puliafito CA. Transpupillary thermotherapy of occult subfoveal choroidal neovascularization in patients with age-related macular degeneration. Ophthalmology 1999;106:1908–1914.

Reiley CE, Lin HC, Yuh DD, Hager GD. A review of methods for objective surgical skill evaluation. Surg Endosc 2010;25(2):356–356.

Reuter MA, Reuter HJ, Engel RM. *History of Endoscopy*. Vol. I–IV. Stuttgart: Max Nitze Museum; 1999.

Ringel F, Ingerl D, Ott S, Meyer B. VarioGuide: a new frameless image-guided stereotactic system—accuracy study and clinical assessment. Neurosurgery 2009;64(5 Suppl 2):365–373.

Robinson RP, Lovell TP, et al. Early femoral component loosening in DF-80 total hip arthroplasty. J Arthroplasty 1989;4(1):55–64.

Romaine-Davis A. *John Gibbon and his Heart—Lung Machine*. Philadelphia, PA: University of Pennsylvania Press; 1991.

Röntgen WC. *Eine Neue Art fon Strahlen*. Würzburg: Physikal, Institut der Universität; 1895.

Rose EA, Gelijns AC, Moskowitz AJ, Heitjan DF, Stevenson LW, Dembitsky W, Long JW, Ascheim DD, Tierney AR, Levitan RG, Watson JT, Meier P, Ronan NS, Shapiro PA, Lazar RM, Miller LW, Gupta L, Frazier OH, Desvigne-Nickens P, Oz MC, Poirier VL. Randomized evaluation of mechanical assistance for the treatment of congestive heart failure (REMATCH) study group. N Engl J Med 2001;345(20):1435–43.

Rosen J, Brown JD, Chang L, Sinanan MN, Hannaford B. Generalized approach for modeling minimally invasive surgery as a stochastic process using a discrete markov model. IEEE Trans Biomed Eng 2006;53(3):399–413.

Rosen J, Brown JD, De S, Mika N, Hannaford SB. Biomechanical properties of abdominal organs in vivo and postmortem under compression loads. ASME J Biomed Eng 2008;130(2).

Rosen J, Hannaford B, MacFarlane M, Sinanan MN. Force controlled and teleoperated endoscopic grasper for minimally invasive surgery-experimental performance evaluation. IEEE Trans Biomed Eng 1999;46(10):1212–1221.

Rosen J, Hannaford B, Satava RM, editors. *Surgical Robotics-Systems, Applications, and Visions*. 1st ed. Springer; 2011.

Rossi S, Fornari F, Pathies C, Buscarini L. Thermal lesions induced by 480 kHz localized current field in guinea pig and pig liver. Tumori 1990;76:54–57.

Rubinsky B, Onik G, Mikus P. Irreversible electroporation: a new ablation modality–clinical implications. Tech Canc Res Treat 2007;6(1):37–48.

Rutherford EE, Tarplett LJ, et al. Lumbar spine fusion and stabilization: hardware, techniques, and imaging appearances. Radiographics 2007;27(6):1737–1749.

Ryan RW, Wolf T, Spetzler RF, Coons SW, Fink Y, Preul MC. Application of a flexible CO(2) laser fiber for neurosurgery: laser-tissue interactions. J Neurosurg 2010;112(2):434–443.

Ryhanen J, Niemi E, Serlo W, Niemela E, Sandvik P, Pernu H, Salo T. Biocompatibility of nickel-titanium shape memory metal and its corrosion behavior in human cell cultures. J Biomed Mat Res 1997;35:451–457.

Salib P. Orthopaedic and traumatic skeletal lesions in ancient Egyptians. J Bone Joint Surg 1960;44-B(4):944–947.

Saliba W, Cummings JE, Oh S, et al. Novel robotic catheter remote controlsystem: feasibility and safety of transseptal puncture and endocardial catheter navigation. J Cardiovasc Electrophysiol 2006;17:1102–1105.

Saliba W, et al. Atrial fibrillation ablation using a robotic catheter remote control system-atrial fibrillation ablation using a robotic catheter remote control system. J Am Coll Cardiol 2008;51:2407–2411.

Sanborn TA, Sleeper LA, Bates ER, Jacobs AK, Boland J, French JK, Dens J, Dzavik V, Palmeri ST, Webb JG, Goldberger M, Hochman JS. Impact of thrombolysis, intraaortic balloon pump counterpulsation, and their combination in cardiogenic shock complicating acute myocardial infarction: a report from the SHOCK Trial Registry. SHould we emergently revascularize Occluded Coronaries for cardiogenic shocK? J Am Coll Cardiol 2000;36(3 SupplA):1123–9.

Sangiorgio SN, Ebramzadeh E, et al. Effects of dorsal flanges on fixation of a cemented total hip replacement femoral stem. J Bone Joint Surg 2004;86-A(4):813–820.

Sangra M, Clark S, Hayhurst C, Mallucci C. Electromagnetic-guided neuroendoscopy in the pediatric population. J Neurosurg Pediatr 2009;3(4):325–330.

Şapçi T, Şahin B, Karavus A, Akbulut UG. Comparison of the effects of radiofrequency tissue ablation, CO2 laser ablation, and partial turbinectomy applications on nasal mucociliary functions. Laryngoscope 2003;113:514–519.

Sarkar S, Sales KM, Hamilton G, Seifalian AM. Addressing thrombogenicity in vascular graft construction. J Biomed Mater Res B Appl Biomater 2007;82(1):100–8.

Sarmiento A. Functional fracture bracing: an update. Instr Course Lect 1987;36:371–376.

Sarmiento A. Distal tibia fracture: opinion: nonoperative treatment. J Orthop Trauma 2006;20(1):75.

Satava RM. Historical review of surgical simulation—a personal perspective. World J Surg 2008;32:141–148.

Sauvaget C, Nagano J, Hayashi M, Spencer E, Shimizu Y, Allen N. Vegetables and fruit intake and cancer mortality in the Hiroshima/Nagasaki Life Span Study. Br J Cancer 2003:689–694.

Sawyer PN, Pate JW, Weldon CH. Relations of abnormal and injury electropotential differences to intravascular thrombosis. Am J Physiol 1953;175:108–116.

Scerrati E. Laser in situ keratomileusis vs. laser epithelial keratomileusis (LASIK vs. LASEK). J Refract Surg (Thorofare, N.J.: 1995) 2001;17:S219–S221.

Schawlow AL, Townes CH. Infrared and optical masers. Phys Rev 1958;112:1940–1949.

Schillinger M, Exner M, Mlekusch W, et al. Fibrinogen predicts restenosis after endovascular treatment of the iliac arteries. Thromb Haemost 2002;87:959–965.

Schlötzer-Schrehardt U, Viestenz A, Naumann G, Laqua H, Michels SM, Schmidt-Erfurth U. Dose-related structural effects of photodynamic therapy on choroidal and retinal structures of human eyes. Graefes Arch Clin Exp Ophthalmol 2002;240:748–757.

Schmidt-Erfurth U, Miller JW, Sickenberg M, Laqua H, Barbazetto I, Gragoudas ES, Zografos L, Piguet B, Pournaras CJ, Donati G, Lane A-M, Birngruber R, van den Berg H, Strong HA, Manjuris U, Gray T, Fsadni M, Bressler NM. Photodynamic therapy with verteporfin for choroidal neovascularization caused by Age-related macular degeneration: results of retreatments in a phase 1 and 2 study. Arch Ophthalmol 1999;117:1177–1187.

Schneider P. *Endovascular Skills: Guidewire and Catheter Skills for Endovascular Surgery.* 3rd ed. Informa Health Care; 2008.

Schoen FJ, Levy RJ. Calcification of tissue heart valve substitutes: Progress toward understanding and prevention. Ann Thorac Surg 2005;79:1072–1080.

Schraff S, Derkay CS, Burke B, Lawson L. American society of pediatric otolaryngology Members' experience with recurrent respiratory papillomatosis and the Use of adjuvant therapy. Arch Otolaryngol Head Neck Surg 2004;130:1039–1042.

Schulz AP, et al. Results of total hip replacement using the Robodoc surgical assistant system: clinical outcome and evaluation of complications for 97 procedures. Int J Med Robot Comput Assist Surg 2007;3(4):301–306.

Schwartz M, Doron A, Erlich M, Lavie V, Benbasat S, Belkin M, Rochkind S. Effects of low-energy He-Ne laser irradiation on posttraumatic degeneration of adult rabbit optic nerve. Lasers Surg Med 1987;7:51–55.

Scott-Conner CEH, editor. Fundamentals of laparoscopy, thoracoscopy, and GI endoscopy. In: *The SAGES Manual.* Springer; 2006.

Seibel B. *Phacodynamics: Mastering the Tools and Techniques of Phacoemulsification Surgery*. New Jersey: Slack Incorporated; 2005.

Seiler T, Holschbach A, Derse M, Jean B, Genth U. Complications of myopic photorefractive keratectomy with the excimer laser. Ophthalmology 1994;101:153–160.

Seiler T, Kaemmerer M, Mierdel P, Krinke H-E. Ocular optical aberrations after photorefractive keratectomy for myopia and myopic astigmatism. Arch Ophthalmol 2000;118:17–21.

Sekiguchi Y, Funami K, Funakubo H. Deposition of NiTi shape memory alloy thin film by vacuum evaporation. In: *Proceedings of 32nd Meeting of Japan Society of Materials*. 1983. p 65–67.

Senders CW, Navarrete EG. Laser supraglottoplasty for laryngomalacia: are specific anatomical defects more influential than associated anomalies on outcome? Int J Pediatr Otorhinolaryngol 2001;57:235–244.

Shepp LA, Logan BF. The Fourier reconstruction of a head section. IEEE Trans Nucl Sci 1974;21(3):21–43.

Shields CL, Shields JA, DePotter P, Kheterpal S. Transpupillary thermotherapy in the management of choroidal melanoma. Ophthalmology 1996;103:1642–1650.

Shields CL, Shields JA, Perez N, Singh AD, Cater J. Primary transpupillary thermotherapy for small choroidal melanoma in 256 consecutive cases: outcomes and limitations. Ophthalmology 2002;109:225–234.

Shimoda K, Wang TC, Townes CH. Further aspects of the theory of the maser. Phys Rev 1956;102:1308–1321.

Shirota T, Yasui H, Matsuda T. Intraluminal tissue-engineered therapeutic stent using endothelial progenitor cell-inoculated hybrid tissue and in vitro performance. Tissue Eng 2003;9(3):473–485.

Shoham M, Burman M, Zehavi E, Joskowicz L, Batkilin E, Kunicher Y. Bone mounted miniature robot for surgical procedures: concept and clinical applications. IEEE Trans Robot Automat 2003;19(5):893–901.

Shoham M, Lieberman IH, Benzel EC, et al. Robotic assisted spinal surgery-from concept to clinical practice. Comput Aided Surg 2007;12(2):105–115.

Shung KK. *Diagnostic Ultrasound: Imaging and Blood Flow Measurements*. Boca Raton (FL): Taylor & Francis Group; 1996. Chapter 3.

Shung KK, Cannatta JM, Zhou Q. High frequency ultrasound transducers and arrays. In: Safari A, Akdoğan EK, editors. *Piezoelectric and Acoustic Materials for Transducer Applications*. New York: Springer; 2008. Chapter 21.

Siegman AE. (2009) Laser history. Available at http://www.stanford.edu/~siegman/stanford_laser_history.pdf. Accessed 2012.

Simon CJ, Dupuy DE, Iannitti DA, Lu DSK, Yu NC, Aswad BI, Busuttil RW, Lassman C. Intraoperative Triple Antenna Hepatic Microwave Ablation. AJR 2006;187: W333–W340.

Simon CJ, Dupuy DE, Mayo-Smith WW. Microwave ablation: principles and applications. Radiographics 2005;25(Suppl 1):S69–S83.

Simone C, Okamura AM. (2002). Modeling of needle insertion forces for robot-assisted percutaneous therapy. In: Proceedings ICRA '02. IEEE International Conference on Robotics and Automation, 2002, Volume 2. p 2085–2091.

Slaughter MS, Rogers JG, Milano CA, Russell SD, Conte JV, Feldman D, Sun B, Tatooles AJ, Delgado RM, Long JW, Wozniak TC, Ghumman W, Farrar DJ, Frazier OH. Advanced heart failure treated with continuous-flow left ventricular assist device. N Engl J Med 2009;361:2241–2251.

Smith A. Full-field breast tomosynthesis. Radiol Manage 2005:25–31.

Smith AD, et al. *Smith's Textbook of Endourology*. Shelton, CT: People's Medical Publishing House USA; 2006.

Solazzo SA, Ahmed M, Liu Z, Hines-Peralta AU, Goldberg SN. High-power generator for radiofrequency ablation: larger electrodes and pulsing algorithms in bovine ex vivo and porcine in vivo settings. Radiology 2007;242(3):743–750.

vanSonnenberg E, McMullen WN, Solbiati L, editors. *Tumor ablation: principles and practice*. New York: Springer; 2005.

Sosis MB. Evaluation of five metallic tapes for protection of endotracheal tubes during $CO2$ laser surgery. Anesth Analg 1989;68:392–393.

Spaide RF, Laud K, Fine HF, Klancnik JM Jr, Meyerle CB, Yannuzzi LA, Sorenson J, Slakter J, Fisher YL, Cooney MJ. Intravitreal Bevacizumab Treatment of Choroidal Neovascularization Secondary to Age-Related Macular Degeneration. Retina 2006;26(4):383–390.

Spanier TB, Chen JM, Oz MC, Stern DM, Rose EA, Schmidt AM. Time-dependent cellular population of textured-surface left ventricular assist devices contributes to the development of a biphasic systemic procoagulant response. J Thorac Cardiovasc Surg 1999;118(3):404–13.

Spence EH. The ROBODOC clinical trial: a robotic assistant for total hip arthroplasty. Orthop Nursing 1996;15(1):9–14.

Spence DE, Kean PN, Sibbett W. 60-fsec pulse generation from a self-mode-locked Ti:sapphire laser. Opt Lett 1991;16:42–44.

Stefánsson E. The therapeutic effects of retinal laser treatment and vitrectomy. A theory based on oxygen and vascular physiology. Acta Ophthalmol Scand 2001;79:435–440.

Steger AC, Lees WR, Walmsley K, Bown SG. Interstitial laser hyperthermia: a new approach to local destruction of tumours. BMJ 1989;299:362–365.

Stein BS, Kendall AR. Lasers in urology: II. Laser Ther Urol 1984;23:411–416.

Stepan LS, Levi DL, Carman GP. A thin film nitinol heart valve. J Biomech Eng 2005;127(6):915–918.

Stokowski G, Steele D, Wilson D. The use of ultrasound to improve practice and reduce complication rates in peripherally inserted central catheter insertions: final report of investigation. J Infus Nurs 2009;32(3):145–155.

Storz. n.d. KARL STORZ Endoskope—welcome to the world of endoscopy. Available at http://www.karlstorz.de/cps/rde/xchg/karlstorz-en/hs.xsl/index.htm. Accessed 2010.

Strüber M, Sander K, Lahpor J, Ahn H, Litzler PY, Drakos SG, Musumeci F, Schlensak C, Friedrich I, Gustafsson R, Oertel F, Leprince P. HeartMate II left ventricular assist device; early European experience. Eur J Cardio Thorac Surg 2008;34(2):389–94.

Tabuse K. A new operative procedure of hepatic surgery using a microwave tissue coagulator. Arch Jpn Chir 1979;48:160–172.

Talacchi A, Turazzi S, Locatelli F, Sala F, Beltramello A, Alessandrini F, Manganotti P, Lanteri P, Gambin R, Ganau M, Tramontano V, Santini B, Gerosa M. Surgical treatment of high-grade gliomas in motor areas. The impact of different supportive technologies: a 171-patient series. J Neurooncol 2010;100(3):417–426.

Taneri S, Feit R, Azar DT. Safety, efficacy, and stability indices of LASEK correction in moderate myopia and astigmatism. J Cataract Refract Surg 2004a;30:2130–2137.

Taneri S, Zieske JD, Azar DT. Evolution, techniques, clinical outcomes, and pathophysiology of LASEK: review of the literature. Surv Ophthalmol 2004b;49:576–602.

Tatli S, Acar M, Tuncali K, Morrison PR, Silverman S. Percutaneous cryoablation techniques and clinical applications. Diagn Interv Radiol 2009; Dec 8.

Tektran. Principles of electrosurgery. Tektran 1997. p 1–33.

Thanopoulos BV, Rigby ML, Karanasios E, Stefanadis C, Blom N, Ottenkamp J, Zarayelyan A. Transcatheter closure of perimembranous ventricular septal defects in infants and children using the Amplatzer perimembranous ventricular septal defect occluder. Am J Cardiol 2007;99(7):984–989.

The European Stroke Organization Executive Committee (ESO), ESO Writing Committee. Guidelines for management of ischaemic stroke and transient ischaemic attack. Cerebrovasc Dis(Basel, Switzerland) 2008;25(5):457–507.

The National Institute of Neurological Disorders and Stroke (NINDS). Tissue plasminogen activator for acute ischemic strokeThe National Institute of Neurological Disorders and Stroke rt-PA Stroke Study Group. New Engl J Med 1995;333(24):1581–1587.

Thielmann S, Seibold U, Haslinger R, Passig G, Bahls T, Jörg S, Nickl M, Nothhelfer A, Hagn U, Hirzinger G. (2010). MICA-A new generation of versatile instruments in robotic surgery. In: IROS 2010; Taipei, Taiwan.

Thompson KP, Ren QS, Parel JM. Therapeutic and diagnostic application of lasers in ophthalmology. Proc IEEE 1992;80:838–860.

Tilney NJ. Invasion of the Body: Revolutions in Surgery. Cambridge (MA): Harvard University Press; 2011.

Torricelli P, Giavaresi G, Fini M, Guzzardella GA, Morrone G, Carpi A, Giardino R. Laser biostimulation of cartilage: in vitro evaluation. Biomed Pharmacother 2001;55:117–120.

Toyokuni S. Reactive oxygen species-induced molecular damage and its application in pathology. Pathol Int 1999;49:91–102.

Tucker RD, Voyles CR. Laparoscopic electrosurgical complications and their prevention. AORN Journal 1995;62(1):49–71.

Turek PJ, Malloy TR, Cendron M, Carpiniello VL, Wein AJ. KTP-532 laser ablation of urethral strictures. Urology 1992;40:330–334.

Tychsen L, Hoekel J. Refractive surgery for high bilateral myopia in children with neurobehavioral disorders: 2. Laser-assisted subepithelial keratectomy (LASEK). J AAPOS// 2006;10:364–370.

Uhl E, Zausinger S, Morhard D, Heigl T, Scheder B, Rachinger W, Schichor C, Tonn J. Intraoperative computed tomography with integrated navigation system in a multidisciplinary operating suite. Neurosurgery 2009;64(5 Suppl 2):231–240.

USNRC, United States Nuclear Regulatory Commission. NRC Regulations (10 CFR) Part 20—Standards for Protection Against Radiation. http://www.nrc.gov/reading-rm/doc-collections/cfr/part020/. Accessed 2011 Sep.

U.S. Census Bureau. 2007. North American Industry Classification System. Available at http://www.census.gov/eos/www/naics. Accessed 2012.

Utley DS, Koch RJ, Egbert BM. Histologic analysis of the thermal effect on epidermal and dermal structures following treatment with the superpulsed CO_2 laser and the Erbium:YAG laser: An in vivo study. Lasers Surg Med 1999;24:93–102.

Valdés PA, Fan X, Ji S, Harris BT, Paulsen KD, Roberts DW. Estimation of brain deformation for volumetric image updating in protoporphyrin IX fluorescence-guided resection. Stereotact Funct Neurosurg 2010;88(1):1–10.

Valobra G, Fiandesio D. Short history of electrocautery and diathermocautery]. Minerva Medica 1968;59(15):829–831.

Van Kampen CL, Gibbons DF. Effect of implant surface chemistry upon arterial thrombosis. J Biomed Mater Res 1979;13:517–541.

Veith FJ, Marin ML, Cynamon J, Schonholz C, Parodi J. 1992: Parodi, Montefiore, and the first abdominal aortic aneurysm stent graft in the United States. Ann Vasc Surg 2005;19:749–751.

Vender JR, Miller J, et al. Effect of hemostasis and electrosurgery on the development and evolution of brain tumor surgery in the late 19th and early 20th centuries. Neurosurgical Focus 2005;18(4):1–7.

Verdonschot N, Huiskes R. Cement debonding process of total hip arthroplasty stems. Clin Orthop 1997a;(336):297–307.

Verdonschot N, Huiskes R. The Effects of Cement-Stem Debonding in THA on the Long-Term Failure Probability of Cement. J Biomech 1997b;30(8):795–802.

Verdonschot N, Huiskes R. Surface roughness of debonded straight-tapered stems in cemented THA reduces subsidence but not cement damage. Biomaterials 1998;19(19):1773–1779.

Verdonschot N, Tanck E, et al. Effects of prosthesis surface roughness on the failure process of cemented hip implants after stem-cement debonding. J Biomed Mater Res 1998;42(4):554–559.

Vernick DM. A comparison of the results of KTP and CO_2 laser stapedotomy. Otol Neurotol 1996;17:221–224.

Vilaseca-González I, Bernal-Sprekelsen M, Blanch-Alejandro J-L, Moragas-Lluis M. Complications in transoral CO_2 laser surgery for carcinoma of the larynx and hypopharynx. Head Neck 2003;25:382–388.

Vogl TJ, Helmberger TK, Mack MG, Reiser MF, editors. *Percutaneous tumor ablation in medical radiology*. Springer-Verlag Berlin Heidelberg: Germany; 2008.

Vogl TJ, Straub R, Zangos S, Mack MG, Eichler K. MR-guided laser-induced thermotherapy (LITT) of liver tumours: experimental and clinical data. Int J Hyperther 2004;20(7):713–24.

Voorhees JR, Cohen-Gadol AA, et al. Battling blood loss in neurosurgery: Harvey Cushing's embrace of electrosurgery. Journal of Neurosurgery 2005;102(4):745–752.

Wall J, Gertner ME, et al. *Energy Transfer in the Practice of Surgery*. New York: Springer; 2008. p 2345–2354.

Walluscheck KP, Bierkandt S, Brandt M, Cremer J. Infrainguinal ePTFE vascular graft with bioactive surface heparin bonding. First clinical results. J Cardiovasc Surg 2005;46:425–430.

Walsh JT, Deutsch TF. Pulsed CO_2 laser tissue ablation: measurement of the ablation rate. Lasers Surg Med 1988;8:264–275.

Walsh JT, Flotte TJ, Anderson RR, Deutsch TF. Pulsed CO_2 laser tissue ablation: effect of tissue type and pulse duration on thermal damage. Lasers Surg Med 1988;8:108–118.

Ward GE. Electrosurgery. The American Journal of Surgery 1932;17(1):86–93.

Waring GO III, Lynn MJ, McDonnell PJ, PERK Study Group. Results of the prospective evaluation of radial keratotomy (PERK) study 10 years after surgery. Arch Ophthalmol 1994;112:1298–1308.

Washburn BR, Diddams SA, Newbury NR, Nicholson JW, Yan MF, Jrgensen CG. Phase-locked, erbium-fiber-laser-based frequency comb in the near infrared. Opt Lett 2004;29:250–252.

Waterhouse RB. *Occurence of Fretting in Practice and Its Simulation in the Laboratory. Materials Evaluation Under Fretting Conditions*. West Conshohocken (PA): American Society for Testing and Materials; 1982. p 3–16.

Watterson JD, Girvan AR, Beiko DT, Nott L, Wollin TA, Razvi H, Denstedt JD. Ureteroscopy and holmium:YAG laser lithotripsy: an emerging definitive management strategy for symptomatic ureteral calculi in pregnancy. Urology 2002;60:383–387.

Weber AL. History of head and neck radiology: past, present, and future. Radiology 2001;218:15–24.

Whipple TL. From mini-invasive to non-invasive treatment using monopolar radiofrequency: the next orthopaedic frontier. Orthop Clin N Am 2009;40(4):531–535.

Wies C. The history of hemostasis. The Yale Journal of Biology and Medicine 1929;2(2):167–168.

Williams GA. 25-, 23-, or 20-gauge instrumentation for vitreous surgery. Eye 2008;22: 1263–1266.

Willis CE, Solvis TL. The ALARA concept in pediatric CR and DR: dose reduction in pediatric radiographic exams—a white paper conference executive summary. Radiology 2005:343–344.

Wilms G, Baert AL. The history of angiography. J Belg De Radiol 1995;78(5):299–302.

Wolff J. Ueber die innere Architecktuer der Knochen und ihre Bedeutung fuer die Frage vom Knochen-wachstum. Virchows Arch Path Anat Physiol 1870;50(389).

Wood EW, Loomis AL. The physical and biological effects of high-frequency sound-waves of great intensity. Philosophical Magazine 1927;4(22):417–436.

Woolson ST, Maloney WJ. Cementless total hip arthroplasty using a porous-coated prosthesis for bone ingrowth fixation. 3 1/2-year follow-up. J Arthroplasty 1992;7(Suppl):381–388.

Wu MH. Fabrication of nitinol materials and components. In: *Proceedings of the International Conference on Shape Memory and Superelastic Technologies, Kunming, China*. 2001. p 285–292.

Wu MP, Ou CS, et al. Complications and recommended practices for electrosurgery in laparoscopy. The American Journal of Surgery 2000;179(1):67–73.

Wyman A, Duffy S, Sweetland HM, Sharp F, Rogers K. Preliminary evaluation of a new high power diode laser. Lasers Surg Med 1992;12:506–509.

Xu LC, Siedlecki CA. Effects of surface wettability and contact time on protein adhesion on biomaterial surfaces. Biomaterials 2007;28:3273–3283.

Yamashita Y, Hosono Y, Itsumi K. Low-attenuation acoustic silicone lens for medical ultrasonic array probes. In: Safari A, Akdoğan EK, editors. *Piezoelectric and Acoustic Materials for Transducer Applications*. New York: Springer Science + Business Media, LLC; 2008. p 161–177. Chapter 8.

Yang DL, Xu QW, Che XM, Wu JS, Sun B. Clinical evaluation and follow-up outcome of presurgical plan by Dextroscope: a prospective controlled study in patients with skull base tumors. Surg Neurol 2009;72(6):682–689.

Yannuzzi LA, Slakter JS, Kaufman SR, Gupta K. Laser treatment of diffuse retinal pigment epitheliopathy. Eur J Ophthalmol 1992;2:103–114.

Yoo AC, Gilbert GR, Broderick TJ. Military robotic combat casualty extraction and care. In: Rosen J, Hannaford B, Satava R, editors. *Surgical Robotics: Systems, Applications, and Visions*. New York: Springer; 2010. p 13–32.

Youjun L, Aike Q, Qun N, Xiaoyong Y. Thermal Characteristics of Microwave Ablation in the Vicinity of an Arterial Bifurcation. IFMBE Proceedings. World Congress on Medical Physics and Biomedical Engineering. Imaging the Future Medicine; 2006 Aug 27–Sep 1; COEX Seoul, Korea; 2006.

Zabriskie N. *The Operating Microscope and Surgical Loupes. Basic and Clinical Science Course: Basic Principles of Ophthalmic Surgery. Basic Principles of Ophthalmic Surgery*. San Francisco (CA): American Academy of Ophthalmology; 2006.

Zack's Equity Research. 2011. Medical Devices Industry Outlook–April 2011. Available at www.zacks.com. Accessed 2012.

Zahn EM, Hellenbrand WE, Lock JE, McElhinney DB. Implantation of the melody transcatheter pulmonary valve in patients with a dysfunctional right ventricular outflow tract conduit: early results from the U.S. Clinical trial. J Am Coll Cardiol 2009;54:1722–1729.

Zayhowski JJ, Dill C III. Diode-pumped passively Q-switched picosecond microchip lasers. Opt Lett 1994;19:1427–1429.

Zehetner J, DeMeester SR. Treatment of Barrett's esophagus with high-grade dysplasia and intramucosal adenocarcinoma. Expet Rev Gastroenterol Hepatol 2009;3(5):493–8.

Zetkin M, Schaldach K. *Lexikon der Medizin: Die umfassende Enzyklopädie*. 1. Aufl ed. Fackelträger-Verlag; 2005.

Zhuo R, Miller R, Bussard KM, Siedlecki CA, Vogler EA. Procoagulant stimulus processing by the intrinsic pathway of blood plasma coagulation. Biomaterials 2005;26:2965–2973.

Zinder DJ. Common myths about electrosurgery. Otolaryngology—Head and Neck Surgery 2000;123(4):450–455.

Zornig C, Mofid H. *NOTES Cholecystectomy: Transvaginal Hybrid and Single-Site Multiple-Port Approaches*. Tuttlingen: EndoPress; 2009.

Zorov DB, Filburn CR, Klotz L-O, Zweier JL, Sollott SJ. Reactive oxygen species (Ros-induced) Ros release. J Exp Med 2000;192:1001–1014.

Medical Devices: Surgical and Image-Guided Technologies, First Edition.
Edited by Martin Culjat, Rahul Singh, and Hua Lee.
© 2013 John Wiley & Sons, Inc. Published 2013 by John Wiley & Sons, Inc.